Water and Wastewater Engineering

A major challenge for the twenty-first century is to provide safe and adequate drinking water to everyone. Preventing pollution of the environment due to rampant and untreated discharges of wastewater is another challenge for most developing countries, including India. The water–energy connection is also being recognized as another growing challenge. The design of water and wastewater treatment facilities must be environmentally sensitive, energy efficient and sustainable into the future.

Conceived as a textbook for undergraduate and graduate students who need to understand the basic concepts and design principles related to water and wastewater engineering, this book begins with an introduction to water resources and the need for water and wastewater treatment. This is followed by an evaluation of water demand in terms of quantity and quality. Major mass transfer and transformation processes that are necessary for understanding the complexity of water pollution issues and treatment processes are dealt with in detail. Treatment processes that are used in water and/or wastewater treatment are detailed subsequently so that they can be designed by the student. A few examples of specific water treatment requirements are provided to enable the student to choose and apply only relevant treatment processes in their design. Conventional and non-conventional treatment schemes for water and wastewater treatment are covered to complete the overview of treatment processes. Collection, transportation and distribution aspects of drinking water supply systems are covered along with wastewater collection systems. Problems and issues arising from the inadequacies of conventional treatment practices, and potential methods for resolving these problems have also been incorporated into this text. An overview of relevant regulations, Indian and other, is also provided.

Sudha Goel is Associate Professor, Department of Civil Engineering (Environmental Engineering and Management), Indian Institute of Technology Kharagpur, India. She has published more than 50 papers in journals and conferences of national and international repute. Her areas of research include water quality and treatment, environmental impact and risk assessment, and solid and hazardous waste management.

Water and Wastewater Engineering

Sudha Goel

Shaftesbury Road, Cambridge CB2 8EA, United Kingdom

One Liberty Plaza, 20th Floor, New York, NY 10006, USA

477 Williamstown Road, Port Melbourne, VIC 3207, Australia

314–321, 3rd Floor, Plot 3, Splendor Forum, Jasola District Centre, New Delhi – 110025, India

103 Penang Road, #05–06/07, Visioncrest Commercial, Singapore 238467

Cambridge University Press is part of Cambridge University Press & Assessment, a department of the University of Cambridge.

We share the University's mission to contribute to society through the pursuit of education, learning and research at the highest international levels of excellence.

www.cambridge.org
Information on this title: www.cambridge.org/9781316639030

First published 2019

A catalogue record for this publication is available from the British Library

Library of Congress Cataloging-in-Publication data
Names: Goel, Sudha, author.
Title: Water and wastewater engineering / Sudha Goel.
Description: Cambridge, United Kingdom ; New York, NY, USA : Cambridge University Press, 2019. | Includes bibliographical references and index.
Identifiers: LCCN 2019016364 | ISBN 9781316639030 (paperback : alk. paper)
Subjects: LCSH: Water--Purification. | Sewage--Purification. | Water-zsupply.
Classification: LCC TD430 .G63 2019 | DDC 628.1--dc23
LC record available at https://lccn.loc.gov/2019016364

ISBN 978-1-316-63903-0 Paperback

Additional resource for this publication at www.cambridge.org/9781316639030

Contents

Preface

'Water and Wastewater Engineering' is a core course in undergraduate programs in civil engineering. The course objective is to ensure that a student is able to evaluate different water resource options for their sustainability, quantity and quality, and to design appropriate municipal water supply and wastewater systems. These water supply systems will necessarily require sourcing of water, its collection, treatment and distribution. Wastewater generated within these systems has to be treated in treatment plants so that it can be reused or disposed of on land or in water bodies after achieving discharge standards. Wastewater reuse has become an extremely important topic these days due to severe water scarcity in many parts of the world, including India.

This textbook covers all aspects of municipal water and wastewater systems and is designed for a one-semester course. Prior to designing water and wastewater treatment plants, it is necessary to identify and develop an appropriate water source. For this, the student must be familiar with different types of water resources: surface water and groundwater, and concepts related to their quantity and quality. These are covered in the first and second chapters of this book. Fundamental concepts from chemistry, microbiology, and chemical engineering are covered in the first part of the book (Chapters 3 and 4) as these are necessary for understanding water quality issues, and designing water and wastewater systems. The second part of the book includes the design of conventional water treatment plants with unit processes like aeration, sedimentation, coagulation, filtration, and disinfection; design of conventional wastewater treatment plants with unit processes like screening, sedimentation, biological processes, activated sludge process and trickling filters or biofilters, sludge treatment, and disposal; water distribution methods; wastewater collection, reuse, and disposal options; and non-conventional treatment strategies for removal of specific pollutants like fluoride, arsenic, nitrate, and natural organic matter.

This textbook started as a spin-off of an online course of the same name. However, several topics that could not be covered in the online course are also included in the textbook. The text material has been expanded and the number of problems increased. Solutions to all problems are provided. Some of the problems will require the use of spreadsheets or other software for graphing and calculations. A word about notation in this book: * symbolizes multiplication in MS Excel and has been retained in all equations instead of 'x'. The online course can be accessed at the following link: http://www.ide.iitkgp.ernet.in/Pedagogy_view/example.jsp?USER_ID=52.

Several pedagogical features have been incorporated in the book, including learning objectives, study outline, and study questions. Learning objectives help the students identify what the outcome of their study should be, while the study outline provides a concise summary of what is important. Several photographs and schematic diagrams are included along with graphical solutions to problems to help the student visualize concepts and solve problems. The book covers only theoretical and empirical principles as they are applied in the field. The practical 'nuts and bolts' of engineering cannot be provided by this textbook! The student should always bear in mind that what is done in practice, i.e., what works, does not always seem to be compatible with theory, i.e., the how and why of what works and what does not work. Research and development is all about bridging the gap between theory and practice.

Finally, while every effort has been made to eliminate typographical and other mistakes from the book, the reader is encouraged to point these out by writing to the author at the following address: sudhagoelcup@gmail.com.

Acknowledgments

As mentioned in the preface, this book is a spin-off of an online course that was created under a National Mission Project on Education through Information and Communication Technology, sponsored by the Ministry of Human Resource Development, Government of India. The author is grateful to the principal investigators of this project (Professor Anup Kumar Roy and Professor Bani Bhattacharya) for giving her an opportunity to develop an online course. Several students were part of developing the online course and include Aashay Arora, Abhishek Ashish, Akhilesh Yadav, Allen Dan Babu, Ankit Surekha, Hiray Kunal Satish, Manas Kansal, Manoj Kumar Mondal, Neelesh Agrawal, Rohit Rout, Prateek Kumar, Shaikh Elias, and Syed Salman Hyder.

Other students who have contributed to this book long after the above-mentioned project was completed include Abhishek Singhal, Rahul Meena, Tandra Mohanta, Ved P. Ranjan, Kruttika Apshankar Kher, and Naseeba Parveen. Experimental data for several problems were obtained from students and their work is referenced at relevant points in the book. The author is grateful to all these students for their contributions to this book. Colleagues and staff at IIT Kharagpur and in other institutions have also supported this endeavor and their help is gratefully acknowledged.

Last but not least, the author is grateful to her family, friends, and teachers who have supported her through all these years.

Abbreviations

AL	aerated lagoon
AODC	acridine orange direct cell count
APHA	American Public Health Association
ASP	activated sludge process
ATAD	auto thermal aerobic digestion
AWWA	American Water Works Association
BCM	billion cubic meters
BFR	brominated fire retardants
BOD	biochemical oxygen demand or biological oxygen demand
BODu	ultimate biochemical oxygen demand
CBOD	carbonaceous biochemical oxygen demand
CEA	Central Electricity Authority
COD	chemical oxygen demand
CPCB	Central Pollution Control Board
CSO	combined sewer overflow
CSTR	continuously stirred tank reactor
CWC	Central Water Commission
CWS	continuous water supply
DAF	dissolved air flotation
DBPs	disinfection by-products
DDT	dichloro-diphenyl-trichloroethane
DF	demand factor
DNA	deoxyribose nucleic acid
DO	dissolved oxygen
DOC	dissolved organic carbon
DW	drinking water

ED	electron donor
ED	electrodialysis
EDR	electrodialysis reversal
FAO	Food and Agriculture Organization
FICCI	Federation of Indian Chambers of Commerce and Industry
FSS	fixed suspended solids
GI	galvanized iron
GW	groundwater
HAA	haloacetic acids
HAN	haloacetonitriles
HPC	heterotrophic plate count
IS	Indian Standards
ISO	International Organization for Standardization
IWS	intermittent water supply
Lpcd	liters per capita per day
LUST	leaking underground storage tank
MBR	membrane bioreactor
MLD	million liters per day
MLSS	mixed liquor suspended solids
MLVSS	mixed liquor volatile suspended solids
MSL	mean sea level
NBOD	nitrogenous biochemical oxygen demand
NOM	natural organic matter
NTU	nephelometric turbidity units
PCP	personal care products
PF	peaking factor
PFR	plug flow reactor
RBC	rotating biological contactor
RNA	ribose nucleic acid
SAR	sodium absorption ratio
SBR	sequencing batch reactor
SEM	scanning electron microscope
SF	solids flux
SOC	synthetic organic compounds (compounds)
SOP	synthetic organic polymers
SS	steady-state

SVI	sludge volume index
SW	surface water
TDS	total dissolved solids
TEA	terminal electron acceptor
TEM	transmission electron microscope
TF	trickling filter
TFS	total fixed solids
THM	trihalomethanes
ThOD	theoretical oxygen demand
TKN	total Kjeldahl nitrogen
TOC	total organic carbon
TOX	total organic halogen
TS	total solids
TSS	total suspended solids
TVS	total volatile solids
uPVC	unplasticized polyvinyl chloride
UV	ultraviolet
VC	viable cells
VLOM	village level operation and maintenance
VOC	volatile organic compounds
VSS	volatile suspended solids
WHO	World Health Organization
WQI	water quality index

Symbols and Dimensions
(Mass, Length and Time–MLT system where possible)

a	activity
A	area, L^2
A	specific light absorbance, dimensionless
A/V	= a = specific surface area, 1/L
A_p	projected area or cross-sectional area of particle in flow direction, L^2
b	endogenous decay coefficient, 1/T
C	concentration, M/L^3
C_d	coefficient of drag, dimensionless
D	dispersion coefficient
D_e	eddy diffusion coefficient
D_m	molecular diffusion coefficient, L^2/T
e	electron charge, $1.60219 * 10^{-19}$ Coulombs
E_a	activation energy for a reaction, kJ/mol
F	flow rate for fire-fighting or fire demand, L^3/T
F/M	food to microorganism ratio, kg BOD_5/kg MLVSS-d
G	velocity gradient, 1/T
h	elevation or height, L
h_f	head loss through filter, L
I	current, amperes
I	impermeability factor or runoff coefficient (ratio of runoff to rainfall)
I	ionic strength, M/L^3
k	Boltzmann constant, $1.38066*10^{-23}$ J/degree Kelvin
K	hydraulic conductivity or coefficient of permeability, L/T
k	maximum substrate utilization rate per unit mass of microbes, mg substrate/mg cells-time, M/M-T
k	reaction rate constant, units vary with reaction order

k_d	deoxygenation constant, 1/T
K_L	overall mass transfer coefficient, M/T
k_o	oxygenation or reaeration constant, 1/T
K_{ow}	octanol-water partitioning coefficient
K_s	half-velocity constant, M/L^3
L	length, L
L_0	ultimate carbonaceous BOD (CBOD) and L_t = ultimate CBOD at time t
M	molality of a solution, moles/L
n	any number
n	Manning's coefficient or coefficient of roughness
N	number of microbes or cells/L
N_A	Avogadro's number, $6.02205*10^{23}$ molecules/mol
P	population, persons; P_0 = population at t = 0; P_s = saturation population in the logistic model
P	power or pressure
Q	flow rate, L^3/T or heat flux, Joules/cm^2-s
q	hydraulic loading rate or surface overflow rate, L/T
R	electrical resistance, ohms
R	ideal gas constant, 8.314 J/mol–K
R	rainfall intensity, L/T
r	rate of change, M/T
R_0	maximum instantaneous growth rate in the logistic model, 1/T
Re	Reynolds number, dimensionless
r_H	hydraulic radius, L
S	growth limiting substrate concentration in solution, M/L^3
S	slope or hydraulic gradient, or drop in head or head loss per unit length = $-h_L/L$, length of pipe, L/L
T	temperature
t	time, T
t_c	critical time,
v	velocity, L/T
V	volume of a solution, L^3
v_s	settling velocity of particle, M/T
X	increment or mass or mass fraction or biomass or cell concentration
Y	maximum yield coefficient, dimensionless

Z	charge of ion
γ (gamma)	activity coefficient, dimensionless
η (eta)	porosity (% of total volume) or Coulombic efficiency, %
Θ (theta)	temperature correction factor or normalized time, i.e., t/τ
κ^{-1}(1/kappa)	double layer thickness, L
μ (mu)	dynamic viscosity
Π (pi)	osmotic pressure
ρ (rho)	density of water or other materials, M/L^3
τ (tau)	V/Q = design hydraulic residence time, T
Ø (phi)	sphericity of the particle, dimensionless

PART - I

Concepts Related to Water

'Clean water and sanitation' is one of the seventeen Sustainable Development Goals set by the United Nations General Assembly in 2015 and is to be achieved globally by 2030. Access to safe drinking water is considered a basic human right and more than 1 billion people in the world continue to suffer due to water-related diseases, resulting in many deaths.

Achieving this goal will require designing appropriate water supply and wastewater treatment systems. The first part of this book provides background information that is essential for understanding and designing these systems. The first step in the design of water supply systems is to identify a sustainable water source for the community. These water sources can be surface water or groundwater and are described in the first chapter. Methods for withdrawal of water from these sources are included. The extent to which these sources are contaminated will determine the degree of treatment that is necessary.

The next important step in the design of water supply systems is to determine the quantity and quality of the water that is needed. Water demand varies both temporally and spatially. Temporal variations on a daily and seasonal basis need to be accounted for in design. Spatial variations in water demand are mainly due to the size of the population that is to be served. Accurate forecasting of future populations is an extremely important step in determining the size and design of both water supply and wastewater systems. Factors that influence water demand are addressed in Chapter 2. Water quality requirements have become more stringent over time necessitating higher levels of treatment. Major water quality parameters of importance in water supply and wastewater systems are described with examples in Chapter 3.

Finally, basic principles of reaction kinetics and reactor design are described in Chapter 4. These principles are necessary for understanding the transport and transformation of contaminants within water and wastewater treatment systems and are applied to the design of treatment processes which are covered in the second part of the book.

CHAPTER 1

Water Resources

Learning Objectives

- *Identify and classify water resources*
- *Describe the hydrologic cycle*
- *Differentiate between surface water and groundwater resources*
- *Describe confined and unconfined aquifers*
- *Derive Darcy's law and apply it to relevant groundwater scenarios*
- *Determine yields, permeabilities, hydraulic conductivities and flow velocities of groundwater through confined and unconfined aquifers*
- *Describe different types of wells, their methods of construction and conditions for their use*
- *Identify sources of groundwater contamination and their transport in the subsurface*
- *Identify potential sources of contamination for surface water bodies*
- *Classify surface water bodies based on CPCB standards*
- *Calculate pollutant or nutrient loading to a water body*
- *Design and describe different types of surface water intakes and wells*
- *Identify sources of contamination, their constituents and concentrations in surface water bodies*
- *Describe and design infiltration wells and galleries and the conditions under which they can be used*

Freshwater is only 3 percent of the total water on the planet, and includes groundwater which is almost 30 percent of this, while surface water is a meager 0.3 percent. Glaciers and icecaps comprise the remaining, at slightly less than 70 percent of the total, freshwater resources of the planet.[1]

1 http://www.unwater.org/statistics_res.html.

The movement of water in any of its three states (solid, liquid, or gaseous) through different parts of the planetary environment is termed the **hydrologic cycle** and is shown in Figure 1.1

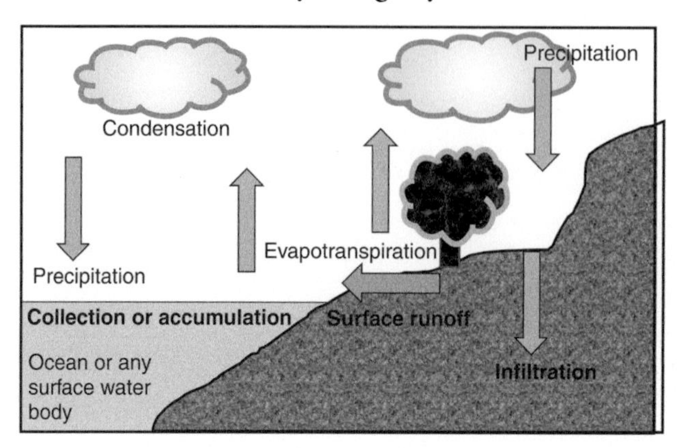

Figure 1.1 Hydrologic cycle and its various components (see color plate).

It can be broadly defined in terms of six major processes:[2]

Precipitation is the sum total of all water that falls over land and surface water bodies including rivers, seas, and oceans and includes rain, ice, snow, hail, and sleet.

Condensation is that part of water vapor in the atmosphere that condenses on particles to form clouds. When these clouds are supersaturated, the water vapor is released in the form of precipitation.

Evapotranspiration is the release of water to the atmosphere from various water bodies by evaporation and from plants during the process of transpiration or respiration.

Surface runoff is defined as the amount of water that flows over land towards any surface water body. Any and all surface water bodies from small ponds and lakes to seas and oceans can serve as sinks for surface runoff. Some of these water bodies are transient (seasonal) while others are permanent.

Infiltration is the amount of water that percolates or infiltrates into the subsurface.

Accumulation or ***collection*** of water is the sixth major process and is the **sink** for all inputs of water. Water is collected in various water bodies: surface or subsurface. Surface water bodies can be oceans, seas, rivers, lakes, and ponds, while groundwater aquifers accumulate water that infiltrates through the subsurface. The latter process is also known as *groundwater recharge*.

As per the last assessment of the hydrologic cycle in India (1993), India receives 4000 billion m^3 of precipitation (including snow) every year. Accessible surface water and rechargeable groundwater comprise 1869 billion m^3 and of this only 60 percent is usable due to topographical and other constraints (Ministry of Water Resources 2002). Based on CWC data, annual evaporation losses in India are estimated to range from 5 to 6 percent (Central Water Commission 2006). Per capita availability of water in India is estimated to be 1820 m^3/y and can be compared to data for other countries as shown in Table 1.1

2 http://cwc.nic.in/newwebsite_aug07/main/webpages/kids/welcome.html.

Table 1.1 Country-wise estimates of per capita water availability (Central Water Commission 2006)

Country	Per capita water availability, m^3/y
India	1,820
USA	8,902
China	2,215
Brazil	40,855
Australia	18,162

Historically, cities and towns flourished along rivers due to easy access and long-term availability of water. However, the last few decades have seen an increasing shift towards the use of groundwater in India even along rivers due to better and more dependable water quality, easy access to groundwater (GW) and deteriorating quality of surface waters (SW). Access to groundwater has become easy due to improved ability to dig deeper tubewells at extremely low costs. On the other hand, storage of surface water, especially by construction of dams and reservoirs, is now becoming a difficult proposition for technical, social and economic reasons. In a survey of 171 students in August 2017, 34 percent of the students came from towns that were completely dependent on surface water sources while 22 percent were using, only, groundwater in their hometowns. The remaining (44 percent) were using both types of water resources, groundwater and surface water in their hometowns. These data are indicative of a major shift from reliance on surface water to groundwater.

Estimates of total water withdrawals in the country in 2010 are provided in Table 1.2 (Panikkar 2012).

Table 1.2 Estimates of total water withdrawals in India in 2010 (Panikkar 2012)

Total water withdrawals	Amount	Units	Percent of total withdrawal
Per inhabitant	627	m^3/y	
Surface water and groundwater withdrawal	7,61,000	10^6 m^3/y	
Withdrawals by use			
Irrigation and livestock	6,88,000	10^6 m^3/y	90.4
Industrial use	56,000	10^6 m^3/y	7.36
Municipal use	17,000	10^6 m^3/y	2.23
Withdrawals by source			
Primary surface water	396	km^3/y	52.0
Primary groundwater	251	km^3/y	32.98
Reused agricultural drainage water	114	km^3/y	14.98

Groundwater is characterized by relatively constant water quality in comparison to surface waters. Water in shallow wells in unconfined aquifers tends to vary with seasons in terms of water level and quality. However, water in deep wells in confined aquifers varies only over the long-term, i.e., it is not subject to seasonal variations. Changes are observed only over decades or more. Spatial variations are also due only to changes in hydro-geologic conditions.

A few decades ago, developing GW sources (investigating, building and operating) was considered more expensive than using SW sources. Several test and monitoring wells were required to determine aquifer yield and water quality. Further, pumping requirements can also be high leading to greater net expenditure in such cases. However, treatment requirements for GW are generally much less than SW sources due to higher water quality making GW a better proposition in most cases.

Other salient differences between groundwater and surface water resources include the following:

1. Groundwater is often considered a non-renewable resource. The term 'mining' of GW is used when there is no likelihood of its replenishment.
2. Pollution of GW resources is more difficult to mitigate than pollution of SW resources.
3. Long-term effects of pollution of GW can be a bigger issue compared to SW pollution. In general, if discharge of wastewater to a SW body is eliminated, the surface water body will recover naturally in a few years. On the other hand, if a GW resource is contaminated and even if the source of contamination is eliminated, the GW resource is never going to recover naturally and completely.
4. Treatment Costs: It is more difficult and more expensive to pump and treat contaminated GW than to restore SW quality.

1.1 GROUNDWATER

Groundwater (GW) is defined as water that has percolated downward (or infiltrated) from the ground surface through the soil and subsurface rocks (Henry and Heinke 1996). A schematic defining the various subsurface zones is shown in Figure 1.2.

Unsaturated zone (vadose zone): The vadose zone is the subsurface region where all the pore spaces or voids between the soil particles contain both air and water. This water is not usable because it is not possible to draw or pump out. However, it can sustain vegetation.

The vadose zone occurs above the water table and the capillary fringe, and the fluid pressure in this zone is less than atmospheric pressure (Freeze and Cherry 1979). This negative pressure head also influences hydraulic conductivity and moisture content in this zone. Hydraulic heads in this zone are measured with tensiometers.

Saturated zone: The saturated zone is that part of the subsurface where all voids are filled with water, i.e., completely saturated. This is the zone below the water table in an unconfined aquifer. The fluid pressure is greater than or equal to atmospheric pressure in confined aquifers and equal to atmospheric pressure in unconfined aquifers. High water yields are possible in this zone. Hydraulic conductivity in this zone is constant and is not influenced by the pressure head.

Capillary fringe: The transition zone between the first two zones or regions is defined as the capillary fringe. Water rises through the pore spaces by capillary action from the saturated to the unsaturated zone.

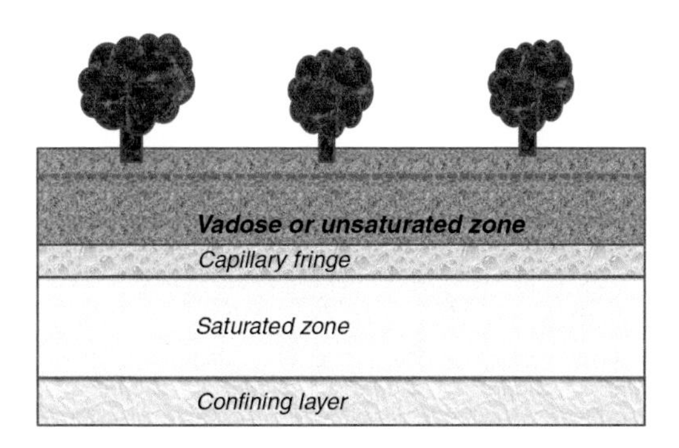

Figure 1.2 Schematic of the subsurface and its zones.

Groundwater can be withdrawn or pumped out of geologic formations called aquifers. Various terms associated with groundwater are shown in Figure 1.3 and described here.

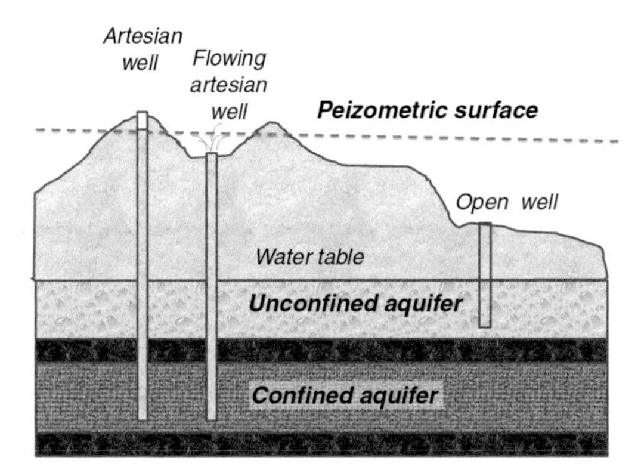

Figure 1.3 Schematic of wells and aquifers.

An **aquifer** is a saturated geologic layer that is permeable enough for water to flow through easily. There are two major types of aquifers:

Confined aquifers are those where water flows through two confining layers called *aquicludes* or *aquitards*. Water is at greater than atmospheric pressure and can rise to a level higher than the local water table. This results in natural artesian wells or springs which can flow without pumping. The potentiometric surface (or peizometric surface) is the plane drawn at the level to which water rises in an artesian well, i.e., the pressure head in the well.

An *unconfined aquifer* sits atop a confining layer and its upper surface is the water table. Water is under atmospheric pressure in an unconfined aquifer and defines the water table level. Wells in these strata are called gravity wells.

Aquiclude or *aquitard* is a relatively impermeable layer or confining layer that restricts the movement of GW.

1.1.1 Quantifying Groundwater Flow

Flow in the subsurface can be described by Darcy's law. *Darcy's law* states that flow (Q) per unit area (A) is proportionate to the hydraulic gradient (–dh/dL), as shown in Figure 1.4. Area considered is perpendicular to the direction of flow.

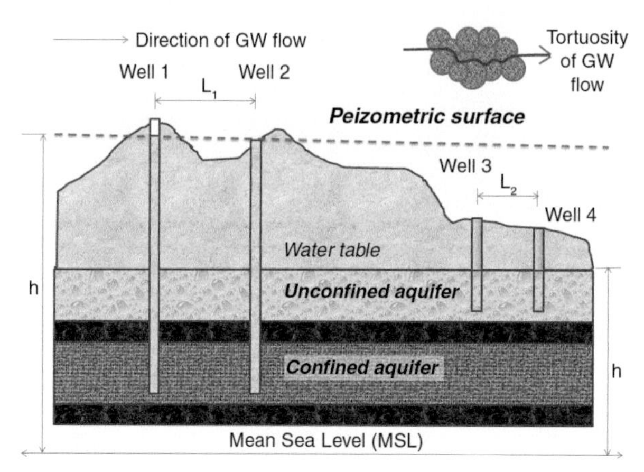

Figure 1.4 Hydraulic gradient in confined and unconfined aquifers.

Hydraulic gradient is the slope of the water table (unconfined aquifer) or piezometric surface (confined aquifer) = dh/dL, where h = height of the water table in an unconfined aquifer or the piezometric surface in a confined aquifer. Height is measured with respect to a datum which is generally the mean sea level (MSL), and L is the distance between two wells.

The constant of proportionality (K) is defined as the *hydraulic conductivity* or *coefficient of permeability* and may change with location and direction of flow in any aquifer.

$$Q/A = K^*(-dh/dL) \quad 1.1.1$$

If K is constant throughout the aquifer, the aquifer is termed *homogeneous*. If K changes with location in the aquifer, that aquifer is termed *heterogeneous*. If K is constant regardless of flow direction then the aquifer is *isotropic*, while if K changes with flow direction then the aquifer is *anisotropic*.

Darcy velocity (v) is v = Q/A. This velocity does not represent 'real' GW velocities at the particle-level. Velocities at the particle level, i.e., the average linear velocities, are much greater than Darcy velocity due to the tortuosity of groundwater flow.

Porosity (η, eta) is the ratio of void volume to total volume. It determines the amount of water stored in an aquifer. Hydraulic conductivity or permeability is not always proportionate to porosity as shown in Table 1.3.

$$\text{Average linear velocity } (v') = v/\eta \quad 1.1.2$$

Other important terms that are used to quantify and characterize groundwater flow are:

Specific yield or effective porosity is the percent volume of water that can be drained from an unconfined aquifer. Particle size determines permeability and specific yield; smaller particle size leads to greater surface tension and lower permeability or specific yield.

Storage coefficient is the volume of water that can be drained from a confined aquifer.

Table 1.3 Range of values for different aquifer materials (Masters 1998; Freeze and Cherry 1979)

	Hydraulic conductivity, m/s	Porosity, percent	Specific yield (percent)
Gravel	10^{-3} to 1	25 to 40	22
Clean sand	10^{-6} to 10^{-2}	25 to 50	25
Silt or clay	10^{-9} to 10^{-5}	35 to 50	3
Sandstone	10^{-10} to 10^{-6}	5 to 30	8
Limestone, dolomite	10^{-9} to 10^{-6}	0 to 20	2
Shale	10^{-13} to 10^{-10}	0 to 10	2

PROBLEM 1.1.1

Three monitoring wells have the following coordinates and heads: W1 at [50, 50] has a head of 200 m; W2 at [50, 100] has a head of 210 m; W3 at [80, 75] has a head of 220 m. Sketch the well field and determine the magnitude and direction of the hydraulic gradient.

Solution

This problem can be solved either graphically or analytically as shown in the following figure.

Graphical method

Step 1: Sketch the well field on graph paper as shown below:

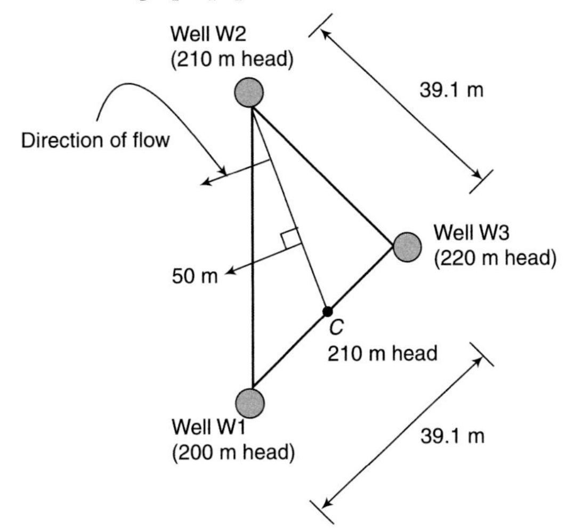

Step 2: Equipotential lines (lines with the same head) have to be drawn through the well field to determine the hydraulic gradient and direction of flow. From the figure it is obvious that there is a point with 210 m head on the line joining wells W1 and W3 and this point is the midpoint of this line which is point C. Joining W2 with C results in one equipotential line and all other equipotential lines are parallel to it by definition.

Step 3: Direction of flow of water is perpendicular to these equipotential lines. Hydraulic gradient is determined by taking the difference in heads of two adjacent equipotential lines (dh) and dividing it by the distance between the two equipotential lines (dL). These distances can be read directly from a graph paper.

Hydraulic gradient (dh/dL) = (210 – 200)/18.5 = 0.54

Direction of flow (or hydraulic gradient) is from higher head to lower head and from graph it is 202 degrees (where 0 degrees corresponds to due east).

Analytical method based on coordinate geometry

The coordinates and heads of the wells are:

W1: (50, 50) with 200 m Head

W2: (50, 100) with 210 m Head

W3: (80, 75) with 220 m Head

The distance between W1 and W2 is 50 m. The distance between W2 and W3 is calculated as:

$$d(W2, W3) = ((50 - 80)^2 + (100 - 75)^2)^{1/2}$$
$$= 39.1 \text{ m}$$

Similarly d(W1, W3) = 39.1 m

The well field is as shown below:

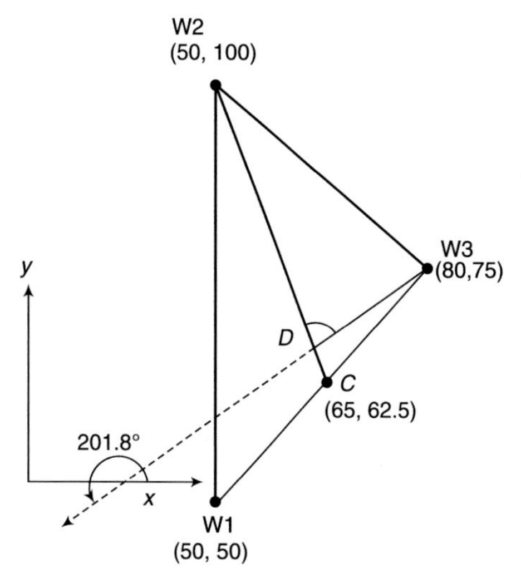

Line connecting the well with highest head (W3) and lowest head (W1) is divided into equal parts and a point is marked where the head corresponds to 210 m, i.e., the head of the intermediate well (W2).

In this case, the required point is the mid-point of line connecting W1 and W3 with coordinates [(50+80)/2, (50+75)/2] or [65, 62.5]. A line is drawn connecting W2 and this mid-point. This is an equipotential line with head of 210 m. Groundwater flow will be in the direction perpendicular to the line towards decreasing head as shown in the figure.

Hydraulic gradient = Change in head / Horizontal distance

We have to find the distance between the equipotential line and W3.

The equation of the line connecting C and W2 is:

$$(y - 100)/(x - 50) = (100 - 62.5)/(50 - 65)$$

$$\equiv 5x + 2y - 450 = 0$$

From coordinate geometry, the distance D of a point (m, n) from any line $Ax + By + C = 0$ is given by:

$$D = |Am + Bn + C| / (A^2 + B^2)^{1/2}$$

Thus, distance of W3 (80, 75) from the line joining C to W2 is given by $5x + 2y - 450 = 0$ is

$$|5*80 + 2*75 - 450| / (5^2 + 2^2)^{1/2} = 18.6 \text{ m}$$

Thus hydraulic gradient = (210 – 200)/18.6 = 0.5376

The slope of the line $5x + 2y - 450 = 0$ is (–Coeff. of x) / (Coeff. of y) = –5/2.

Thus slope of a line perpendicular to this line is (–1)/(–5/2) = 2/5.

Thus the gradient is in a direction which makes $180 + \tan^{-1}(2/5) = 201.80°$ with x axis.

PROBLEM 1.1.2

If hydraulic conductivity is 6.52 m/d and the hydraulic gradient is 0.025, determine Darcy velocity. Further, assume the aquifer is made of fine sand with a porosity of 40 percent, determine the average linear velocity.

Solution

Darcy velocity = Q/A = K(dh/dL) = 6.52 m/d * 0.025 = 0.163 m/d or 1.88×10^{-6} m/s

For a porosity of 40 percent, the average linear velocity = (0.163/0.4) m/d or 4.7×10^{-6} m/s

PROBLEM 1.1.3

Porosity of media is known to be independent of grain size if we assume that all grains are equal-sized with some radius r and are perfectly spherical. If that is true what accounts for the differences in porosity observed in the field.

Solution

We will first consider a simple cubic lattice and assume that all particles are completely spherical. In a simple cubic lattice, the particles are located only at the corners of the cube and touch each other along the edge. Thus, in a unit cell only 1 particle is present (8 × 1/8 of each particle)

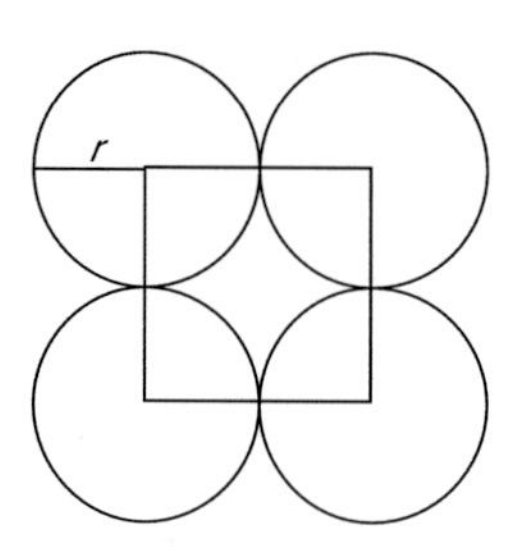

From the figure given above we can see that size of the unit cell is $2r \times 2r \times 2r$

Volume of the unit cell = $(2r)^3$

As the number of particles present is only one, therefore volume of particle inside the unit cell is = $\frac{4}{3}\pi r^3$

$$\text{Void ratio} = \frac{\textit{Volume of voids}}{\textit{Volume of particle}} = \frac{8r^3 - \frac{4\pi r^3}{3}}{\frac{4\pi r^3}{3}} = 0.90$$

$$\text{Porosity = void volume/ total volume} = \frac{8r^3 - \frac{4\pi r^3}{3}}{8r^3} = 1 - (\pi/6) = 0.47 \text{ or } 47 \text{ percent.}$$

Thus it can be seen that porosity or void ratio for this arrangement does not depend on the size of the particle.

The simple cubic lattice in this problem is a highly unstable arrangement and can not exist under natural conditions. A detailed comparison of 6 different ways of packing equal sized spheres (including the above which is the easiest to analyze but impossible in practice) is provided in a paper by Graton and Fraser (1935). The most compact or 'tightest' packing arrangement for equal-sized spheres is 'rhombohedral' and the corresponding porosity is 25.95 percent or 26 percent. As particles become irregular, i.e., they deviate from sphericity, or when the media is heterogeneous, i.e., when all particles are not of the same size, porosity is further reduced and tends towards zero. This accounts for the differences in permeability and hydraulic conductivity for different materials and different grain sizes and shapes.

PROBLEM 1.1.4

Compare the surface area to volume ratio for a cube of side w with a sphere of diameter w.

Solution

For a simple cube of side w, the total volume = w^3 and surface area = $6w^2$

Therefore, the surface area (SA)/volume (V) ratio = $6/w$

For a simple sphere of diameter w, i.e., $r = w/2$

Surface area (SA) = $4\pi r^2$ and volume = $4/3\pi r^3$

Therefore, the surface area (SA)/volume (V) ratio = 6/w

These calculations illustrate the impact of size on the surface area/volume ratio, also known as specific surface area. As size decreases, the surface area per unit volume increases. There are several examples in the environment where this principle is applicable. Smaller-sized particles provide a larger specific surface area for adsorption of contaminants as compared to larger-sized particles. The small size of bacteria and other microorganisms serves as an advantage since the large specific surface area allows greater uptake of nutrients from the environment by absorption.

1.1.2 Design and Construction of Wells

Recent reports show that groundwater is now the main source of water supply in India for irrigation. A CWC report shows that wells are the water source for 61.6 percent of the net irrigated area in India (CWC 2015). A recent survey using remote sensing (satellite imagery of 500 m resolution) showed that 62 percent of the irrigated area was under minor irrigation projects, i.e., groundwater, reservoirs and tanks and only 38 percent was under surface water irrigation (major projects) (Thenkabail et al. 2012).[3]

Wells can be categorized in various ways based on different criteria. Some examples are provided here.

Pressure head

The pressure head is used to categorize wells as gravity wells or artesian wells. Gravity wells are defined as wells where the water is under atmospheric pressure and therefore, has to be pumped out. Wells where the water pressure is higher than atmospheric pressure are known as pressure wells or artesian wells and are shown schematically in Figure 1.3. Often, the peizometric head is sufficient for these well waters to be used directly without pumping.

Depth of well

Shallow wells are the most common wells especially in rural India and tap unconfined aquifers. These are either dug wells or hand pumped wells. Water quality in these wells is highly variable and ranges from poor to very high quality water. Tubewells or deep wells and borewells are used to tap confined aquifers. Tubewells can be cost- and energy-intensive in terms of capital and operating costs but are fairly common in urban and peri-urban communities. Their high outputs with constant quality over long periods of time are sufficient to offset capital and operating costs.

Type of construction

Dug wells are the oldest types of wells that can be dug manually and are often simply lined with bricks with open joints. The bottoms of these wells are open and water seeps through the sides through the open joints. The water levels in these wells represent the local water table.

3 http://www.iwmigiam.org/info/GMI-DOC/GIAM-India-Stats.pdf.

Hand-pumped wells can be dug wells which have been covered with a concrete slab or drilled wells. The most common hand pumps used in India and in other developing countries are the Mark II and Mark III types which were developed for use in remote rural locations. Mark III pumps are an improved version of Mark II pumps, which were developed in 1991 with the sole objective of providing ease in operation and maintenance, i.e., village level operation and maintenance (VLOM) (Colin 1999). They have open top cylinders with riser pipes of 50 or 65 mm diameter. The riser pipes are made of either galvanized iron (GI) or unplasticized polyvinyl chloride (uPVC). These hand pumps can tap water up to a depth of 45 m for uPVC and 60 m with GI providing a discharge of 800 to 900 L/h.[4]

Driven wells are generally small diameter wells that are easy to construct and often abandoned when not needed. Driven wells are generally used to dewater construction sites. A sharply pointed perforated pipe or pipe with well screen is driven into the ground, i.e., hammered into the ground or lowered by water jet.

Drilled or bored wells: Tubewells are the best and most common examples of drilled or bored wells. Different types of drilling methods can be used to construct these wells and are summarized here.[5] Augers or drill bits are used to drill into the ground and different types of augers are used depending on ground conditions, as shown in Figure 1.5. The excavated material is collected by the auger and raised to the ground surface where it is emptied before being lowered into the ground again. The helical auger allows excavated material to be lifted to the ground surface due to the spiral structure of the auger without raising the auger itself. Drilling or boring can be done manually, mechanically, or using animal power.

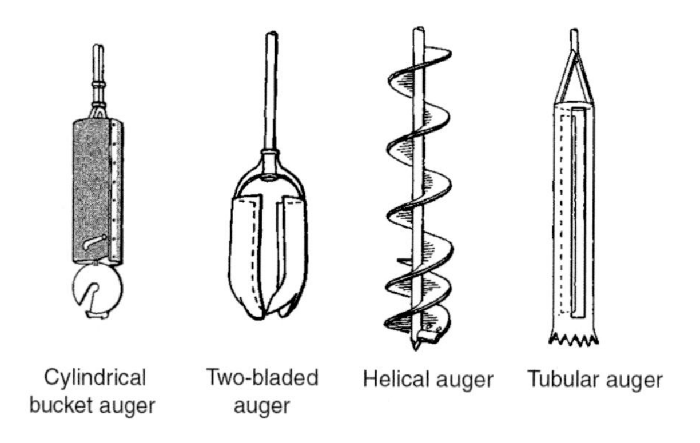

Figure 1.5 Earth augers used for constructing wells (FAO 2012).

Drilling methods

Percussion drilling: In this method, the drilled hole is kept full of water and a combination of mechanical and hydraulic processes are used for excavating the subsurface (FAO 2012). A chisel-

4 http://spanpumpsindia.com/wp-content/themes/betheme/product_pdf/india%20mark%20III.pdf.

5 http://www.fao.org/docrep/X5567E/x5567e06.htm#TopOfPage.

edged cutting bit is attached to the bottom of a string of drill pipe. The hollow bit has inlet ports a small distance above its cutting edge. During drilling, the drill pipe is alternately raised and dropped. Pressure due to the impact of the cutting bit in the bottom of the hole and the inertia of the water cause a mixture of water and cuttings to enter the inlet ports of the cutting bit. This causes the already full drill pipe to overflow. A check valve in the cutting bit prevents the mixture of water and cuttings from flowing out of the ports when the drill stem is raised. The cuttings may be settled out from the water in a pool or barrel after the mixture overflows from the drill pipe and the water can then be recycled. Hydraulic percussion is limited to drilling through relatively fine materials, since coarse materials will not rise to the surface through the drill pipe. This method has been used to drill to depths of more than 900 meters (3,000 feet) in alluvial areas where neither hard formations nor coarse materials were encountered.

The percussion method is versatile and allows all types of materials to be penetrated. However, in very hard stone, progress is slow. While this method is frequently associated with large, motorized, truck-mounted equipment, it can be successfully scaled down and used with manpower or small engines.

Core drilling: This method is used for drilling in rocky, hard formations using a rotating hollow pipe with a drill bit or cutter. The cut material rises in the hollow pipe as the cutter or bit advances and is removed periodically (Duggal 2007).

Rotary drilling: There are two types of rotary drilling methods – direct rotary and reverse rotary recirculation method (Duggal 2007). A drilling fluid (generally mud slurry) is pumped into a hollow pipe and forced out through apertures in the drill bit bottom. As more material is loosened, the slurry rises and is pumped out and discharged into a settling basin. This method is suitable for soft rock and unconsolidated formations up to depths of 150 m.

The reverse rotary recirculation method is a modification of the above where the drilling fluid is pumped into the outer annulus of the pipe carrying the drilling bit. The slurry of loosened material is pumped out from the hollow drilling pipe and discharged to a basin for settling. Large diameter bores of 600 mm can be prepared rapidly using this method.

1.1.3 Darcy's Law and Pumped Wells

When water is pumped from a well, regardless of whether it is tapping a confined or an unconfined aquifer, the water level in the well drops to a level below the local water table in an unconfined aquifer and below the piezometric surface if the well is tapping a confined aquifer. This drop in water level in the well is defined as the *drawdown* in the well. As shown in Figure 1.6, if the height of the water table is h, the height of water in the pumped well will be less than h during pumping, some value h_1. The hydraulic gradient created by pumping a well will define the *radius of influence*, R of the well as shown in the following figure. An observation (or monitoring) well or piezometer located on the perimeter of this circle will show no drawdown while all wells within the radius of influence will show some drawdown with respect to the piezometric surface as shown in the sectional elevation view of the well. The drawdown in these wells is inversely proportional to their distance from the pumped well, i.e., as the distance from the pumped well increases, the drawdown in the monitoring well decreases. Pumping of a well creates an inverted cone with its axis coincident to the central axis of the well; this cone is called the *cone of depression*.

Since water flows into a pumped well radially, Darcy's law has to be applied for this radial flow where the well is assumed to be a cylinder with radius r_w and therefore, the hydraulic gradient is dh/dr where h = the height of the piezometric surface at any location within the radius of influence of the well and r = distance of that location from the center of the pumped well.

Therefore, $Q/A = K(dh/dr)$

The surface area of a cylinder of any radius r within the radius of influence is $2\pi rh$,

therefore, $Q = K^* 2\pi rh\ (dh/dr)$

The above differential equation can be integrated for any two points, r_1 and r_2 corresponding to pressure heads, h_1 and h_2 within the radius of influence.

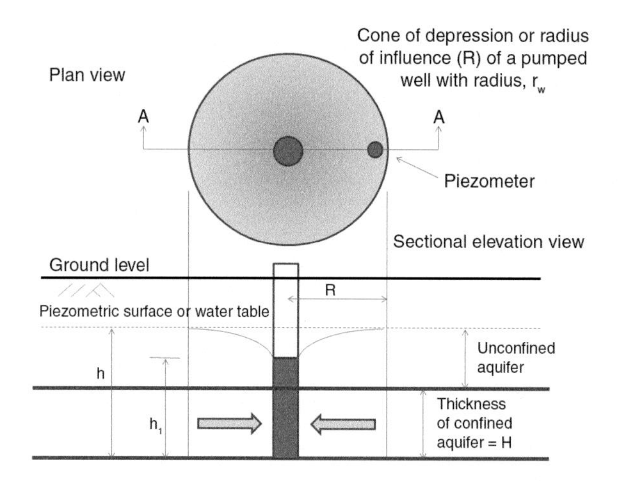

Figure 1.6 Radius of influence of a pumped well in a confined or unconfined aquifer (see color plate).

This results in the following equation for an unconfined aquifer (Masters and Ela 2008):

$$Q = \frac{\pi K(h_1^2 - h_2^2)}{\ln\left(\frac{r_1}{r_2}\right)} \qquad 1.1.3$$

For a confined aquifer, the derivation is similar to the one above except that the entire thickness of the aquifer (H) is available for pumping water and

$$Q/A = K(dh/dr) \text{ or } Q = K^*(2\pi rH)\ (dh/dr)$$

Integrating this differential equation leads to

$$Q = 2\pi KH\left[\frac{(h_1 - h_2)}{\ln\frac{r_1}{r_2}}\right] \qquad 1.1.4$$

PROBLEM 1.1.5

A confined aquifer of 50 m depth is pumped by a fully penetrating well at a steady rate of 4000 m^3/d for a long enough time to be at steady-state. Drawdown at an observation well 1 which is 20 m away from the pumped well is 5 m and drawdown at a second observation well 200 m away is 0.5 m. Find the hydraulic conductivity of the aquifer.

Solution

Q, Flow rate = K*A*(dh/dL) = 4000 m^3/d

Hydraulic gradient (dh/dL) = $(h_2 - h_1)/(L_2 - L_1)$

h_1 = elevation of piezometric surface in well 1 which is L_1 (20 m) away from pumped well = (50 – 5) = 45 m

h_2 = elevation of piezometric surface in well 2 which is L_2 (200 m) away from pumped well = (50 – 0.5) = 49.5 m

Therefore, hydraulic gradient (dh/dL) = (49.5 – 45)/(200 – 20) = 0.025

Using Equation 1.1.2 for Q results in

Q = 2π*50*K*(49.5 – 45)/[ln(200/20)] = 613.97K

Solving for K, given that Q = 4000m^3/d, results in hydraulic conductivity, K = 6.52 m/d

PROBLEM 1.1.6

Calculate discharge from a tubewell tapping a confined aquifer of 25 m thickness. Drawdowns at two observations wells W1 and W2 at distances of 10 and 100 m from the pumping well are 2.0 m and 0.2 m, respectively. Assume hydraulic conductivity (m/d) of the aquifer is 150 m/d.

Solution

For the two observation wells W1 and W2, corresponding heads h_1 and h_2 are calculated as:

h = 25 – 2 = 23 m and h_2 = 25 – 0.2 = 24.8 m.

H = 25 m, r_1 = 10 m and r_2 = 100 m

$Q = 2\pi KH(h_1 - h_2)/\ln(r_1/r_2) = 18419\ m^3/d$

PROBLEM 1.1.7

Calculate discharge from a tubewell 0.6 m diameter with a drawdown of 3.0 m in an unconfined water-bearing stratum 15 m thick. Permeability may be taken as 1500 L/d and radius of influence as 300 m.

Solution

Head at the well, h_1 = (15 – 3) = 12 m and head at the boundary of the radius of influence, h is 15 m. Corresponding radii are r_1 = 0.3 m for the pumped well and r = 300 m for the radius of influence.

$$Q = \pi K(h_1^2 - h^2)/\ln(r_1/r) = 55257.24 \text{ L/d} = 55.26 \text{ m}^3/\text{d}$$

PROBLEM 1.1.8

A shallow open well serves as the only water source for a small village. It is highly contaminated in terms of turbidity, coliforms, and organic matter. What measures can be suggested to improve this water source?

Solution

The main reasons for turbidity, coliforms and organic matter in the water are the open structure of the well, inputs of fecal matter (coliforms and organic matter) from birds and unhygienic practices in withdrawing and using well water. Organic matter and turbidity are also mainly due to transport of materials by wind, and leaf fall.

A few measures that can be taken to improve well water quality are:

a. Since the well is open, it first needs to be covered with a concrete slab.
b. A hand pump with a screen at the bottom can be used to extract water from the well.
c. The well water can be filtered through a fine fabric or sand filter.
d. Potassium permanganate tablets should be added to the drawn well water to remove coliforms and make the water potable.
e. If electricity and a submersible pump are available, the water from the well can be extracted into a basin and then chlorinated.
f. A small-scale treatment plant can be set up to remove turbidity.
g. Water quality analysis should be conducted regularly to ensure that the water meets acceptable drinking water standards provided in Appendix A.

1.1.4 Groundwater Pollution

GW pollution is a major concern where it is the main drinking water (DW) source. Pollution can occur due to infiltration or leaching of pollutants from surface or subsurface structures. These pollutants may be geogenic, i.e., natural in origin or anthropogenic. Examples of geogenic pollutants are arsenic [As], fluoride [F], and high total dissolved solids (TDS, measured as electrical conductivity). Major anthropogenic contaminants include nitrates, pesticides, petroleum derivatives, and heavy metals.

Pollution sources can be *point* or *non-point* sources. Point sources of pollution are small, localized areas where pollutants are discharged. Pollutant discharges from end-of-pipe, leaks and spills are examples of point sources. Non-point sources are widespread sources of pollution such as fertilizer and pesticide applications on fields.

The list of anthropogenic sources of contamination includes leaking underground storage tanks (LUSTs) that contain oil or petroleum products, inorganic compounds like nitrates which are formed when fertilizers are applied to agricultural fields, pesticide application, and various industrial activities.

When pollutants, especially anthropogenic ones, are released into groundwater they can travel with the groundwater, they can be adsorbed to aquifer solids and retarded in their flow, and/or they may undergo various physical, chemical and biological reactions.

Some processes that occur in the subsurface and need to be accounted for in understanding contaminant fate and transport are:

Diffusion which occurs in response to a concentration gradient, and

Dispersion which occurs due to velocity gradients or differences in flow velocities of discretized volumes of fluid, i.e., deviations from ideal laminar flow conditions.

Hydrodynamic dispersion includes dispersion and diffusion and results in the spreading out of a plume in all three directions thereby causing the smearing of plume boundaries and mixing of the contaminant in a greater volume. As flow conditions approach ideal plug flow (PF), effects of hydrodynamic dispersion are reduced and smearing of the plume boundary is reduced.

Tracer studies are generally conducted to compare the behavior of contaminants with respect to the behavior of groundwater.

Conservative contaminants or compounds like chloride flow at the same velocity as groundwater. A *conservative* compound is one whose chemical identity is not altered by changes in its conditions; these compounds are generally used as tracers in groundwater studies. Examples of commonly used tracers include chloride, nitrate, some dyes, or radioactive compounds like deuterium and tritium. Other contaminants that have a tendency to adsorb to aquifer solids do not travel at the speed of groundwater. This results in *retardation* of contaminant velocity relative to groundwater. Retardation factor, R, can be defined for various contaminants. For chloride, which is a conservative tracer, R is 1. R is much greater than 1 for contaminants like perchloroethylene (PCE), carbon tetrachloride (CTET), dichlorobenzene (DCB), hexachloroethene (HCE) (Masters 1998). R is not constant over time and based on field data appears to increase over time until steady-state (SS) values are reached (Masters 1998).

An example of the fate and transport of contaminants released from leaking underground storage tanks (LUSTs) or from pesticides from field applications is shown in Figure 1.7. Underground storage tanks containing oil, petroleum or other chemical compounds tend to leak due to corrosion or other structural damages. When the contents of the tank are released into the subsurface, they may volatilize, they may adsorb to the solids, or flow through the void spaces and enter the aquifer. These compounds may also be transformed chemically or biologically in the subsurface resulting in the generation of other compounds that may be more or less toxic than the parent or original compounds. Volatile compounds are released into the atmosphere where they can affect public health due to inhalation if they are toxic. The remaining compounds that are released into groundwater may be ingested along with the water or may remain in adsorbed form in the subsurface.

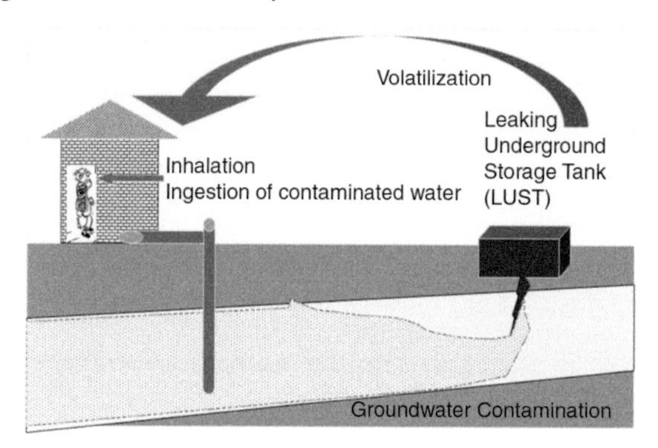

Figure 1.7 Groundwater contamination from leaking underground storage tanks (LUST) (see color plate).

PROBLEM 1.1.9

List the possible sources of pollution (anthropogenic) of groundwaters. Classify as point and non-point sources.

Solution

Some sources of groundwater pollution:

1. Urban runoff (non-point)
2. Underground storage tanks (point)
3. Septic systems (point)
4. Landfills (point or non-point depending on size of landfill)
5. Fertilizer applications (non-point)
6. Industrial facilities (point or non-point depending on size of facility)
7. Agricultural chemicals processing and handling facilities (point or non-point depending on size of facility)
8. Pipelines and sewers (point, assuming leakage at fixed points)
9. Stormwater drains (non-point if unlined)

PROBLEM 1.1.10

For Problem 1.1.1, oil was detected at well W3. A village depends on water from well W1. Assume a hydraulic conductivity of 6.51 m/d, and determine the Darcy velocity. Calculate the time it will take for the oil to show up in the village well. Assume a retardation factor of 2 for oil.

Solution

Darcy velocity v = K*(dh/dL) = 3.52 m/d

For a retardation factor of 2

Contaminant velocity = 1.76 m/d

1.2 SURFACE WATER

Any water body that is open to the atmosphere can be termed a surface water resource. These resources include rivers, streams, lakes, reservoirs, and ponds. Generally, surface waters originating in the hills or mountains are of high quality while those in the plains tend to have poorer quality, mainly due to the greater burden of silt and surface runoff. Treatment is always essential when water quality is poor or when there are significant variations in water quality. Heavy rains and runoff lead to high turbidity due to the high silt content leading to significant variations in water quality in surface water bodies. Surface waters are also prone to contamination from accidental spills or open discharges of untreated or partially treated municipal and industrial wastewaters.

Lakes and reservoirs often have less variable water quality than rivers and streams. Thermal stratification occurs in all surface water bodies and is discussed in detail in Chapter 4. It can lead to dissolved oxygen (DO) depletion in the lower layers resulting in massive fish kills after the ice thaws in temperate and colder regions. Anoxic or reducing conditions in the lower layers can lead to solubilization of metals like iron (Fe) and manganese (Mn). Taste and odor problems may increase due to anoxic conditions and lead to the release of gases like carbon dioxide, methane, ammonia, and hydrogen sulfide (H_2S). Thermal stratification and turnover (or complete mixing) happen only in lakes in temperate and colder regions. Turnover cannot occur in tropical regions with low elevations where surface water never freezes or its temperature never falls below 4 °C.

Eutrophication is another major concern for standing water bodies and leads to accelerated degradation in water quality over time. Algal blooms due to high carbonate concentration, nutrients and temperature can contribute to higher turbidity, alkalinity, taste and odor, pH and lower DO levels. These issues are addressed in Chapter 4. Water quality data for some of the major rivers, lakes and groundwater in India can be obtained from Central Pollution Control Board's (CPCB) website: www.cpcb.nic.in.

Rivers

Major *perennial rivers* of North India are either snow-fed or glacier-fed rivers, especially those of the Indo-Gangetic plains, while most major perennial rivers of the South, i.e., the Deccan peninsula, have their sources in artesian springs located in the hills. Examples of perennial rivers of North

India include the Ganga, Yamuna and Brahmaputra, all of which originate in the Himalayan region. Tributaries to these rivers include both perennial and non-perennial rivers. Major perennial rivers of the Deccan peninsula include Narmada, Krishna, Kaveri and Godavari.

Non-perennial rivers: Non-perennial rivers are mostly monsoon-dependent rivers that rely only on surface runoff from the catchment or watershed region. Generally, water does not last throughout the year in these rivers. Examples are Sabarmati in Gujarat, Ib – a tributary of the Mahanadi in Odisha, and many coastal rivers of Southern India.

The largest cities in India like Delhi and Kolkata derive their water supply from perennial rivers like Yamuna and Hooghly (a distributary of the Ganga), respectively. Mumbai and Chennai use water from reservoirs created to dam (store) monsoon flows of perennial or non-perennial rivers.

The principal advantage of using a perennial river as a source of water supply is the large quantity of water available for supply throughout the year. In general, river water quality deteriorates as the water moves from source to sink. Surface water tends to have higher concentrations of natural organic matter (dissolved solids) and suspended particles like clay and silt, compared to groundwater. This results in greater turbidity in surface waters as well. Surface waters are generally softer than groundwater, i.e., they have lower *hardness* due to lower concentrations of calcium (Ca) and magnesium (Mg). Ca and Mg are often found in very high concentrations in groundwaters.

Lakes and Ponds

Lakes and ponds generally have their sources in springs or surface runoff from their watersheds. Examples include Nakki Lake in Mount Abu, Lake Tamzey in Sikkim and Dal Lake in Kashmir. These are natural storage basins formed where there is low-lying land surrounded by hilly or mountainous terrain. Mumbai and New York are examples of mega-cities that depend on upland lakes for their water supplies. Toronto is another large metropolis that depends entirely on pumping water out of Lake Ontario for its needs. Water quality in these surface water bodies varies spatially and temporally.

Impounding Reservoirs

An impounding reservoir can be defined as an artificial lake created by the construction of a dam across a valley containing a water course (river or stream) (Duggal 2007). The objective is to impound or store a portion of the stream-flow so that it may be used for water supply over a long period of time. The reservoir essentially consists of three parts (i) a dam to hold back water, (ii) a spillway through which excess stream-flow can be discharged, and (iii) a gate chamber containing valves for regulating the flow of water from the reservoir. Examples of impounding reservoirs in India include Sardar Sarovar across River Narmada and Krishna Raja Sagara across River Kaveri.

1.2.1 Surface Water Intakes

Intakes are structures constructed in or for a surface water source that are used to withdraw water from the source. This water is subsequently discharged into an intake conduit, which conveys it to the water works system. All intakes consist of a conduit, screens with perforations to screen the influent, and gates and valves for flow regulation.

Many different intake structures are used for withdrawing water depending mainly on the nature of the water source. Some of the major intake structures that are used for withdrawing surface waters include:

1. Infiltration galleries or horizontal wells
2. Infiltration wells
3. River intakes
4. Reservoir intakes
5. Canal intakes

Important design considerations for these intake structures are listed in Table 1.4. Previously, design of intake structures was based on water supply considerations only. However, recent guidelines for maintaining environmental flows (e-flows)[6] now need to be included in the design of intake structures.

Table 1.4 Design considerations for water intake structures (based on Morris 2015)

Design criteria	For water supply	For aquatic habitat
Structural strength and stability	Ability to withstand floods and accidents	Structure should not lead to bank erosion or channel instability
Sediment accumulation and transport	Sediment should not accumulate at the intake (water quality should be maintainable over long-term)	Coarse materials should be transported and not inhibit e-flows
Migration paths	Not applicable	The intake should not impede migration of organisms in upstream or downstream direction, e.g., salmon
Minimum flows	Adequate withdrawals should be available during all seasons (water supply should be reliable)	Minimum in-stream or e-flows should be maintained to avoid disruption of ecosystems
Economic considerations	Should be cost-effective	Not applicable

1.2.1.1 *Infiltration Galleries or Horizontal Wells*

Infiltration galleries are horizontal wells constructed at shallow depths (6 to 9 m) along river banks (>15 m away) to collect clean river water after natural filtration through the river bank.[7]

6 E-flows are minimum flows to be maintained in terms of quantity, quality and timing in a river or stream to sustain freshwater and estuarine ecosystems and the human livelihoods and well-being that depend on these ecosystems (Wikipedia 2015).

7 In some cases noted in the literature, the water drawn maybe from the local aquifer adjacent to the river.

Wells are constructed in open cut with masonry walls and roof slabs and have widths of about 1 m, depths about 2 m, and lengths ranging from 10 to 100 m (Duggal 2007). Porous or perforated lateral drain pipes are used for collecting water as shown in Figure 1.8.

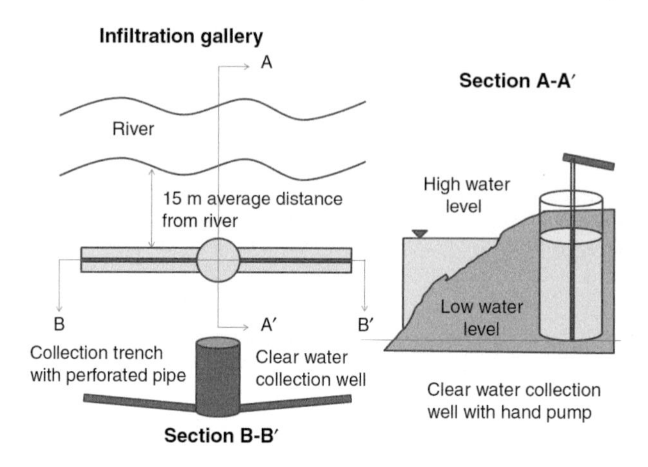

Figure 1.8 Infiltration gallery alongside a river.

Similar to the derivation for an unconfined aquifer, discharge from an infiltration gallery can be derived and results in the following equation (Garg 2001):

$$Q = K*L*(H^2 - h^2)/2R \tag{1.2.1}$$

where

Q = pumping rate or flow rate, 50 L/min

K = hydraulic conductivity, m^3/min

L = length of infiltration gallery, m

R = radius of influence of pumping well, m

H = depth of water in gallery, m

h = depth of water in pumped well = depth – drawdown, m

1.2.1.2 *Infiltration Wells*

Infiltration wells are shallow wells constructed in series along river banks to collect river water seeping through the bank soil or river bed (Garg 2001; Duggal 2007). The objective here is to obtain naturally filtered river water. These wells are constructed in brick masonry with open joints and are generally covered at the top and open at the bottom. In some cases, the well may have a porous plug at the bottom. In another modification, radial pipes with strainers are placed horizontally from the interior of a large jack well (3 – 6 m) and used to obtain naturally filtered river water from mid-stream (Duggal 2007).

Kharagpur city's water supply is provided for by infiltration wells tapping River Kangsabati. Major considerations in the design of these wells are to ensure the following:

1. The top of the well should be below the minimum water level so that water is available throughout the year. Also, the structure should not impede navigation and should be strong enough to withstand damage in case of collisions or high flood water levels.
2. Water in an infiltration well is river water and under dry conditions, it can be groundwater drawn from the river bed if the river bed is saturated with groundwater. (Refer Figure 1.9)

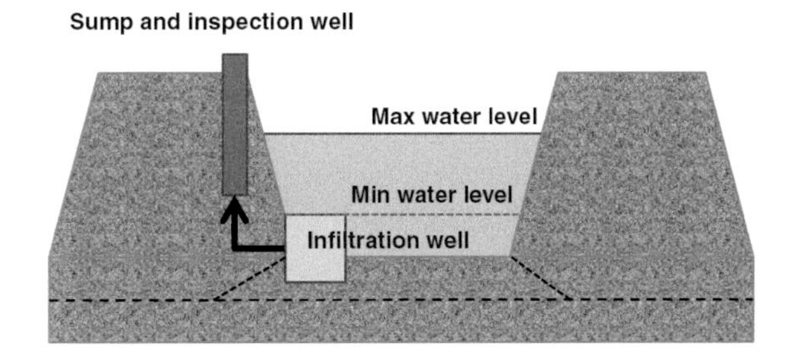

Figure 1.9 Infiltration well with base below river bed. Dashed, black line represents groundwater table and its mounding below the river bed.

1.2.1.3 *Reservoir Intakes*

These intakes are located within the body of a masonry dam or along the toe (upstream end) of an embankment dam (Duggal 2007). Several inlets are provided at different levels to take care of water level fluctuations. It is essential to ensure that inlet or intake pipes are provided below the lowest recorded water level to ensure year-round water supply. These intake pipes are covered with screens to prevent the entry of large debris or solids. The inlet pipes are connected to a single main pipe for collecting and conveying water. A supporting concrete structure or pump house is generally built around these pipes and therefore, no water is present inside the pump house. Thus, the intake can be conveniently inspected and operated from the pump house.

1.2.1.4 *Intake Towers*

Intake towers are constructed on rivers or reservoirs where there are large fluctuations in water levels. These concrete structures have openings called ports controlled by gates which also help in regulating the quality of water. A foot bridge usually connects the tower to the shore or dam. They are of two types (Duggal 2007):

(a) Wet Intake Towers

In these towers, there is a circular shell made of concrete and a vertical shaft connected to the intake conduit. Entry ports exist in the outer concrete shell and gate controlled ports in the shaft (which is directly connected to the intake conduit) to regulate the flow. The intake conduits may lie on the river bed or below the river bed. The water from the intake is taken directly to the treatment plant if it is located at a lower elevation or pumped to the treatment plant.

(b) *Dry Intake Towers*

It is similar to the wet intake towers except that in wet towers, water first enters the tower through the ports and then into the intake conduit through gated openings whereas in dry intake towers, gated entry ports directly draw water into the intake conduit. If the entry ports are closed in these towers, additional buoyant forces will act on the structure. So these intake structures must be made of heavy materials. They offer the advantage of drawing water from a specific level of the river or reservoir by adjusting the gates.

1.2.1.5 River Intakes

River intakes are simple in design and generally consist of a pipe with a perforated screen at the inlet end. The screen is essential for preventing the entry of large solids into the water supply system. A barrier with buoys or nets is provided around the screened pipe to prevent collisions with animals or vehicles like boats, etc. Major factors that should be accounted for in designing a river intake include:

1. Depth of intake: Intake should be located so that it is underwater throughout the year. Minimum and maximum depth of the river water should be monitored prior to designing a river intake.
2. Location of intake: For greater accessibility and to avoid obstructing navigation, intakes are usually located close to the bank while maintaining adequate flow and water levels at all times.
3. Water quality: Screens have to be provided to ensure that no large solids enter the system.
4. Reliability of water supply is important to ensure that adequate water is available throughout the year.
5. Structural strength: The intake should be sufficiently strong and well-constructed to ensure that it can withstand fluctuations in water velocities, changes in water quality, and accidental collisions.
6. Economical: The cost of constructing the intake and maintaining it is a major factor in determining its design.

A typical river intake has the following components:

(1) Inlet well

(2) Inlet pipe

(3) Jack well

The inlet well/collector well is usually circular and located away from the bank. Its location is chosen such that its top is above the river high flood level (HFL) and water supply is possible even during low flows. It is made of concrete or masonry and is connected to the river bank by a foot bridge. Bar screens regulate the entry of water into the intake wells. The intake pipe (made of reinforced cement concrete or RCC) connects the inlet well to the jack well/sump well. The pipe is of gravity type and is located well above the bottom of the intake well to prevent entry of silt. The jack well houses pumps (centrifugal or turbine type) which lift the water and feed it into the mains.

Other types of river intakes are:

(a) Cross-weir intake

In these intakes, a weir is built across the river. The rise in water level permits water to be drawn via a canal diversion as shown in Figure 1.10 or through a suction pipe which leads to the pumps. They are generally used in narrow rivers and where there are no significant fluctuations in water levels.

(b) Side weir intake

Side weir intakes are usually used where water is diverted from the mainstream flow. A chamber is built in the river with bar screen to prevent entry of floating debris. These intakes are cheap and easy to construct; the banks should be steep and stable and intake siting should ensure that even low flows can be collected.

Figure 1.10 Cross-weir intake with canal diversion. The water level is fixed by the weir height, i.e., some damming of water is possible in this structure. (a) Micro-hydroelectric power projects in Ererte, Ethiopia[8] (b) and in Gobecho,[9] Ethiopia (see color plate).

(c) Floating Pontoon intake

Where there are large fluctuations in river water levels, cross-weir intakes and side-weir intakes are uneconomical and of little use. In floating pontoon intakes, buoyancy tanks are used to help the intake to float. Other components of the intake include pumps and anchors/tethers. It should be anchored properly to withstand the impact of floods, boats, and floating debris.

(d) Piled Crib intake

It is generally constructed where there is little fluctuation in the water level of the river. A suction pipe from the pumps goes to a point in the river where a strong foundation is available. The pipe has a bell shaped mouth which is covered with an iron mesh. A wooden crib partially/wholly filled with riprap (rocks) surrounds and supports the pipe. The water from the suction pipe goes into wet wells and from there to the supply.

8 https://energypedia.info/wiki/Micro_Hydro_Power_(MHP)_-_Ethiopia,_Ererte.

9 https://energypedia.info/wiki/Micro_Hydro_Power_(MHP)_-_Ethiopia,_Gobecho_I.

1.2.1.6 Canal Intakes

It consists of a brick masonry chamber built on the bed and on a part of the canal bank. It has a side opening. A coarse screen covers the opening of the intake and prevents entry of silt and large solids. Entry into the conduit is through a fine screen and is regulated by the use of gate valves. As the intake obstructs a part of the canal, there is an increase in velocity and therefore, both the upstream and downstream ends must be protected against scouring.

PROBLEM 1.2.1

Design an infiltration gallery for a population of 5,000 people and a per capita demand of 200 L/d. A test gallery of length 6 m and radius of influence of 100 m was pumped. The depth of water in the gallery was 2.0 m, the pumping rate was 50 L/min, and the drawdown during pumping tests was 1.0 m.

Solution

Based on the derivation shown previously,

$Q = K*L*(H^2 - h^2)/ 2R$ where

Q = pumping rate or flow rate, 50 L/min

K = hydraulic conductivity, m^3/min

L = length of infiltration gallery, m = 6 m

R = radius of influence of pumping well, m = 100 m

H = depth of water in gallery, m = 2 m

h = depth of water in pumped well, m = depth –drawdown = 2 – 1 = 1 m

Therefore, $K = Q*2R/[L*(H^2 - h^2))] = 0.56$ m/min

Where Q required = 5,000 people x 200 L/person-day = 10,000 m^3/d = 0.69 m^3/min

Therefore,

Minimum length required = 83.33 m. Therefore, design length can be taken as 100 m

1.2.2 Surface Water Contamination

A large number of contaminants find their way into surface water bodies either from point sources or non-point sources. Point sources of contamination or pollution include wastewater discharges from municipal sewers or from industries and agricultural activities. Non-point sources are mainly surface runoff. Surface runoff from agricultural areas can include high concentrations of fertilizers and pesticides while those from urban areas include high concentrations of organic and inorganic contaminants. Examples of point and non-point discharges are shown in Figure 1.11.

For more than the last 100 years, wastewater discharge standards in most countries were based on the concentrations of individual contaminants or contaminant groups such as chemical oxygen demand (COD), biochemical oxygen demand (BOD) and suspended solids (SS).

In India, general and specific wastewater discharge standards apply to treated wastewaters. General discharge standards are provided in Appendix D while specific industrial discharge standards can be downloaded from CPCB's website. There has been a paradigm shift in the last couple of decades in countries like the US, where discharges of individual contaminants are now regulated based on the total mass loading of the contaminant to a river on a daily basis. Total maximum daily loads (TMDL) are calculated for individual watersheds and individual contaminants so as to regulate the maximum amount of pollutant being discharged to a water body while ensuring that water quality criteria are met.[10] This requires identification of major pollutant sources to a river or water body and reduction in the total amount of contaminant discharged by these major sources. Use based classification of surface water bodies is based on a few parameters and is provided in Appendix B.

Figure 1.11 A single point source (large drain hole) discharge of municipal wastewater and multiple drainage points (small drain holes) for discharge of surface runoff and infiltrate from the paved areas above the river embankment are clearly visible in this photo. Location: River Hooghly, Kolkata (see color plate).

PROBLEM 1.2.2

List the possible sources of pollution of surface waters. Classify as point and non-point sources.

Solution

Sources of surface water pollution:

1. Runoff from fields containing pesticides and insecticides or excess nutrients like N, P, and K (non-point)
2. Runoff from lawns/parking lots (non-point)
3. Untreated municipal wastewater discharges (point)
4. Municipal water and wastewater treatment plant discharges (point)

10 https://www.epa.gov/tmdl.

5. Runoff from dairy, poultry activities and animal feedlots (non-point or point depending on area)
6. Industrial wastewater discharges (point)
7. Sediments, debris from surface runoff (non-point)

PROBLEM 1.2.3

Many rivers in India are severely polluted due to multiple sources of pollutants being released into them. Major sources of pollutants are point discharges of municipal and industrial wastewaters and non-point discharges of surface runoff.

a. The following data were obtained from regular monitoring of one particular river: average annual flow 1 m^3/s; wastewater discharges constituted 80 percent of the total average flow in the river. The wastewater discharge standards for heavy metals like Hg and Cd are 0.01 and 2 mg/L for discharge to inland rivers (CPCB).[11] Calculate the pollutant load to the river during one year assuming the wastewater meets discharge standards.

b. Nutrients like nitrates and phosphates result in further degradation of river water quality. The average nitrate concentration in the wastewater discharges was 5.0 mg/L and that of phosphate was 1.0 mg/L. Determine the mass loading of nutrients in the river on a daily and yearly basis.

Solution

a. Average river flow = 1.00 m^3/s = 86400000.00 L/d

   Average wastewater flow = 0.80 m^3/s = 69120000.00 L/d

   Concentration of Hg in wastewater = 0.01 mg/L

   Concentration of Cd in wastewater = 2.00 mg/L

   Total load to the river in one day, Hg = 0.69 kg/d

   Total load to the river in one year, Hg = 252.29 kg/y

   Total load to the river in one day, Cd = 138.24 kg/d

   Total load to the river in one year, Cd = 50457.60 kg/y

b. Concentration of nitrates = 5.00 mg/L

   Concentration of phosphate = 1.00 mg/L

   Total mass of nitrates in river = 432.00 kg/d or in one year = 157680.00 kg/y

   Total mass of phosphates in river = 86.40 kg/d or in one year = 31536.00 kg/y

1.3 SOURCE WATER PROTECTION

A major issue in designing water supply systems especially for surface water bodies is to provide long-term source water protection so as to make the supply sustainable in terms of quantity and

11 http://cpcb.nic.in/.

quality. Effective source water protection programs help to reduce water treatment requirements and costs (AWWA 1990).

Popular measures usually taken for source water protection are listed below:

1. Prevention of wastewater discharges into surface water bodies is necessary for ensuring water quality for downstream users, i.e., riparian rights have to be protected.
2. In many countries, buffer zones are generally created around the water source. This is easily done around lakes, ponds and reservoirs where a zone with vegetation is maintained to ensure that surface runoff discharging into the source carries a minimal burden of soil, dirt, microbes, nutrients, and other contaminants. River water quality can be protected by collection and treatment of point and non-point sources of wastewater and creation of embankments that prevent direct discharges of wastewater to the river.
3. Further, sources are protected by preventing all anthropogenic activities that are likely to result in degradation of water quality. For example, areas around water sources are zoned as eco-sensitive in many countries and industrial and commercial activities are generally not allowed in such areas.

1.3.1 Groundwater Protection

Groundwater may be contaminated due to several reasons as mentioned in Problem 1.1.9. Source protection is based on eliminating sources of pollution and preventing direct discharges of contaminants on land or into surface water bodies. All non-point and point sources of pollutant discharges like urban runoff, return flows from agricultural fields which contain high levels of fertilizers, and pesticides should be collected and treated instead of being discharged over land. Septic tanks, cess pits, and open dumps of solid waste should be eliminated and replaced with properly engineered systems for collection, treatment and disposal of wastewater and solid waste. Stormwater drains should be lined in urban and semi-urban areas.

Table 1.5 Sources of groundwater pollution and possible remedial measures

S. No.	Sources of Pollution	Remedial Measures
1	Urban runoff	Collection, and treatment of discharges where possible
2	Leaking underground storage tanks (LUST), pipelines and sewers	Regular monitoring and repair of leaking tanks
3	Septic systems, cess pits, open dumps of solid waste	Properly engineered systems are needed in all cases to prevent infiltration of leachate into subsurface; such systems should be eliminated or replaced with better systems.
4	Stormwater drains	Should be lined in urban and semi-urban areas; water may require some treatment prior to discharge over land or into surface water body
5	Agricultural chemicals processing and handling facilities; fertilizer applications; industrial facilities	Collection and treatment prior to discharge over land or into surface water body

1.3.2 Surface Water Protection

The most important source protection measure for surface water bodies is elimination or reduction of all point discharges: municipal or industrial. This implies that all these discharges need to be treated to the level of relevant standards. Non-point sources of pollution should be collected and treated where possible, prior to discharge into surface water bodies. Use of chemicals like fertilizers, and pesticides should be done in a controlled manner and minimized as far as possible to avoid groundwater and surface water pollution. Urban runoff, including that from animal feedlots and similar activities, should be collected and treated to minimize entry of emerging contaminants like pesticides, pharmaceuticals like antibiotics, and heavy metals.

Study Outline

Hydrologic cycle

The movement of water in any of its three states: solid, liquid, or gaseous through different parts of the planetary environment is termed the **hydrologic cycle.**

It can be broadly defined in terms of six major processes:[12] precipitation, condensation, evapotranspiration, surface runoff, infiltration, and accumulation or collection.

Water resources

There are two types of water resources: surface water and groundwater.

Water withdrawals can be characterized by their uses or by their sources.

Groundwater

Groundwater is defined as water that has percolated downward (or infiltrated) from the ground surface through the soil and subsurface rocks.

Quantifying groundwater flow

Darcy's law states that flow per unit area, perpendicular to the direction of flow, is proportionate to the hydraulic gradient.

Hydraulic conductivity or *coefficient of permeability* may change with location and direction of flow in any aquifer.

Porosity (η, eta) is the ratio of void volume to total volume of aquifer material.

Design and construction of wells

Wells can be categorized based on:

Pressure head: gravity or artesian wells

12 http://cwc.nic.in/newwebsite_aug07/main/webpages/kids/welcome.html.

Depth of well: shallow wells such as dug wells or hand-pumped wells versus deep (tubewells)

Type of construction: percussion drilling, rotary drilling, and core drilling. Rotary drilling methods include direct rotary and reverse rotary recirculation method.

Groundwater pollution

GW pollution may be geogenic, i.e., natural in origin, or anthropogenic.

Examples of geogenic pollutants are arsenic [As], fluoride [F], and high total dissolved solids (TDS).

Major anthropogenic contaminants include nitrates, pesticides, petroleum derivatives, and heavy metals.

Surface water

Any water body that is open to the atmosphere can be defined as a surface water body.

Surface water intakes

Intakes are structures constructed in or for a surface water source which are used to withdraw water from the source.

Various types of intakes can be used for different types of surface water sources such as infiltration galleries and wells, reservoir and river intakes, wet or dry intake towers, and canal intakes.

Surface water contamination

Contaminants enter surface water bodies from point sources or non-point sources.

Point sources of contamination or pollution include wastewater discharges from municipal sewers or from industries and agricultural activities. Non-point sources are mainly surface runoff.

Wastewater discharge standards in most countries were based on the concentrations of individual contaminants or contaminant groups such as chemical oxygen demand (COD), biochemical oxygen demand (BOD), and suspended solids (SS).

A relatively new method for controlling pollution of surface water bodies is limiting the total maximum daily loads (TMDL) for individual contaminants and individual watersheds to ensure that water quality criteria are met.

Source water protection

Incorporation of source water protection programs is necessary for ensuring long-term sustainability of water resources in terms of quantity and quality.

Good source water protection programs also help to minimize treatment requirements and costs.

Study Questions

1. Identify the source of water in your community. Determine the methods used for withdrawing water from this source.
2. Differentiate between SW and GW resources.
3. Differentiate between confined and unconfined aquifers.
4. Define the following terms: saturated and unsaturated (vadose) zones in the subsurface, capillary fringe, aquitard, artesian well, peizometric surface, water table.
5. What are the different methods used for tapping groundwater?
6. State Darcy's law and provide a few examples of its application.
7. What is groundwater recharge?
8. How does groundwater get contaminated? Describe with at least one example.
9. When is surface water preferred to groundwater?
10. List different water intakes and where they can be used.
11. Why is source water protection essential in water resource management?

CHAPTER 2

Water Demand

Learning Objectives

- *Determine water requirements for different uses, communities, and future populations*
- *Define consumptive and non-consumptive uses of water*
- *Cite appropriate standards for different uses of water and determine minimum water requirements*
- *Describe how scale affects quantification of water use*
- *Describe differences in water use and supply in rural and urban communities in India*
- *Model available population data to forecast future populations in different communities*
- *Justify the use of different population forecasting methods for different conditions*
- *Quantify seasonal and daily variations in water demand for design of water supply systems*

Water is the most important requirement for sustaining life. Design of water supply systems requires that the water source be adequate in terms of quantity, quality and sustainability. Design of wastewater systems requires knowledge of water quantities used and treatment standards to be achieved. This chapter deals with the different uses of water and how much needs to be supplied at a minimum.

Figure 2.1 shows the flow of water through water supply and wastewater systems for any given location, e.g., a city, town, or small community. Water is extracted from a groundwater or surface water resource and is taken for treatment to a water treatment plant. The degree of treatment provided depends on the quality of the raw water. After treatment, the water is sent into the distribution system from where it reaches individual consumers. These consumers may be single households, apartment buildings, offices, industries or factories, commercial or business establishments, or any other place where clean water is required.

After use, the discarded water is termed 'wastewater' and in general, it goes into drains in sewered communities. The wastewater in sewerage systems is taken for treatment to a wastewater treatment plant. After treatment, the wastewater is discharged on land or in the sea, or it can be reused for horticulture, irrigation, industrial uses, and cleaning activities depending on the degree of treatment provided. Where drainage or sewerage systems are not present, the wastewater may be drained into cess pits, septic tanks, or reused after providing some treatment.

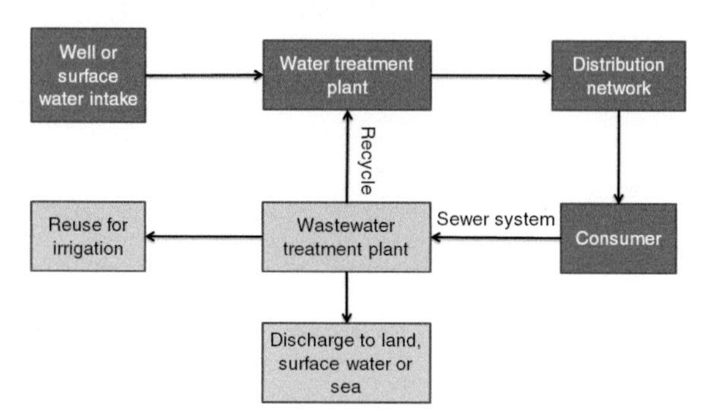

Figure 2.1 Flow of water through a water supply and wastewater system for any urban center (see color plate).

2.1 DESIGN OF WATER SUPPLY SYSTEMS

The most important task in the design of water supply systems is identification of an appropriate source of water. Some of the main criteria that are used to determine whether a water source should be developed include:

1. *Quantity*: A water source that can provide large quantities of water over a long period of time is a better source for water supply systems than a source with less quantity. Another major factor affecting water demand is the presence of sewerage systems. Unsewered communities, which are the norm in rural India, require much less water than sewered communities that are mostly urban.

2. *Quality*: If a water source is of better quality compared to another, cost of providing this water after treatment will be much less. Therefore, the better quality source becomes a better option for development. Quality requirements are also dependent on water use, i.e., not all water uses require the same quality of water. For example, drinking water use requires the highest water quality for health reasons; while gardening or horticultural uses of water do not require that the water should meet drinking water standards.

3. *Location of the source*: Source should preferably be located close to the consumers; this improves access and reduces costs of development, operation, and maintenance.

4. *Cost of development, collection and distribution*: Economic considerations are a major factor in determining which source can be developed for long-term supplies.

5. ***Sustainability*:** How long will the source last? What population size has to be served now and in the future? Will the quality of water be adequate over the long-term? These are the questions that need to be answered to determine the sustainability of a water supply.

Most systems are designed for a long-term design period which ranges from a minimum of 10 years to 30 or more years. Often, execution of systems design is in a phased manner if the design period is more than a decade. Current and future populations and their water requirements need to be estimated *a priori* for designing adequate and effective water supply and wastewater systems.

In a nutshell, the main steps in the design of any water supply system are:

1. Identify need (water demand) in terms of QUANTITY and QUALITY
2. Identify SOURCES that can fulfil water needs. Criteria for source selection include:
    a. Quantity
    b. Quality
    c. Location
    d. Cost of development, collection, and distribution
    e. Sustainability
3. Create and implement Source Protection Programs.

2.2 WATER USES AND REQUIREMENTS

As mentioned before, water can be abstracted, extracted or withdrawn from groundwater or surface water resources. Water withdrawals are often reported in the literature and refer to the amount of water extracted from a source. The water withdrawn is either consumed or returned after use and the water balance can be defined as: Withdrawals = consumption + returns (Masters 1998).

The schematic shown in Figure 2.2 refers to a surface water body like a river, but is equally applicable to all resources like lakes, ponds, reservoirs, or groundwater. The only difference between groundwater and surface water is that the return flow from a groundwater source may go to a surface water body or it can be used to recharge the local aquifer.

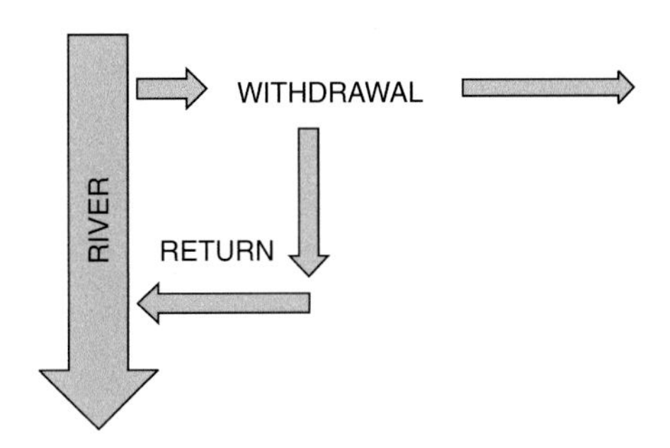

Figure 2.2 Water withdrawals and returns.

Withdrawal = water extracted from surface or groundwater bodies

Consumption (or consumptive uses) = water used but not returned. Examples of consumptive uses include drinking, cooking, evaporation, transpiration, and irrigation.

Returns (non-consumptive uses) = water returned to water body after use and available for future use. Examples include water used for cooling in thermal power plants, industries, water used for cleaning, washing, and bathing.

Besides consumptive and non-consumptive uses of water, water usage can be classified based on **offstream** or **instream** water uses. **Offstream** water uses include the withdrawal or diversion of water from a surface or groundwater source; quality is an issue in almost all off-stream water uses. Some examples of common off-stream water uses are listed below:

- ***Domestic and residential uses*:** These water uses include daily water requirements like drinking, cooking, bathing, cleaning, gardening, toilet use, and other domestic activities. Activities like gardening, drinking, and cooking are consumptive uses of water since there is no return flow; while cleaning, bathing, and toilet uses are non-consumptive uses of water where the return flow first goes to the drain and then to the sewer.
- ***Industrial uses*:** Different industries have very different water requirements and these are summarized in Table 2.1. Some industrial uses are consumptive where water is used in the industrial process/manufacture of products while those like cooling and house-keeping are non-consumptive uses of water.
- ***Agricultural or horticultural uses*:** Almost all agricultural uses of water are consumptive since there is little return flow from fields and gardens.
- ***Energy development uses*:** Large amounts of water are used in thermal power plants. As shown in Table 2.1, they are the single largest industrial users of water especially for steam generation and as cooling water. These uses are mainly non-consumptive and the only consumptive uses are for house-keeping use.

Table 2.1 Water use and wastewater generation by different industries in India, 2004[13]

Industry	Annual water consumption, million m^3	Water consumption, %
Thermal power plants	35,157	87.86
Engineering	2,020	5.05
Pulp and paper	906	2.26
Textiles	830	2.07
Steel	517	1.29
Sugar	195	0.49
Fertilizer	74	0.18
Others	314	0.78
Total	40,013	100.00

13 http://www.idfc.com/pdf/report/2011/Chp-18-Industrial-Water-Demand-in-India-Challenges.pdf.

Instream water uses are those which do not require diversion or withdrawal from the water sources such as:

- Habitat maintenance and improvement, i.e., eco-conservation or maintenance of environmental flows (e-flows) are all instream uses of water. Maintaining e-flows is now mandatory for all major water bodies to ensure that there is sufficient water in the stream year-round for aquatic systems to survive and thrive. Maintenance of good water quality is a major issue for these uses.
- Recreational uses: Aquatic systems are often the focal point of recreational, tourist, religious, and various other cultural activities. In general, these activities do not require extraction of water and are considered instream water uses. These activities are heavily dependent on maintenance of good water quality.
- Navigation: Flowing streams serve as transportation channels and navigation is a major service provided by surface water bodies. Water quality is generally not considered a major issue for these uses.
- Fisheries and aquaculture: A major commercial activity or instream use of water is for production of fish and other aquatic organisms. Water quality is a major issue and any damage to water quality can bring these commercial activities to an end.
- Hydroelectric power production is one of the most important instream water uses. Hydro-electric power is one of the major energy sources in countries like India as shown in Figure 2.3. Approximately 14 percent of the installed electricity generation capacity in India is derived from hydro-electric power (Central Electricity Authority 2017).

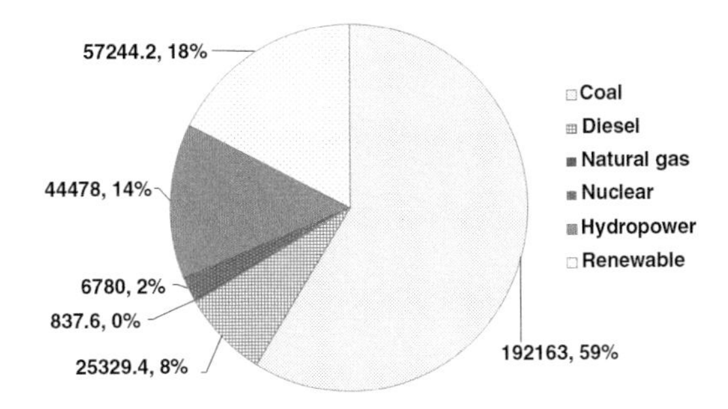

Figure 2.3 Total installed electricity generation capacity in India (as of 31 March 2017).

In summary, quality is dependent on water use, i.e., not all water uses require the same water quality. Water use depends on the nature of the activity for which it is used, location, and many other factors.

2.2.1 Factors Affecting Water Use

Several factors affect water use and are discussed here in brief:

1. **Population:** Total water consumption is directly proportionate to population size and is determined by multiplying the per capita consumption by the total population, i.e., Q = total water to be supplied = P*Per capita water demand (L/person-day), where P = population. Per capita requirements are also related to population size and are discussed in Sections 2.3 and 2.4. As the total population increases, per capita water consumption also increases due to increase in water demand for common civic services like fire-fighting, cleaning and recreation.
2. **Nature of activities:** Water requirements vary with activities such as residential, commercial, industrial and so on. Indian standards for water requirements for major activities like factories, hospitals, hostels, schools and offices are listed in Table 2.2.

Table 2.2 Minimum water requirements for different activities (buildings)[14]

S. No.	Type of activity or building	Minimum requirement in Lpcd (liters per capita day)
1	Factories where bathrooms are required to be provided	45 per head
	Factories where no bathrooms are required to be provided	30 per head
2	Hospitals (including laundry) per bed	
	No. of beds not exceeding 100	340 per head
	No. of beds exceeding 100	455 per head
	Nurses homes and medical quarters	135 per head
3	Hostels	135 per head
4	Offices	45 per head
5	Restaurants	70 per seat
6	Hotels	180 per bed
7	Cinemas, concert halls and theatre	15 per seat
8	Schools	
	Day schools	45 per head
	Boarding schools	135 per head
9	Railway and bus stations	45 per head
	where bathing facilities are provided	70 per head
	where no bathing facilities are provided	25 per head
10	International and domestic airports	70 per head

14 IS 1172: 1993.

3. **Climatic conditions:** Precipitation, evaporation losses, requirements for gardening, and cleaning, can fluctuate with season and location which are climatic factors. For example, people often bathe two times a day in summer but not in winter. Water requirements for gardening are higher in summer compared to the monsoon season when none is required.

4. **Water supply and sewerage systems:** Another major factor is provision of sewerage systems. Rural areas in many countries including India do not have sewerage systems and per capita water consumption is drastically lower in these areas than in urban centers or areas with sewered communities. Further, modern 'flush' systems consume much higher amounts of water than systems without flush or sewerage systems which result in extremely low water use. Minimum requirements for different communities, with and without sewerage systems, are listed in Table 2.3.

Table 2.3 Minimum water supply required for different communities[15]

Population	Minimum water supply, Lpcd
<20,000 without flushing system and supply through standpost	40
<20,000 without flushing system and supply through house service connection	70 to 100
20,000 to 1,00,000 with full flushing system	100 to 150
>1,00,000 with full flushing system	150 to 200

5. **Economic conditions:** In general, as economic conditions improve, there is greater development of infrastructure and resource consumption. For example, average per capita water consumption in poor countries is much less than that in richer countries. Water use in India is compared with that in the USA in Table 2.4. The same principle can be applied at the individual level where people in higher income groups use more water than people in lower income groups. Per capita consumption of water was found to be proportionate to income in a study of metropolitan cities in India (Shaban and Sharma 2007). Per capita consumption in high income groups was found to be 100 Lpcd while that in slum areas was found to 81.9 Lpcd.

6. **Cultural conditions:** Significant quantities of water are used for cultural or religious or ritual activities.

15 IS 1172: 1993.

Table 2.4 Minimum water requirements in India and USA

Use	IS 1172-1983, Liter/capita-day	IS 1172-1993, Liter/capita-day	AWWA 1999, US Gallon/capita-day
Drinking	4.5	5	
Cooking	49.5	5 (2.5 %)	
Bathing		75 (37.5 %)	12.8 (18.5 %)
Washing clothes	22.5	25	15 (21.7 %)
Washing utensils		15	10 (14 %)
Cleaning homes		15	10.9 (15.7 %)
Gardening		15	
Flushing toilets	22.5	45	18.5 (26.7 %)
Total	100	200	67.2 (100 %)

7. **Nature of industries and commerce:** Major water users include large-scale industries and commercial enterprises. A list of all major industrial water users in India is provided in Table 2.1 along with their estimated water consumption. Major industrial users of water in India are the thermal power plants followed by the engineering sector, pulp and paper mills, and textile mills. In a survey of per capita water use conducted in the big cities of India, the highest per capita water use was observed for Jamshedpur at 203 Lpcd (Asian Development Bank 2007). The most probable reason for the higher than average per capita water consumption in Jamshedpur is the presence of a large iron and steel manufacturing industry.
8. **Environmental protection and quality issues:** Uses of water for public health and improving environmental quality, greater public awareness, and demand for higher quality water are factors that also influence water usage.
9. **Conservation** and reuse of available resources is another factor that has gained importance in recent years. Water conservation practices like use of 'no water' flush toilets or small flush tanks of 10 L capacity rather than old flush tanks of 40 L capacity are examples of water conservation measures. Use of grey water from kitchens and other cleaning areas for use in horticulture is another popular example of water conservation practices [Note: Detergents and high salt concentrations in grey water can be detrimental to the growth of some plants]. Rain water harvesting is another conservation measure that has gained popularity in recent times.
10. **Cost of water:** In many areas of the world, and more so in Indian cities, a water fee or tax is charged to the consumer and is often part of the annual property tax or rental fees. In most cases it is a flat or fixed charge. Water usage is affected by the method of charging for water and the price of water. This is often a major issue for industries and commercial enterprises (FICCI 2011). As the cost of water increases, water usage decreases.
11. **Method of charging or billing:** A water fee or tax is often imposed either as a flat fee or based on the amount of water used if water meters are available. In metered systems, if rate

or price is significant, water conservation by consumers is generally higher. With flat or fixed rates, the consumer has no incentive for conserving water, and wastage is higher!

12. **Pressure in distribution system:** As pressure increases, quantity of flow in the system increases and usage is higher. Further, losses due to leaks also increase.
13. **Supply period:** Water is supplied either intermittently or continuously. Intermittent supply systems are common in countries like India (Asian Development Bank 2007) whereas continuous supply systems are generally used in developed countries like the US.
    a. **Intermittent:** Losses are higher even if the consumer is conservation-conscious. For example, taps are left open inadvertently, or water is stored in excess of need and thrown when service is resumed.
    b. **Continuous:** Losses are less even though net consumption may be higher. Only 17.9 percent of the 2734 metropolitan households surveyed in 7 Indian cities were getting 24 hour water supply (Shaban and Sharma 2007). The remaining households had intermittent water supplies.

    Domestic water consumption depends on adequacy of water supply and not on the supply period (Andey and Kelkar 2009). Intermittent water supply resulted in lower water consumption in two of the four cities surveyed, and was less than that consumed under continuous water supply schemes. However, where water demands were met adequately, i.e., supply under both schemes was approximately 500 Lpcd, there were no differences in water consumption in all four cities.

Water consumption in Indian cities ranges from 75 Lpcd in Bhopal to 203 Lpcd in Jamshedpur (Asian Development Bank 2007). Domestic water consumption alone was found to range from 77 Lpcd in Hyderabad to 115.6 Lpcd in Kolkata in a survey of 7 Indian cities (Shaban and Sharma 2007). Only 65.4 percent of the total 2734 households surveyed in this study were getting adequate water, i.e., >100 Lpcd.

2.2.2 Calculating Water Demand

Water demand is the amount of water to be supplied to a community (this can be a village, town or city). Water demand can be calculated for a community using the following thumb rules (Duggal 2007):

- Domestic requirement – 50 percent of total
- Industrial/commercial requirement – 20 to 25 percent of total
- Public uses – 10 percent of total
- Losses and leaks – 15 percent of total
- Fire demand is to be calculated based on empirical formulae provided

Besides providing adequate quantities of water, water supply systems have to be designed to provide appropriate flow rates as well. The water supply system should be designed to provide 'coincident draft' which is the water required during a day of maximum demand and a coincidental fire event. Design flow rate is calculated based on the following formula where F = fire demand:

Total flow required in the system = Q_{max} (daily) + F

Fire-flow requirements have two important aspects: flow rate and duration of flow to be provided. This is additional storage required in case of a conflagration when high flow rates are needed, while the net quantity used may be small depending on the duration of use. Storage for fire-fighting can range from a minimum of 2 h up to 24 h.

Various formulae are available for calculating fire demand. Most formulae are based on population size or nature of construction materials used in the building (Duggal 2007; Garg 2001; Sincero and Sincero 1996):

$$F = 3.7*10^{-3}*C*A^{0.5}, \text{ where}$$

F is the flow rate for fire-fighting (m^3/s), C is a dimensionless coefficient related to type of construction material: 0.8 for non-combustible, 0.6 for fire-resistive construction, A is total floor area excluding the basement (m^2).

Formulae based on population size include the following where Q = fire demand in L/min (lpm) unless stated otherwise and P = population in thousands.

1. National Board of Fire Underwriter's Formula:

   $$Q = 1020\ P^{0.5}\ (1-0.01P^{0.5}) \text{ (in US Gallons/min)}$$

2. Kuichling formula:

   $$Q = 3182\ P^{0.5}$$

3. Ministry of Urban Development, India formula: $Q = 100*P^{0.5}$
4. Freeman formula:

   $$Q = 1136.5\ (P/5 + 10)$$

5. Indian Standard (IS 9668-2010) formula: 1800 Lpm is to be provided for every 50,000 population and an additional 1800 Lpm for each 1 lakh population that is greater than 3 lakhs. For towns with population < 1 lakh, total requirement should be doubled. The fire reserve or storage should be maintained for a minimum of 2 h. In high-risk areas or where there are high-rise buildings, there should be a static tank of 2,20,000 liters capacity for every 1 km^2 area.

2.2.3 Factors Affecting Water Losses

A large amount of the water supplied that is lost is categorized as 'unaccounted for water'. This includes leaks and losses from the system for various reasons that are described below:

- **Leakage at joints and corrosion of pipes**
- **Pressure in distribution systems:** higher pressure leads to higher losses during leakage
- **Type of supply:** Intermittent versus continuous water supply systems; intermittent supply leads to fewer leakage losses. However, there is more wastage of water in many cases.
- **Metering:** Unaccounted water loss is easy to monitor where water supplies are metered since leaks can be detected and fixed. However, the cost of metering can be prohibitive and in general, water is not metered in India.

- **Unauthorized connections** are reduced where supply is metered; it is easy to detect illegal connections where supply is metered.
- **Negligence:** Negligent activities like leaving the taps open when not in use also leads to losses. These activities are also common in the absence of metering. Where water is metered, losses due to negligence are lower. Lack of awareness among people about water conservation leads to greater losses as well.

The percentage of water supplied that remains unaccounted for in India ranges from 12 to 60 percent in large Indian cities (Asian Development Bank 2007).

PROBLEM 2.1.1

A. List all possible uses of water and classify them as consumptive or non-consumptive.

B. Similarly list all possible domestic uses of water and classify them as consumptive and non-consumptive.

Solution

	Overall uses of water	
1	Drinking, cooking, gardening	Consumptive
2	Washing clothes/utensils, toilet flushing, bathing	Non-consumptive
3	Fire-fighting	Consumptive
4	Public uses, e.g., fountains, parks and gardens, public toilets and washrooms, street cleaning, etc.	Both
5	Industries/factories	Both
6	Hydroelectric power generation	Non-consumptive
7	Recreation	Non-consumptive
8	Navigation	Non-consumptive
9	Fisheries, eco-preservation	Non-consumptive
10	Institutional and commercial cleaning	Non-consumptive

Domestic Uses of Water	
Drinking	Consumptive
Washing clothes	Non-consumptive
Washing utensils	Non-consumptive
Cooking	Consumptive
Bathing	Non-consumptive
Flushing toilets	Non-consumptive
Gardening	Consumptive
Culturing aquatic life (aquariums)	Non-consumptive

PROBLEM 2.1.2

What standards are applicable in determining acceptable drinking water quality? Should all water treatment plants meet the same standards?

Solution

Standards applicable in India for acceptable drinking water quality are provided in IS 10500, 2012. All water treatment plants do not have to meet the same standards since the purpose of treatment may not always be drinking water supply. Water treatment standards applicable for industries, irrigation, and other uses will be necessarily different from those for drinking water.

PROBLEM 2.1.3

A new campus community is to be set up for an initial student population of 5000 students and 500 faculty members, with 1000 supporting staff (assume an average family size of four). The community will be expanded in the future. It needs to have the following types of buildings: residences, offices, classrooms, hospital (one of 100 bed capacity), hostels (10 buildings of 500 capacity), restaurants (10 with 50 seats each), shops (50), schools (6 from nursery to class 12), and other establishments like banks, post office, etc. Determine the immediate water requirements (minimum based on IS: 1172-1993, 1983) for this campus. Keep in mind additional requirements for leakage, fire demand, and horticulture.

Solution

S No	Description	Requirement per head/unit (liters/day)	Total Units/ person	Total Requirement (liters/ day)	Explanation
1	Hostels	135	5,000	6,75,000	Total 10 hostels, 500 students per hostel
2	Restaurants	70	500	35,000	10 restaurants each with a capacity of 50 seats
3	Hospital	340	100	34,000	1 hospital with 100 beds
4	Schools	45	3,000	1,35,000	Assuming a family size of four with each family having 2 children. Total 1,500 families and hence 3,000 children
5	Residences	135	6,000	8,10,000	Total 1,500 families and therefore 6,000 residents (4 residents per family)

(*Contd.*)

S No	Description	Requirement per head/unit (liters/day)	Total Units/ person	Total Requirement (liters/ day)	Explanation
6	Offices	45	1,500	67,500	Offices for 1,500 people (500 faculty and 1,000 supporting staff)
7	Classrooms	45	5,000	2,25,000	For 5,000 students living in the campus; requirement taken equal to that for schools
8	Fire demand	3,600 lpm	2 × 60 = 120 min	4,32,000	As per IS: 9668-2010, fire demand for population < 50,000 = 3,600 lpm to be maintained for minimum two hours
			Total	24,13,500	

However, we have to account for losses (due to bad plumbing, leaky mains, unaccounted for connections, etc.) and water requirements for horticulture. Assuming 15 percent and 25 percent of the total water requirement for losses and horticulture respectively, **24,13,500** is 60 percent of the total water requirement. Thus, the total water requirement is 24,13,500 x 100/60 = **40,22,500 liters/day ≈ 4.0 MLD.**

PROBLEM 2.1.4

An underdeveloped village with the same population, i.e., greater than 11,000, versus the planned community mentioned in the above problem will have different water needs. What is the difference in requirement and why?

Solution

Water requirement for the planned community as calculated in the above problem

= **4.0 MLD**

The population used for the planned community was 5,000 + 6,000 = 11,000.

The water requirement for underdeveloped village of population 11,000 may be taken as 80 Lpcd. Therefore, total water requirement is:

80 × 11,000 = 8,80,000 liters/day ≈ **0.9 MLD**

For an underdeveloped village, we have considered only the domestic water requirement.

Hence, the difference is 4.0 M – 0.9 M = **3.1 MLD**

The difference in requirement is due to absence of sewerage systems, fire demand and losses are not included, there is an absence of industrial/commercial buildings and the standard of living of the rural community is lower compared to the campus community.

PROBLEM 2.1.5

A general assumption in India is that intermittent water supplies (IWS) result in lower water consumption compared to continuous water supplies (CWS). In a study in 4 Indian cities, water consumption under two supply schemes was monitored and the data are provided in a paper by Andey and Kelkar (2009). What can you conclude from these data sets?

Solution

1. The assumption holds true for only one of the 4 cities – Ghaziabad. A 30 percent increase was observed under CWS since demand was not satisfied under IWS.
2. Jaipur has adequate storage facilities while Nagpur has long hours of IWS (16 h, close to CWS) and switching to CWS resulted in no significant difference in usage in these two cities.
3. Panaji, similarly, showed no significant differences in water consumption under the two schemes.

PROBLEM 2.1.6

For a city with 10 lakhs population, determine the fire demand to be provided based on Indian standards (IS 9668-2010).

Solution

Since 1800 liters/min (lpm) are to be provided for every 50,000 population, and an additional 1800 lpm for every one lakh above 3 lakhs population, the total fire demand is

$$1800*10*10^5/50000 + (10 - 3)*1800 = 27 * 1800 = 48{,}600 \text{ liters/min}$$

This water should be available for at least 2 hours, therefore

$$48{,}600 \text{ lpm}*60 \text{ min/h } *2 \text{ h} = 5.832 \text{ million liters}$$

2.3 CITY-LEVEL AND HIGHER SCALES OF WATER USE

Water demand or use is a function of the scale at which the calculations are done. This can be illustrated by examining water use at the individual level, household level, institutional level, regional or national levels as shown in Table 2.5. For example, personal water use by an individual is less than that used by the household since many activities in the home are common activities and not just personal. These are generally categorized as 'domestic' or residential water use. At the city level, additional water uses would include municipal services like cleaning of public areas, and fire-fighting, industrial and commercial activities, civic requirements for cleaning, aesthetic or recreational activities. Similarly, moving from the city level to the state or regional level, water uses will increase to further include large-scale industrial and commercial activities, and irrigation.

Table 2.5 Per capita water requirement in India and scale of use

Scale	Estimated per capita requirement, L/capita-day or Lpcd	Remarks
Individual	100	Personal requirements only
Households	150 to 200	Additional uses like cleaning of common areas, gardening, cooking, and washing
City or community level	200 to 500 or more	Additional uses like recreational areas around water bodies, fountains, street cleaning, horticulture, fairs and festivals, and fire demand
Regional or national	>1,000	Additional uses like irrigation, and energy development

If only domestic or residential water uses are computed, per capita water usage is 262 Lpcd in the US and 135 to 200 Lpcd in India. Per capita water use in developed versus developing countries can be compared at the national level as noted below.[16] Here the US is representative of developed countries while India is representative of developing countries.

Developed (e.g., US) = 1280 gal/cap-d = 1280 x 3.785 L/US gallon = 4845 L/cap-d

Developing (e.g., India) = 609 m^3/cap-year = 1669 L/cap-d.

Table 2.6 provides estimates of water requirements in India for different water uses or sectors at the national level.

Table 2.6 Estimated water requirements in India for 2010[17]

Sectors or uses of water	Water requirements, BCM[18]	Percent use
Irrigation	550	78.35
Domestic	42.5	6.04
Industries	37	5.27
Power	18.5	2.63
Navigation	7	1.00
Ecology	5	0.71
Evaporation losses	42	5.98
Total	702	100.00

16 Based on USGS estimates and National Commission for Integrated Water Resources Development Plan, India, 1999.

17 Based on (UNICEF 2013) in National Commission for Integrated Water Resources Development Plan, 1999.

18 BCM = billion cubic meters.

2.4 URBAN AND RURAL WATER SUPPLIES IN INDIA

Full-scale water supply to individual homes is almost entirely an urban phenomenon. Most rural areas in India have community systems (public water supply systems) if any. Examples include common open wells where water is drawn by a pulley and bucket system, or hand pumps. In some cases, where rural communities can afford it, they have river and canal intakes and pumps to deliver water to their homes and fields.

Non-community systems are privately controlled water supply systems like private wells, borewells, lakes, and ponds. Educational institutions with residential campuses and industrial estates are examples of non-community water supply systems since they make their own water supply arrangements.

IIT Kharagpur's campus lies within the municipal limits of Kharagpur city. IIT Kharagpur has its own arrangement for transporting water from River Kangsabati or Kasai (about 5 km away from campus). Campus supply is 9 lakh gallons (UK gallons) a day which is approximately 4 MLD. If the number of residents is estimated to be 15,000 to 20,000 in 2018, per capita water consumption is about 200 to 267 Lpcd.

2.5 POPULATION FORECASTING[19]

Design of any water supply system requires an estimate of the population to be served at present and the population to be served during its intended period of service. Hence, forecasting future populations is necessary and generally, the first step in the design of any water or wastewater treatment scheme.

Forecasting of the future population of any city is done by statistical analysis of the pattern of population growth of the city considering future possibilities of economic development, migration and other related factors. Where no population data for a city exist, population can be forecast based on studies of other cities with similar features.

Various methods for forecasting population of a city or a region are available. A few are briefly explained here.

19 Forecasting is based on an analysis of past and present data to predict future outcomes while prediction may or may not be based on an analysis of the past. All forecasts are predictions but all predictions are not forecasts.

2.5.1 Arithmetic Progression Method

This method is used when population is found to increase by a fixed amount each year such as in areas where development is planned or controlled. Examples would include residential academic campuses, or industrial estates. Population at the end of *n* increments (increments of time may be annual or decadal or any other specified period of time) can be calculated as follows (Garg 2001):

$$P = P_0 + n*X \qquad 2.5.1$$

P = Population at the end of *n* increments of time

P_0 = Population at time $t = 0$

X = fixed increment in population, generally annual or decadal (fixed amount or constant, not a variable)

2.5.2 Exponential Method

This method assumes exponential growth or a first-order rate of increase in population. The population may be forecast based on the following equation:

$$P = P_0 e^{r*t} \qquad 2.5.2$$

where P is the present population, r is the annual population growth rate (1/unit time or in this case, 1/year) and P is the population at the end of *t* years. The exponential growth model is often the best-fit for populations of developing countries like India.

2.5.3 Geometric Progression Method

This method is similar to the exponential method, except that it is not a continuous function like the exponential function but a discrete function. The assumption is that annual percentage growth is a constant that occurs over equal time periods. The population at the end of *t* years (or decades) may be written as

$$P = P_0 (1 + r/100)^t \qquad 2.5.3$$

Where P_0 is the population at time $t = 0$ and P is the population after t years with *r* being the annual (or decadal or other increments of time) percentage rate of increase per year (or per decade), i.e., geometric mean of the percent increase in population.

2.5.4 Incremental Increase Method

In this method, two terms are added to the last known population: a) last increment in population; and b) average of all increments during the period of data available. All methods are illustrated with a single set of data shown in Table 2.7.

The equation used to forecast future population is

$$P = P_0 + nX + \{n(n+1)/2\}*Y \qquad 2.5.4$$

where n = number of increments in population data

X = average increase in population for all increments

Y = average incremental increase in population

2.5.5 Logistic Curve Method

The logistic curve method is based on the assumption of exponential growth, where the rate of exponential growth is maximum initially (R_0) and slows down as the population approaches the carrying capacity of its environment. The curve is S-shaped and is called a logistic curve. The logistic curve is generally accepted as the best description of the growth of natural populations and is used in many fields such as ecology, microbiology, and population demographics. However, it is important to remember that human societies do not always grow naturally since policy interventions, natural catastrophes and wars can destroy natural patterns of growth.

The difference between an exponential growth equation and a logistic growth equation is that in the first case, growth rate (*dP/dt*) increases as population increases while in the second case, the growth rate decreases as the population increases. Mathematically, the differential form of the logistic equation can be stated as:

$$\frac{dP}{dt} = rP\left[1 - \frac{P}{P_s}\right] \qquad 2.5.5$$

Where r = exponential growth rate when the population size is far below P_s which is the population at saturation, i.e., the carrying capacity of the environment. The factor in parenthesis is also known as *environmental resistance*.

Unlike exponential growth where r is assumed to be constant, r is not constant in logistic growth. Therefore, two additional terms are defined for t = 0: R_0 which is the initial, instantaneous growth rate and P_0 which is the initial population. These are used to compute r at any time t. The solution to the above equation is:

$$P = P_s/\{1 + b^*\exp[-R_0^*t]\} \qquad 2.5.6$$

$$R_0 = r^*\,[1 - P_0/P_s] \text{ implies } r = R_0/[1 - P_0/P_s] \text{ and } b = (P_s - P_0)/P_0$$

If at $t = t_0$, $P = P_0$, then $t = t^*$ when $P = P_s/2$, the above equation can be solved for P and t^*:

$$P = P_s/\{1 + \exp[-r(t - t^*)]\}$$

$$t^* = 1/r \,\ln\left[\frac{P_S}{P_0} - 1\right] \qquad 2.5.7$$

The logistic growth model applies only to those populations where the growth rate is maximum in the beginning and decreases monotonically over an extended period of time, i.e., the population is growing exponentially at all times but with a decreasing growth rate as P approaches P_s. The graph shown in Problem 2.5.5 is a logistic growth curve. There is no time lag or lag phase in these populations. **Note: The initial part of the 'S' curve is not indicative of a lag phase, it is the nature of the logistic growth curve. This is the fastest growth phase where r is at its maximum.** For populations, where the growth rate varies over time but non-monotonically, e.g., India over the period from 1901 to 2011, it could not be applied. However, the total Indian population has been growing but at a decreasing growth rate since 1971 and fits the logistic growth model for the period 1971 to 2011. Even for the US population where this model was used in a famous, 'first-application' paper (Pearl and Reed 1920), it was not an accurate forecast since many other factors were not accounted for.

2.5.6 Changing Rate of Increase Method

This method is similar to the geometrical progression method except that the rate of increase is assumed to change instead of remaining constant. Unlike the logistic growth curve, the changing rate of increase method is used for non-monotonic changes in growth rates. This method may be useful for forecasting populations of large cities when their population growth rate starts decreasing.

The equation for forecasting population using this method is the same as for geometric growth with only one difference, i.e., r is not a constant:

$$P = P_0 (1+r/100)^t \qquad 2.5.8$$

$r = X - nY$ where

X = last percent increase in population

Y = average percent increase in population

2.5.7 Curve-Fitting or Best-Fit Method

In this method, a population-time curve is fit through previously available data and the best-fit model or equation is chosen based on a reasonable coefficient of regression or coefficient of determination (R^2 value); an R^2 value >0.9 is considered acceptable. The most frequently used models are linear, exponential, and polynomial. The equation of the best-fit curve is then used to extrapolate population figures for a future time.

2.5.8 Ratio Method

In this method, the local population and national population data for the last four to five decades of the town or city for which population is to be forecast are taken into consideration. Ratios of local population to national population are then calculated and a graph of these ratios with time is plotted. The graph is extrapolated to get the ratio for a future point in time and then it is multiplied with the expected national population at that time. Thus, the town's or city's expected population at that point in time can be calculated. This method can be used to forecast relative population growth, e.g., urban versus total or rural versus total.

Example: Suppose the population of a small town during 1950, 1960, 1970, 1980 is as given in Table 2.7 below. Populations in the years 1990 and 2020 is to be forecast using some of the numerical methods described above and the results compared:

Table 2.7 Population data and calculations using different forecasting methods

Year	Population	Arithmetic increase	Incremental increase	Percent increase or geometric increase	Change in percent increase
1950	10,000	-	-	-	-
1960	11,150	1,150	-	11.5	-
1970	12,350	1,200	50	10.7	0.8
1980	13,650	1,300	100	10.5	0.2
Total		3,650	150		1
Average		1,220	75	35.95	0.33
1990		14,870	14,945	18,556	15,038
2020		18,530	18,980	46,696	19,396

1. Arithmetical Progression Method:

   Population (1990) = 13650 + 1220 = 14,870

   Population (2020) = 13650 + 4*1220 = 18,530

2. Geometric Progression Method:

   Population (1990) =13650 + (1+ 35.95/100)*13650 =18,556

   Population (2020) =13650 $[1 + 35.95/100]^4$= 13650*1.51 = 46,696

3. Incremental increase method:

   Population (1990) = 13650 + 1220 + 75= 14,945

   Population (2020) = 13650 + 4*1220 + 4*5*75/2= 19,280

PROBLEM 2.5.1

A master plan for a town with one lakh population in 2009 is to be developed. Use the exponential, geometric, and logistic growth models to forecast populations for a period of 50 years starting from 2009. Minimum time interval should be 10 years, i.e., 5 data points. Average annual rate of increase is assumed to be 2 percent. Assume carrying capacity for logistic growth model of 2.5 lakhs.

Solution

Using the equations provided in the text, populations can be estimated using the three methods as shown below. Plots of each of these growth curves are shown below the table.

Year	Number of years, n	Exponential Population	Geometric Population	Logistic Population
2009	0	1,00,000	1,00,000	1,00,000
2019	10	1,22,140	1,21,899	1,12,203
2029	20	1,49,182	1,48,595	1,24,658
2039	30	1,82,212	1,81,136	1,37,120
2049	40	2,22,554	2,20,804	1,49,343
2059	50	2,71,828	2,69,159	1,61,101
2069	60	3,32,012	3,28,103	1,72,201
2079	70	4,05,520	3,99,956	1,82,496
2089	80	4,95,303	4,87,544	1,91,888
2099	90	6,04,965	5,94,313	2,00,329
2109	100	7,38,906	7,24,465	2,07,813

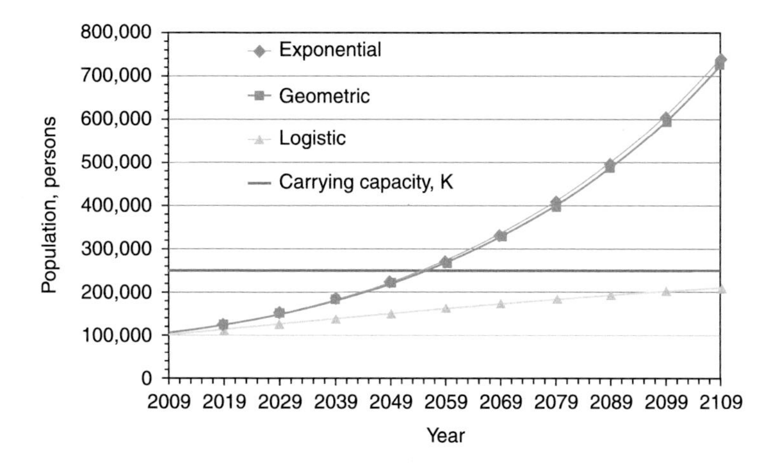

There is very little difference between forecasts made using the exponential and geometric models. These models are good representations of growth in developing countries.

The logistic model behaves very differently from the other two and is a better representation of the growth of natural communities or even growth of organisms like bacteria in pure cultures or mixed cultures. The logistic growth model is also based on exponential growth. However, in the logistic model, the exponential growth rate decreases as the population increases, i.e., growth is greatest at the beginning of the time series. The problem with using the logistic model to forecast the growth of developing economies is the definition of carrying capacity. The carrying capacity is not an absolute or fixed value as assumed in the logistic model and is easily extended (or increased) by adding more infrastructure or improving existing infrastructure.

PROBLEM 2.5.2

Fit the following Census data provided for total population and urban population of India to linear and exponential growth equations and determine the rates of increase for both models. Which equation (model) is a better fit, use it to predict the population in 2051?

Year	Population	Urban population, percent	Urban population
1921	2.51×10^8	11.20	2.81×10^7
1941	3.19×10^8	13.90	4.43×10^7
1961	4.39×10^8	18.00	7.91×10^7
1981	6.83×10^8	23.30	1.59×10^8
2001	1.03×10^9	27.78	2.85×10^8

Census of India, 2001

Solution

Populations have been calculated for linear and exponential growth models in the following table and are plotted in the graph below.

	Total population		Urban population		
Year	Linear	Exponential	Linear	Exponential	Urban population as % of total
2011	1.10×10^8	1.29×10^9	3.30×10^7	4.26×10^8	33.07
2021	2.10×10^8	1.54×10^9	6.30×10^7	5.73×10^8	37.17
2051	5.10×10^8	2.64×10^9	1.53×10^8	1.39×10^9	52.80

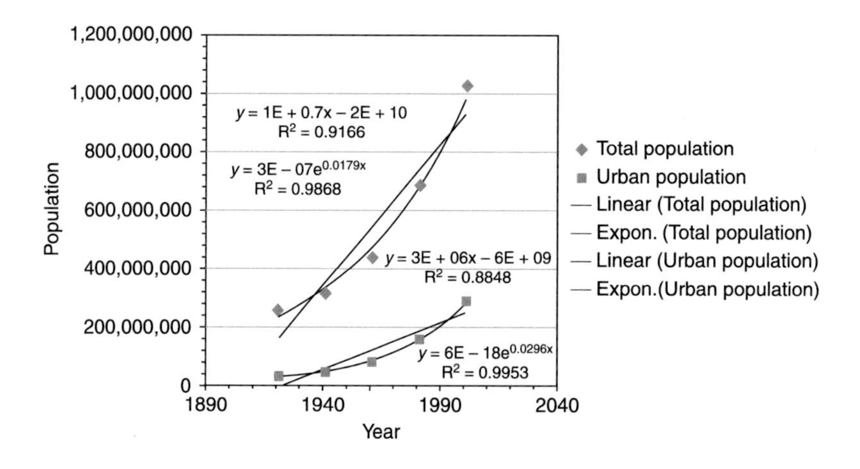

The exponential equation is a better fit than the linear equation based on R^2 values for total and urban population as shown in the graph.

$$y = 1*10^7 x - 2*10^{10}; R^2 = 0.9166$$

$$y = 3*10^{-7} \exp(0.0179x); R^2 = 0.9868$$

$$y = 3*10^6 x - 6*10^9; R^2 = 0.8848$$

$$y = 6*10^{-18} \exp(0.0296x); R^2 = 0.9953$$

Rate of growth (total population) = 0.0179 or 1.79 % annual average

Rate of growth (urban population) = 0.0296 or 2.96 % annual average

Forecast for 2051

Total population = 2.638×10^9 which is 100 %

Urban population = 1.393×10^9 which is 52.80 % of the total population.

PROBLEM 2.5.3

Use the data in the above problem and the remaining population forecasting methods to forecast population growth for the next 50 years. Comment on the differences in results and implications for the country.

Solution

Year	Population	Increase per interval	Incremental increase	Percent increase	Change in % increase
1921	2.51×10^8	-	-		-
1941	3.19×10^8	6.80×10^7	-	0.27	-
1961	4.39×10^8	1.20×10^8	5.20×10^7	0.38	0.11
1981	6.83×10^8	2.44×10^8	1.24×10^8	0.56	0.18
2001	1.03×10^9	3.47×10^8	1.03×10^8	0.51	–0.05
	Total	7.79×10^8	2.79×10^8		0.24
	Average	1.95×10^8	9.30×10^7	0.414	0.08

Arithmetic progression method

Population in 2051 = Population in 2001 + 2.5*Average increase in population

$$= 1.03*10^9 + 2.5*1.95*10^8 = 1.52*10^9$$

Incremental increase method

Population in 2051 = Population in 2001 + 2.5*Average increase in population + (2.5*3.5/2)* Average incremental increase

$= 1.03*10^9 + 2.5*1.95*10^8 + (2.5*3.5/2)*9.3*10^7$

$= 1.03*10^9[1+0.4875+0.407] = 1.03*1.89*10^9 = 1.95*10^9$

Geometric progression method

$$P = P_0 (1+r/100)^t$$

Population in 2051 = Population in 2001*$(1 + 0.414)^{2.5} = 1.03*10^9*[2.37] = 2.45*10^9$

Changing rate of increase method

Population in 2021 = Population in 2001*$[1 + (0.51 – 0.08)] = 1.47*10^9$

Population in 2041 = Population in 2021*$[1 + (0.43 – 0.08)] = 1.98*10^9$

Population in 2051 = Population in 2041*$[1 + (0.35– 0.08)/2] = 2.25*10^9$

The last method can be used to predict the population of those cities which are nearly stable and have a decreasing rate of growth.

Prediction of population using any arithmetic progression method gives a lower value whereas the geometric method gives a higher value. The changing rate of increase method is more accurate and rational as it takes into account the change in the rate of increase for the preceding period.

PROBLEM 2.5.4

A water treatment plant and distribution system is to be built for a city of 1 lakh people. The rate of exponential growth is assumed to be 3 percent. The design period for the plant is 20 years. What is the population that the water works should be built for?

Solution

Using the equation $P = P_0 \exp(rt)$ where P and P_0 are populations at t = 20 years and t = 0 years, respectively; r = 0.03 and t=20 years, determine P.

P = 1.82 lakhs after 20 years assuming exponential growth. Therefore, the water works plant should be designed for a population of 1.82 lakhs and can be built in 5 or 10 year increments or phases depending on availability of funds.

PROBLEM 2.5.5

World population increased exponentially for a very long time until the annual growth rate peaked sometime between 1965 and 1975. Since then the exponential growth rate has been declining steadily. The following population milestones have been declared by the UN: in 1960 the world population was 3 billion, in 1974 the world population was 4 billion, in 1987 it became 5 billion, in 1999 it was 6 billion and it was 7 billion in 2011. It is evident that population growth in these last decades followed the logistic growth model. Determine the growth parameters and predict the next three milestones.

Solution

Step 1: Tabulate year and population and determine dP/dt (calculated as $[P_1 - P_0]/[t_2 - t_1]$) and r as shown.

Time, years	Population	dP/dt	r
1960	3.00×10^9		
1974	4.00×10^9	7.14×10^7	0.020549
1987	5.00×10^9	7.69×10^7	0.017165
1999	6.00×10^9	8.33×10^7	0.015193
2011	7.00×10^9	8.33×10^7	0.012846
2030	7.86×10^9		
2033	8.02×10^9		
2058	9.00×10^9		
2080	9.50×10^9		
2100	9.76×10^9		
2120	9.91×10^9		
2140	1.00×10^{10}		
$P = P_s/(1+b\exp(-R_0 t))$			
$R_0 =$	intercept	0.0302	
$-R_0/P_s =$	slope	-3×10^{-12}	
P_s		1.01×10^{10}	
$b = (P_s - P_0)/P_0$		2.36	

Step 2: Plot a graph of r versus population and determine the equation of the line passing through the data points along with their R^2 value. If the data fits the straight line well, then the logistic model applies clearly.

Step 3: The intercept of the line with the y-axis is R_0 and the slope of the line is $-R_0/Ps$. Therefore, in this problem the intercept or R_0 is 0.0302 or 3.02 percent annual growth rate while the carrying capacity of the planet is 10.1 billion people.

Step 4: With values of R_0 and Ps, b can be determined and used to estimate populations at any time t. Based on these calculations, the next three population milestones will be:

8 billion in 2033, 9 billion in 2058 and 10 billion in 2140. $y = -3*10^{-12}x + 0.0302$

$R^2 = 0.9866$

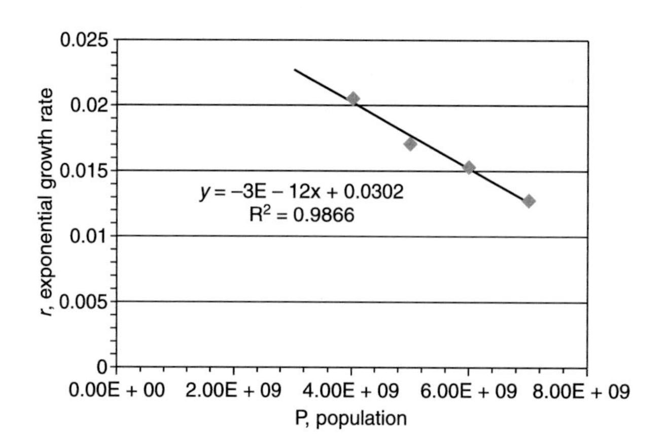

2.6 VARIATIONS IN WATER DEMAND

Water demand varies both spatially and temporally. Spatial variations are related to climatic, economic and cultural factors while temporal variations can be daily, hourly, weekly, or seasonal. The average per capita daily consumption of water is calculated as:

(Quantity of water consumed per year)/(365 x population)

However, water is not consumed at the same rate throughout the year. Fluctuations in the rate of consumption have to be accounted for in determining the amount of water that has to be supplied to a community.

2.6.1 Seasonal and Monthly Variations

The rate of water consumption peaks during summer owing to greater use of water for street and lawn sprinkling, running water coolers, drinking, bathing, and evaporation losses. Consumption goes down in the monsoon and is minimum during the winter mainly due to reduced evaporation losses. Fluctuations in the rate of consumption may be as much as 150 to 200 percent of the average annual consumption. The effect of seasonal variations in water consumption is more pronounced in tropical countries like India. Storage requirements (i.e., the design of large storage reservoirs) increase with increase in seasonal variations in water consumption.

2.6.2 Weekly, Daily and Hourly Variations

Consumption of water on certain days may differ from average daily consumption. Consumption may be higher on Sundays and holidays due to washing of clothes, common areas, car washing, and other weekend activities. It depends on the habits of consumers living in a particular district, and climatic conditions. Daily and hourly variations also depend on whether the area being served is residential, industrial or commercial. In industrial areas, certain industries might consume more water by working in day and night shifts. Certain industries might require water at a particular time or specific season depending on its application. For example, the food processing industry is most affected by seasonal variations while other industries like the paper and pulp industry may operate uniformly year-round. In residential areas, the peak flow hours are generally between 8 a.m. and 12 p.m. and minimum flow hours between 12 a.m. and 4 p.m. Consumption begins to rise from 8 am onwards on weekends and holidays, while it may be as early as 6 p.m. on other weekdays.

On an average, the maximum daily consumption may be 250 percent of the average daily consumption and the maximum hourly consumption may be 200 percent or more of the average hourly consumption of the day. The design of pumps, service reservoirs, mains, etc., depends on these hourly and daily variations. The reader may refer to Figure 2.1 (Duggal 2007) for hourly fluctuations in water consumption on a daily basis.

Demand factors (DF) are calculated based on historical municipal records (Mihelcic and Zimmerman 2010):

$$DF = Q_{event}/Q_{average}$$

Examples of demand factors are:

Maximum day demand = average of all recorded annual max day demand; 1.2 to 3

Minimum day demand = average of all recorded annual min day demand; 0.3 to 0.7

Peak hour demand = average of all recorded annual max hour demand; 3.0 to 6.0

Maximum day of record = highest recorded maximum day demand; <6.0

Peaking factor (PF) is applied to the average daily flow rate to design or size different components of water supply and wastewater systems (Table 2.8):

- $Q_{design} = Q_{average} * PF$

Table 2.8 Design criteria for water and wastewater treatment plants (Mihelcic and Zimmerman 2010)

Treatment process	Water treatment plant	Wastewater treatment plant
Plant hydraulic capacity	Qmax,d* (1.25 to 1.5)	Qmax, instantaneous
Treatment processes	Qmax,d	Qaverage*(1.4 to 3.0)
Sludge pumping	Qmax,d	Qaverage*(1.4 to 2.0)

PROBLEM 2.6.1

A campus community has a high degree of variation in water demand during the academic session versus the vacations. Estimate the average water demand during the academic session when the resident population is 25,000 versus during vacation periods when the resident population is 15,000. This assumes that there are 10,000 residents throughout the year, while the student population is 15,000 of which 5,000 remains on campus during vacations. What peaking factors should be used for designing the water supply for this community.

Solution

Assuming an average water consumption of 500 Lpcd (liters per capita per day),

The average water demand during the vacation period = 500 Lpcd × 15,000 = 75,00,000 L/d

The average water demand during the academic session = 500 Lpcd × 25,000 = 1,25,00,000 L/d

They are in the ratio 3:5.

Wide variations can be observed in the water consumption. Some examples are:

1. **Seasonal Variations**

   More water is consumed in summer compared to winter or monsoon. In this problem, less water is consumed in the vacation period compared to the academic session period.

2. **Daily Variations**

   More water is consumed on Sundays and holidays than on normal working days.

3. **Hourly Variations**

   Water consumption is minimum in the early morning hours, rises to a maximum during the period: 8 a.m. – 11 a.m., decreases in the afternoon, again increases in the evening and finally drops to low values in the late night hours.

   Peaking factors that can be used in the design of water supply systems : $Q_{max,d}$ = 1.25 to 1.5

PROBLEM 2.6.2

Calculate fire demand for a city of 5.0 lakhs population. How much water is unaccounted for or 'losses and leakage'?

Solution

Calculating fire demand for a 5 lakh population using IS 9668, 2010 [5*2 + (5 – 3)]*1800 = 12*1800 = 21,600 Lpm

Total amount of water required for minimum 2 h duration, i.e., static storage = 25,92,000 L

For high risk areas like bazaars, markets, commercial complexes, static water tanks/ km^2 area are provided 2,20,000 L

PROBLEM 2.6.3

You have to design a water supply system for a city with an estimated population of 2.0 lakh persons in the year 2000. Assuming a 3 percent exponential increase in urban population growth, forecast the population in 2025 and 2050 using the exponential increase method. Therefore, the first step in the design is to determine the minimum water requirements for the city in 2025.

Solution

Population in 2000	2,00,000
r =	0.03
Population in 2025	4,23,400
Population in 2050	8,96,338

Determine per capita-day requirements for 2025 for this city

Since 135 L/capita-day are recommended for domestic use for low income groups, we assume double that, i.e., 270 L/capita-day to account for other uses and losses.

Therefore, average water demand = 4,23,400*270 L/capita-d = 11,43,18,001 liters/day or approximately 115 MLD

Fire demand is calculated based on IS 9668, 2010:

For a population of 4,23,400 people, the amount of water required is

1800 * (4,23,400/50,000) + 1800*((4,23,400/1,00,000) – 3) = 17,464 L/min

If the minimum fire duration of 2 hours is designed for, then the static fire storage required is 20,95,632 L.

PROBLEM 2.6.4

A newly planned township is to be set up with a population of 1.0 lakh people. Two sources of water are available: groundwater from a deep confined aquifer and surface water. Groundwater can be pumped on-site while surface water will have to be transported 10 km. What information do you need to decide which source should be developed for water supply for the next 25 years. List the criteria that need to be evaluated for making this decision.

Solution

Various questions need to be answered to determine the suitability of a water source for long-term development.

Criteria for water source selection include the following:

a. ***Quantity*:** How much water is required at present and in the near future - decide design period, determine population at the end of the design period and determine if water quantity will be available throughout the design period?

b. ***Quality*:** What is the source water quality; is it adequate for drinking water purposes; what treatment is necessary and what are the costs of water treatment to achieve acceptable levels?

c. ***Location*:** How far is the source water, what is the cost of installation and transportation?

d. ***Cost of development, collection and distribution*:** What level of infrastructure will be needed to protect, develop, collect, and distribute water? What is the cost of building the required infrastructure?

e. ***Sustainability***: Will the source water be adequate in terms of quantity and quality throughout the design period? Often, quality and quantity deteriorate over time and result in unforeseen costs in the future. Further, there are greater demands than expected with increasing knowledge of contaminants, better treatment methods, and higher water quality requirements. Planning for future requirements for enhancement in quantity and quality is therefore necessary for providing sustainable and acceptable water to the community in the long-term.

Study Outline

Water demand

Water is essential for all activities and the objective of water supply systems is to provide water to all residents ensuring that it is adequate in terms of quantity, quality and is sustainable in the long-term. Water treatment plants are required to treat water to meet drinking water standards. Water that is returned to the system after use is termed wastewater and has to be treated to meet discharge standards.

Design of water supply systems

The first and most important task in designing water supply systems is identification of a sustainable water resource that can meet water demands in terms of quantity and quality. Besides the three factors mentioned above, the location and cost of developing that water resource for water supply systems are other important factors.

Water uses and requirements

Water extracted from a water body (surface or groundwater) is termed withdrawal, while water that is discharged to a water body after use is 'water returned'.

Withdrawal = consumption + return. Water can be used 'in-stream' without withdrawing it from the source or it can be used 'off-stream' after withdrawal.

Water is used in residential, commercial, industrial, institutional, and agricultural activities and for energy development.

Factors affecting water use

Several factors that affect water use are: population, cultural conditions, nature of activities such as residential, commercial, institutional, and municipal, climatic and economic conditions, availability of water supply and sewerage systems, nature of industrial activities, environmental protection and quality, water conservation methods, cost of water and method of billing, pressure required in the water supply system, and the period of water supply.

Factors affecting water losses

Factors affecting water losses in any water supply system are leakages, pressure, system of supply (whether intermittent or continuous), availability of metering, and unauthorized connections.

Scale of use

Total amount of water utilized and per capita water requirement depends on the scale of use. At the individual level, 100 liters per capita per day may suffice but the municipal authority has to provide far more than that to account for water uses that are common and not individual. This results in higher total amount of water utilized as well as higher per capita consumption of water. At the country or regional scale, water requirements for agriculture and energy development result in much higher total and per capita water requirements than at the municipal level. Therefore, the order in which per capita consumption increases with increase in scale of use is:

Individual < city < state < country or regional level

Population forecasting

After identifying the source of water, the next most important task in designing water supply systems is forecasting the population for the service period.

Several methods are used for different situations and include: arithmetic, geometric, exponential, incremental increase, logistic, changing rate of increase, ratio method and finally, curve-fitting method.

Variations in water demand

Water demand fluctuates on an hourly, daily, weekly, monthly, and yearly basis. Design of water supply systems are based on demand factors and peaking factors. These factors can be calculated from historical municipal records if available or by estimates based on experience in other locations.

Study Questions

1. The number of students in a college was 5000 in the year 2000 and remained constant for 10 years. The number of seats in this college was increased to 7000 in 2010 and there is some discussion that the seats will be increased to 10,000 in 2020. What population forecasting model is appropriate for describing this kind of growth in student population? Justify your answer.
2. Explain the factors that affect water use.
3. Explain the factors that affect water losses. What is 'unaccounted for water'?
4. Why does the scale of water use matter in estimates of total and per capita water consumption? Explain with calculations or collect data from the published literature to answer this question.

5. Differentiate between urban and rural water supply scenarios with examples.
6. Describe the different population forecasting methods that are in use and situations where each is applicable.
7. Why does water demand vary over time and location? Explain with examples.
8. What is the difference between peak demand in summer versus winter? Collect data from at least one source to answer this question.
9. What is the difference between daily water demand and demand for fire-fighting?

CHAPTER 3

Water Quality

Learning Objectives

- *Identify pollutants present in water and evaluate their impact on the water quality of surface and groundwater resources*
- *Analyze water samples for basic water quality parameters and describe their significance for public health or the environment*
- *Describe the importance of pH and the factors affecting it*
- *Define ionic strength and calculate it*
- *Define activity coefficients and describe their importance in quantifying chemical equilibrium relationships*
- *Define electrical conductivity and why it can serve as a surrogate for total dissolved solids*
- *Differentiate between different categories of solids in water or wastewater samples based on their measurement methods*
- *Describe the importance of temperature in evaluating the health of an aquatic system, the factors affecting it and water quality parameters affected by it.*
- *Describe the importance of dissolved oxygen levels in water with respect to public health and aquatic systems*
- *Describe the importance of turbidity in evaluating the health of an aquatic system and the factors affecting turbidity*
- *Describe sources of alkalinity in water sources, and their importance in determining water quality*
- *Describe problems associated with hardness, its impact on water quality and water uses.*
- *Describe the importance of color as a water quality parameter*
- *Define and describe different methods for estimating oxygen demand in water, their importance and applications*

- *Calculate ThOD for specific compounds or for a wastewater with known chemical composition*
- *Describe the COD test and its relation to ThOD*
- *Describe the BOD test and its relation to ThOD and COD and to the health of aquatic systems*
- *Describe the use and importance of TOC measurements for aquatic samples*
- *Describe the processes by which synthetic organic compounds enter natural aquatic systems*
- *Evaluate the presence of different elements in water and their importance*
- *Evaluate concentrations of major macronutrients in water and their importance*
- *Define the limiting nutrient in any aquatic system*
- *Evaluate the concentrations of micronutrients in water and their impact on the growth of organisms*
- *Evaluate the presence of toxic heavy metals in water and their impact on public health or environment*
- *List microbiological constituents in natural waters and their importance from a water quality perspective*
- *Differentiate between prokaryotes and eukaryotes*
- *Describe viruses and list some examples of diseases associated with viruses*
- *List common pathogenic bacteria and diseases associated with bacteria in water*
- *Describe algae and their importance in water quality and treatment*
- *Describe lotic or flowing water systems and list water quality parameters of significance in these systems*
- *Derive the equation for the dissolved oxygen sag curve and describe its implications*
- *Describe environmental flows and their importance for lotic systems*
- *Describe standing water bodies and processes that affect them*
- *Explain the process of eutrophication and the factors influencing it*
- *Describe the process of thermal stratification, the factors leading to it and its impact on water quality*

Water quality can be defined in many ways and by many different parameters. Basic water quality parameters include pH, ionic strength, temperature, electrical conductivity or total dissolved solids (TDS), suspended solids, turbidity, alkalinity, hardness, dissolved oxygen and color. Chemical constituents of natural waters include organic and inorganic compounds or elements while biological constituents of water include microbes and other higher organisms. Constituents of water may be harmful or beneficial. The most useful chemical constituent of water is dissolved oxygen. Compounds like nitrogen, phosphate, sulphate and various micronutrients may or may not be useful depending on the location or species of interest. Toxic elements or compounds may also be present in natural waters and include geogenic elements like fluoride and arsenic. Anthropogenic elements or compounds may include synthetic organic compounds like pesticides, polycyclic aromatic hydrocarbons (PAHs), and toxic heavy metals like chromium (Cr), selenium (Se), lead (Pb), and cadmium (Cd).

Pollutant versus contaminant

The two words are often used synonymously and their definitions are subject to some debate. A pollutant is defined as a chemical species that causes **undesirable effects** on the environment or any of its components including living organisms (Nazaroff and Alvarez-Cohen 2004). Undesirable effects include anything that endangers the health of humans and other organisms, endangers the safety of humans or the environment, or causes financial or aesthetic losses. These pollutants may be natural or anthropogenic in origin.

Contaminants are species or substances (physical, chemical or biological) that do not naturally belong to the environment in which they are found, i.e., their concentrations are higher than natural or background levels. For example, discharge of wastewater into a river will result in microbial and organic matter concentrations that are in excess of what is naturally present in the river resulting in contamination of the river water. These contaminants are also pollutants since they have undesirable impacts on human health and the environment.

Arsenic is a geogenic or natural pollutant (but not a contaminant) since it causes adverse health effects on humans and other organisms. DDT which is a potent pesticide is of anthropogenic origin making it a contaminant in all cases where it is detected in the environment and a pollutant since it causes adverse health effects on humans and other organisms.

3.1 GENERAL

Basic water quality parameters include pH, ionic strength, turbidity, total dissolved solids or electrical conductivity, temperature, dissolved oxygen, alkalinity, hardness and color. These parameters are used to evaluate the 'potability' or other uses of water such as recreational and ecological uses, irrigation, or industrial uses.

3.1.1 pH

The most important water quality parameter that determines the usability of water is pH. Acidic water has a sour or 'tart' taste while basic water has a soapy taste. pH is defined as the negative logarithm of the activity of hydrogen ions, i.e., pH = –[log H^+] where [log H^+] is the concentration of protons or hydronium ions.

Hydrogen electrodes are used as an absolute standard for measuring pH. The dissociation of pure water results in the following reaction, i.e., the generation of one proton and one hydroxide anion,[20] and measurements using a hydrogen electrode show that the activity of protons in pure water is 10^{-7} mol/L:

$$H_2O \leftrightarrow H^+ + OH^-$$

The dissociation constant (also known as the thermodynamic ion-product constant) for water is derived based on the above reaction to yield

$$K_w = [H^+] * [OH^-]/[H_2O] = 10^{-14} \text{ at 25 °C and the activity of } H_2O \text{ is 1.0.}$$

K_w values are significantly affected by temperature so that pK_w at 0 °C is 14.943, at 50 °C is 13.262 and at 100 °C is 12.31 (Skoog et al. 2004).

As the ionic strength of a solution increases, so does the value of K_w. For example, K_w at 0.1 M is 1.7×10^{-14} which is 70 percent higher than the value in pure water (Skoog et al. 2004).

Drinking water standards require that the pH value lies between 6.5 to 8.5. pH can be controlled by adding acid or base to water. The addition of various chemicals used in water and wastewater treatment lead to major changes in pH, which need to be controlled to meet regulatory standards. The presence of alkalinity in water helps to buffer a water sample against any significant pH change and is discussed in Section 3.1.8.

Solubility of various metals and other compounds is influenced by pH. Low or acidic pH results in higher solubilities of metals like Fe(II), Fe(III), Mn(II), Al(III) and Cr(III) (Sawyer et al. 2000).

Bronsted acids: These are proton donors (Atkins et al. 2010).

Bronsted bases: These are proton acceptors.

Lewis acids: These are electron pair acceptors.

Lewis bases: These are electron pair donors.

Conjugate acids: When a species gains a proton, it becomes a conjugate acid, e.g., H_3O^+ is the conjugate acid of H_2O.

Conjugate bases: When a species donates a proton, it becomes a conjugate base, e.g., F^- is the conjugate base of HF.

An acid or base is classified as weak or strong depending on the magnitude of the acidity constant. If pKa for an acid is <0, it is a strong acid and dissociates almost completely in water; if pKa for an acid is >0, it is a weak acid and the extent of its dissociation depends on its activity and the ionic strength of the solution. Equilibrium constants for some strong and weak acids are summarized in Table 3.1.

20 The word 'hydroxyl' is used for functional groups [R-OH] or hydroxyl free radicals [OH•].

Table 3.1 Equilibrium constants for some acids

Acid	Equation	pK_a
Hydrochloric acid	$HCl \leftrightarrow H^+ + Cl^-$	–6.3
Sulphuric acid	$H_2SO_4 \leftrightarrow 2H^+ + SO_4^{2-}$	–3 (1st ionization constant)
		1.99 (2nd ionization constant)
Nitric acid	$HNO_3 \leftrightarrow H^+ + NO_3^-$	–1.4
Acetic acid	$CH_3COOH \leftrightarrow H^+ + CH_3COO^-$	4.757
Hypochlorous acid	$HOCl \leftrightarrow H^+ + OCl^-$	7.57
Carbonic acid	$H_2CO_3{}^* \rightleftharpoons H^+ + HCO_3^-$	6.37 (1st ionization constant)
	$HCO_3^- \rightleftharpoons H^+ + CO_3^{2-}$	10.33 (2nd ionization constant)
Phosphoric acid	$H_3PO_4 \rightleftharpoons H^+ + H_2PO_4^-$	2.12 (1st ionization constant)
	$H_2PO_4^- \rightleftharpoons H^+ + HPO_4^{2-}$	7.21 (2nd ionization constant)
	$HPO_4^{2-} \rightleftharpoons H^+ + PO_4^{3-}$	12.32 (3rd ionization constant)
Fluoric acid	$HF \leftrightarrow H^+ + F^-$	3.252

3.1.2 Ionic Strength

Ionic strength (I) is an extremely important water quality parameter that is a measure of the concentrations of all cations and anions present in water; it is a property of the solution and not of individual ions. Ionic strength was first defined by Lewis and Randall in 1921.[21] Mathematically, it is written as:

$$I = \tfrac{1}{2} \sum C_i Z_i^2 \qquad 3.1.1$$

where

C_i = concentration of any ion *i* in moles/L

Z_i = charge of any ion *i*

Ionic strength calculations require complete characterization of the water samples for each individual cation and anion which is best done using instruments like an ion chromatograph. Ionic strength can also be estimated by multiplying total dissolved solids by 2.5×10^{-5} based on Langelier's method (Sawyer et al. 2003).

Another important point to note is that electroneutrality of a solution requires that the sum of the charge concentrations of all cations should be equal to the sum of the charge concentrations of all anions where concentrations are in meq/L for all ions.

For dilute solutions of monovalent ions, the concentration of any ion $[C_i]$ (designated by square brackets) is approximately the same as its activity $\{a_i\}$ (designated by curly brackets), i.e., for solutions with ionic strength <0.01 M (Skoog et al. 2004).

Activity of an ion decreases with increase in ionic strength probably due to an increase in electrostatic attraction between oppositely charged ions. A dimensionless activity coefficient (γ_i) is used to

21 https://en.wikipedia.org/wiki/Ionic_strength.

determine the activity of any given ion and activity corrections are essential for all solutions with I > 0.01 M (Sawyer et al. 2000).

$$\{a_i\} = \gamma[C_i]$$

Several methods are available for estimating activity coefficients for different ions. Four are noted here with constants that are applicable for 25 °C temperature:[22]

Debye–Huckel rule: $\log \gamma_i = -0.5085 * Z_i^2 \sqrt{I}$ for $I < 5 \times 10^{-3}$ M 3.1.2

Extended Debye–Huckel rule: $\log \gamma_i = -0.5085 * Z_i^2 \dfrac{\sqrt{I}}{1 + 3.281\, x_i \sqrt{I}}$ for $I < 0.1$ M 3.1.3

Davies rule: $\log \gamma_i = -0.5085 * Z_i^2 \dfrac{\sqrt{I}}{1 + 3.281\, x_i \sqrt{I}} - 0.3I$ for $I < 0.5$ M 3.1.4

Guntelberg approximation: $\log \gamma_i = -0.5 * Z_i^2 \dfrac{\sqrt{I}}{1 + \sqrt{I}}$ for $I < 0.1$ M 3.1.5

Where x_i is the effective diameter of the hydrated ion X, expressed in nanometers (nm). Effective diameters of different ions are empirical constants and can be obtained from various references (Skoog et al. 2004).

Potentiometric measurements (electrode-based) are linearly related to ion activity but not to ion concentration. As ion concentration increases, its activity in solution decreases resulting in a non-linear relationship with concentration. The theoretical slope of the line for measured electrode potential versus ion activity is 0.0296 (Skoog et al. 2004).

3.1.3 Electrical Conductivity

The electrical conductivity (EC) of water is a measure of the concentration of all ions in it and has units of micro-Siemens/cm. The inverse of electrical conductivity is electrical resistivity (ER) which has units of MΩ-cm (Mega-ohms-cm). Both values are generally reported for 25 °C. It is a direct and inexpensive measure of the ionic strength of a solution unlike the ionic strength calculations noted in the preceding section which require complete characterization of all ions in solution.

Conductivity is also a surrogate parameter for total dissolved solids (TDS). TDS by definition includes all dissolved compounds not just ions. However, in most natural waters, ions are the dominant group compared to neutral compounds. Therefore, conductivity serves as a convenient surrogate parameter for TDS which is a difficult and time-consuming parameter to measure as per standard methodology and can only be measured in the laboratory (APHA, AWWA, WEF 2005). Conductivity, on the other hand, serves as a good approximation that can be measured in the field instantaneously with a portable (pocket-sized) probe. Conductivity and TDS are directly proportionate and a conversion factor is used for a specific water sample. Ideally, the conversion factor for a water sample should be determined by measuring TDS and conductivity independently

22 http://www.aqion.de/site/101.

and deriving correlations between the two parameters. However, in practice, the following equation is generally used:

$$\text{TDS (mg/L)} = A^{*}\text{Conductivity} \qquad (3.1.6)$$

where A = conversion factor ranging from 0.55 to 0.75 (mg-cm/L-μS).

Conductivity is linearly related to ionic strength in dilute solutions with $I < 0.1$ M. This relationship does not work for solutions with high ionic strength such as seawater (Skoog 2004).

Conductivity is also the basis for designating the quality or grade of water depending on their application. The International Organization for Standardization (ISO) and the American Society for Testing and Materials (ASTM) have different types of water and their defining characteristics are noted in Table 3.2.[23] Type 1 or Grade I water is the purest form of water that can be produced in the laboratory or industrially and is used for high-end analytical work with instruments such as atomic absorption spectrometer (AAS) and inductively coupled plasma mass spectrometer (ICP-MS). Other high grades of water include deionized water or water from reverse osmosis, distilled water, and potable water. Distilled water may have EC values of approximately 0.5 μS/cm while potable water should have EC values ranging from 500 to 1000 μS/cm.[24] Seawater has an EC value of 54,000 μS/cm.

Table 3.2 ASTM and ISO standards for different types of water

ASTM D1193-91							
Water type	**pH**	**EC, max**	**ER, min**	**TOC, max**	**Na, max**	**Si, max**	**Cl, max**
Units		**μS/cm**	**MΩ-cm**	**μg/L**	**μg/L**	**μg/L**	**μg/L**
Type 1	-	0.056	18	100	1	3	1
Type 2	-	1	1	50	5	3	5
Type 3	-	0.25	4	200	10	500	10
Type 4	5.0 to 8.0	5	0.2	No limit	50	No limit	50

ISO 3696	**pH**	**EC, max**	**Oxidisable matter oxygen content, max**	**UV Abs @ 245 nm, max**	**Na, max**	**Si, max**	**Cl, max**	**Residue after evaporation at 110 °C, max**
Unit		**μS/cm**	**MΩ-cm**	**μg/L**	**μg/L**	**μg/L**	**μg/L**	**mg/kg**
Grade I		0.1	-	0.001		10		-
Grade II		1	-	0.01		20		1
Grade III	5.0 to 7.0	5	-	-				2

23 https://www.elgalabwater.com/blog/different-types-pure-water-lab-what-you-need-know.

24 http://www.aqion.de/site/130.

3.1.4 Solids

All natural waters contain dissolved compounds and suspended particles which together comprise total solids. These solids are further categorized based on various practical measures.

Total Solids (TS): All solids, dissolved and suspended, are included in total solids. Practically, the difference between the weight of the water sample before and after drying, i.e., evaporating all moisture from the sample, and dividing by the volume of the sample defines TS concentration.

Total solids are further divided into two categories: dissolved and suspended solids as shown in Figure 3.1. The distinction between dissolved and suspended solids varies greatly in the literature. There are no theoretical cut-offs for these particles or solids. The definition or cut-off for the two categories is a matter of operational convenience. For example, if biological particles or solids are to be separated from a sample, the conventional choice of membrane pore size used is 0.45 micron or lower (0.2 micron for sterilization). In most other applications, larger pore size based membranes are used. Choice of filter pore size recommended in Standard Methods (APHA et al. 2005) is anywhere from 1.5 micron to 0.2 micron based on operational requirement or objective. Colloids fall into either category 'operationally' speaking.

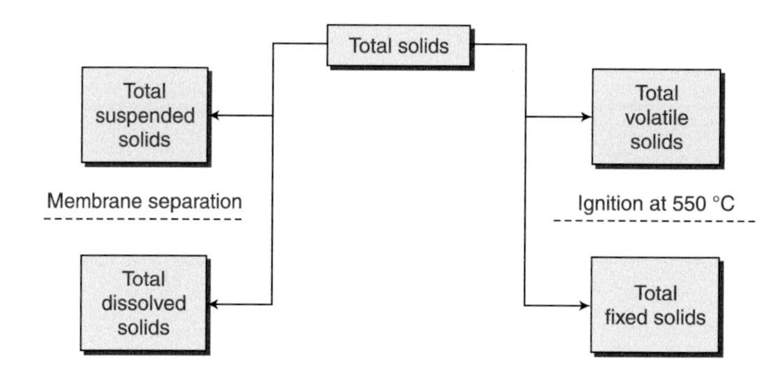

Figure 3.1 Categories of solids present in various environmental samples.

Total Suspended Solids (TSS): For water quality measurement, suspended solids are defined as solids that are separated using a membrane filter of 1.5 or 1.7 micron pore size (APHA et al. 2005).

Total Dissolved Solids (TDS): The solids that pass through the above-mentioned filter, i.e., solids remaining in the filtrate are defined as total dissolved solids.

Total Fixed Solids (TFS): When total solids are burned at 550 °C in a muffle furnace, the residue remaining after burning is defined as TFS.

Total Volatile Solids (TVS): The difference in weight between total solids and total fixed solids is defined as total volatile solids.

Fixed Suspended Solids (FSS): When TSS are burned at 550 °C in a muffle furnace, the residue remaining after burning is defined as fixed suspended solids (FSS). For measuring FSS, the water sample needs to be filtered on ashless (glass fiber) filter paper which is weighed prior to filtration and after filtration.

Weight of filter before burning and after filtration = TSS + clean filter

Weight of filter after burning and filtration = FSS + clean filter

Therefore, weight of filter before burning - weight of filter after burning = VSS

Since TSS = FSS + VSS

Volatile Suspended Solids (VSS): The difference in weight between TSS and FSS defines volatile suspended solids (VSS).

3.1.5 Temperature

Another important water quality parameter that is often overlooked in surveys is temperature. It determines the fate and transport of contaminants in water since the kinetics of all chemical and biological reactions are temperature-dependent. Dissolved oxygen levels, the health and biodiversity of any ecosystem, and the extent and rate of all physical, chemical and biological processes that impact a given system including aquatic systems are all temperature-dependent processes.

Factors influencing water temperature for unpolluted surface water bodies are depth of the water column, incident light intensity, turbidity of water, and ambient temperature. Diurnal variations in light intensity and ambient temperatures will result in corresponding changes in water temperatures. When wastewaters like cooling water are discharged at temperatures often higher than those in the water body, water body temperatures are affected. In some cases, these changes can be severe enough to destroy natural ecosystems since dissolved oxygen concentrations decrease with increase in temperature due to lower saturation limits and higher deoxygenation reaction rates.

Temperature affects water quality parameters like density of water, dissolved oxygen concentrations, solubilities and/or dissolution of various contaminants, growth of organisms, and volatilization of contaminants.

Reaction rates and speciation of various contaminants at equilibrium are dependent on temperature.

Temperatures of water bodies are influenced by ambient air temperatures, incident light energy, wind action and degree of mixing, and generation of heat by various chemical and biological processes.

Three processes by which energy or heat is transferred are:

Conduction: When heat or energy is transferred due to direct physical contact between objects, it is termed conduction. For example, electrical water heaters heat water by conduction when electrical current heats a metal vessel containing water and transfers the heat to the water in the vessel.

Radiation: Radiative transfer of heat or energy occurs even in the absence of a medium. The best example of radiative transfer of heat is sunlight. Sunlight is electro-magnetic radiation that travels through space, i.e., in the absence of a medium, and heats the Earth and other planets.

Convection: Convective transfer of energy occurs when a medium separates the objects. For example, cooking food by steaming is an example of convective transfer of heat or energy.

Fourier's law of heat conduction: *Heat flux by conduction (Q) is proportionate to the temperature gradient (dT/dy), i.e.,*

$$Q = k\, dT/dy$$

where Q = heat flux or heat flow per unit area, $Joules/cm^2$-s

k = thermal conductivity of the material through which heat is transferred, Joules/cm-s-degree K

T = temperature in degree Kelvin

y = distance or thickness of the material through which heat is transferred, cm

3.1.6 Dissolved Oxygen

Oxygen is sparingly soluble in water and dissolved oxygen concentrations are dependent on two major factors in unpolluted waters: temperature and salinity. Saturation concentration of oxygen in pure water is 14.6 mg/L at 0 °C and 0 mg/L salinity at 1 atm pressure. As temperature or salinity increase, saturation concentration of oxygen decreases so that at 40 °C and 40,000 mg/L salinity, DO saturation concentration is 5.2 mg/L.[25]

DO saturation (DO_s) can be estimated based on the following equation where salinity corrections are ignored (Hendricks 2011):

$$DO_s = 14.59 - 0.3955*T + 0.0072*T^2 - 0.0000619*T^3 \quad 3.1.7$$

where T is the temperature in °C or Centigrade.

25 http://water.usgs.gov/software/DOTABLES/.

Aquatic ecosystems are extremely sensitive to changes in DO levels and minimum DO levels recommended for ecosystems in India are 4 mg/L (Appendix B). Some ecosystems especially in temperate and colder regions are more sensitive and recommended DO levels can be as high as 8 mg/L.

Large amounts of biodegradable organic matter are often discharged into aquatic systems in the form of wastewater, surface runoff or leachate from landfills and open dumps. The biodegradation of this organic matter results in the depletion of oxygen in water. Sensitive species which need high oxygen levels are unable to survive in such conditions. At very low DO concentrations or under anoxic conditions where DO = 0 mg/L, very few higher organisms can survive. Anaerobic bacteria are the dominant species found in anoxic systems.

Aquatic chemistry under oxic and anoxic conditions

When suboxic or anoxic conditions prevail, various metals and other compounds are in reduced state. Examples include Fe(II), Mn(II), and As(III). Sulphur containing compounds are converted or present in sulphide (S^{2-}) form. This is usually the case with deep groundwaters that are not in equilibrium with the atmosphere. When such waters are pumped to the surface, they start getting oxygenated due to equilibrium with the atmosphere and the above metals are slowly oxidized to Fe(III), Mn(IV) and As(V).

3.1.7 Turbidity

Turbidity is a very important water quality parameter and indicates the presence of light-scattering substances in water. Turbidity, unlike color, is associated with particles which may be substances like clay and silt, or biological particles like bacteria or algae. Color is associated with the presence of light-absorbing substances including dissolved substances or compounds.

High turbidity levels preclude light from penetrating deep into the water column and result in a shallow photic (light) zone. Photosynthetic organisms are capable of growing only in the photic zone since the darker regions of the water column, i.e., the aphotic zone, cannot sustain their growth. The result is lack of generation of oxygen in the aphotic zone by photosynthetic organisms and therefore, low to zero DO levels in this zone.

A turbidity meter or nephelometer is used for measuring turbidity and is based on measuring the intensity of light either transmitted or scattered by a sample, respectively (Sawyer, McCarty, and Parkin 2000). In a turbidity meter, the incident light is reduced to a single wavelength that is incident on the sample and the intensity of transmitted light passing through the sample is measured. In a nephelometer, light is incident on a sample and the amount of light scattered by the particle suspension or water sample is measured at right angles to the light source and sample, i.e., based on Tyndall effect. The intensity of scattered light is a measure of the sample's turbidity and is measured in nephelometric turbidity units (NTU). In both cases, the instruments are compared to standard solutions of known turbidity for calibration. This is in contrast to a spectrophotometer where the intensity of light absorbed by the solution is measured in terms of absorbance.

Based on Beer–Lambert's law,

Absorbance A $= \log I_0/I = a*C*L$ 3.1.8

where

I = intensity of light leaving the solution,

I_0 = intensity of light incident on the solution,

a = constant for the solution

C = concentration of compound of interest in solution

L = path length of light through solution (or cell or cuvette)

3.1.8 Alkalinity

Alkalinity is a measure of the capacity of water to neutralize acids.[26] Alkalinity in water is due to the presence of hydroxides, carbonates, bicarbonates, silicates, borates, phosphates, sulphides and ammonia (Peavy et al. 1985). These ions are generated by the dissolution of soil and due to equilibrium with atmospheric constituents like carbon dioxide. Phosphates are constituents of wastewaters, fertilizers and pesticides while sulphides and ammonia are discharged along with wastewaters and generated in water by biological processes.

The most common contributors to alkalinity in water are bicarbonates, carbonates, and hydroxides which remove H^+ ions and lower the acidity of the water, i.e., increase its pH. They usually do this by combining with H^+ ions to make new compounds, thereby acting as a buffer. Without this acid-neutralizing capacity, any acid added to a stream would cause an immediate change in the pH. Measuring alkalinity is important in determining a stream's ability to neutralize acidic pollution from rainfall or wastewater. It's one of the best measures of the sensitivity of the stream to acid inputs. Total alkalinity is measured by measuring the amount of acid (e.g., sulphuric acid) needed to bring the sample to a pH of 4.5 (APHA, AWWA, WEF 2005). At this pH, all alkaline compounds in the sample are consumed in the neutralization reaction. The result is reported as milligrams per liter of calcium carbonate (mg/L as $CaCO_3$).

Reactions contributing to alkalinity are listed here:

1. Dissolution of carbon dioxide from atmosphere is governed by Henry's law:

   $[CO_2] = K_H P_g$ where K_H = Henry's constant, for CO_2 = 31.6 atm/M, and P_{CO2} (in 2018) = 410 ppmv = 4.1×10^{-4} atm.

2. Generation of carbonic acid: $CO_{2\,(aq)} + H_2O \rightleftharpoons H_2CO_3$

3. Bicarbonate formation: $H_2CO_3 \rightleftharpoons H^+ + HCO_3^-$ $pK_1 = 6.37$

4. Carbonate formation: $HCO_3^- \rightleftharpoons H^+ + CO_3^{2-}$ $pK_2 = 10.33$

5. Hydroxide formation: $H_2O + CO_3^{2-} \rightleftharpoons OH^- + HCO_3^-$

6. Solubility of calcium carbonate: $Ca^{2+} + CO_3^{2-} \rightleftharpoons CaCO_3$ $pK_{sp} = 8.34$

26 http://water.epa.gov/type/rsl/monitoring/vms510.cfm.

Algae utilize aqueous carbon species as their carbon source for their growth resulting in a decrease in proton concentration and an increase in pH to 9 or 10 (Peavy et al. 1985). Based on the above reactions and the concentration-pH diagram for all three alkalinity-contributing species (hydroxides, carbonates and bicarbonates) a few points can be noted:

1. Hydroxide ion dominates only at pH>11.
2. At pH = 10.33, carbonate and bicarbonate concentrations are equal and hydroxide concentration is negligible in comparison.

If a water sample with alkalinity is titrated using an acid such as sulphuric acid, a pH versus titrant volume graph can be obtained (Sawyer et al. 2003).

1. **Hydroxide alkalinity** is significant only when the sample pH is high, i.e., >11 and is directly determined based on pH and K_w.
2. **Phenolphthalein end-point at pH =8.3:** The first inflection point in the titration curve is determined using a phenolphthalein indicator (P-alkalinity). At the first inflection point of pH = 8.3, almost all carbonates have been converted to bicarbonates and hydroxides have been completely converted to water. Bicarbonates also contribute to alkalinity, therefore, neutralization of carbonates is only half-complete at this point.

   P-alkalinity = hydroxide alkalinity + carbonate alkalinity

   If no carbonates are present, then P-alkalinity = hydroxide alkalinity. If no hydroxides are present, P-alkalinity = ½ carbonate alkalinity.
3. **Bromcresol end-point at pH = 4.5:** The second inflection point is determined in titration using a bromcresol indicator (B-alkalinity) and has an endpoint at pH = 4.5. This represents the essentially complete conversion of bicarbonate to carbonic acid. At this point, there are no alkalinity-contributing species remaining in solution. Therefore, the amount of acid required to titrate a water to pH = 4.5 is the total alkalinity of the water.

   Total alkalinity = amount of acid used to bring sample pH down to 4.5

   For samples containing hydroxide and carbonate alkalinity, the titrant required to bring pH from 8.3 to 4.5 is ½ the carbonate alkalinity.
4. Samples containing only carbonates and bicarbonates and insignificant amounts of hydroxides generally have 8.3 < pH < 11. In that case,

   Bicarbonate alkalinity = total alkalinity – carbonate alkalinity
5. Samples containing bicarbonates alone have pH < 8.3 and bicarbonate alkalinity = total alkalinity.

3.1.9 Hardness

Hardness in water is attributed to the presence of divalent cations like Ca and Mg. Other multivalent cations like Fe, Mn, Sr, Al, etc., may also be present in water but are not generally used to define hardness. The presence of these cations leads to various problems in water supply systems such as:

1. Formation of soap curd (white precipitates of soap) and lack of frothing or foaming that is essential for bringing dirt particles into solution, increased soap requirement due to that and subsequent difficulty in all cleaning activities. Synthetic detergents can reduce but not eliminate this problem.
2. On heating, scale formation or precipitation of hardness-contributing ions, $CaCO_3$ and $Mg(OH)_2$, leads to reduced efficiency of heating elements, and subsequent failure of the appliance.
3. Where groundwater is used and large amounts of hardness-contributing ions, $CaCO_3$ and $Mg(OH)_2$ are present, they precipitate in the form of $CaCO_3$ and $Mg(OH)_2$ and may show up as white flakes or crystals in water.

Hardness is expressed in units of mg/L as $CaCO_3$. The desirable limit for total hardness (TH) is 200 mg/L and permissible limit in the absence of another source is 600 mg/L (BIS 2012).

Total hardness (TH) is defined as the sum of the concentrations of Ca and Mg expressed as mg/L of $CaCO_3$. Hardness that is associated with carbonate and bicarbonate ions is defined as carbonate hardness (CH) while non-carbonate hardness (NCH) is defined as the difference between TH and CH. If the concentrations of carbonate and bicarbonate ions are greater than TH, then TH = CH and NCH = 0. Carbonate hardness is also called 'temporary hardness' since it can be removed by heating.

Note: *For drawing bar graphs of cations and anions and for solving hardness, alkalinity and softening problems, it is important to work with meq/L rather than mg/L. Laws of mass and charge balance are easier to adhere to with these units rather than mg/L.*

3.1.10 Color

Color in any water can be due to the presence of dissolved or particulate matter that may be either organic or inorganic in nature. Organic matter includes dissolved compounds like humic and fulvic acids which give water a light tea-like yellow color and particulate matter like phytoplankton which impart a green or yellow color to water. Inorganic matter includes dissolved metals like iron, manganese and chromium which cause water to have colors ranging from yellow to red. Many of these metals like iron and manganese precipitate slowly as they get oxidized in contact with atmospheric oxygen leading to colored precipitates that are red or black.

PROBLEM 3.1.1

Determine the pHs of 2 milli moles/L (mM) solutions of HCl and NaOH.

Solution

2 mM of HCl = 2×10^{-3} M. Assuming complete dissociation of HCl as shown below, the concentration of protons is also 2×10^{-3} M:

$$HCl \leftrightarrow H^+ + Cl^-$$

Taking the negative logarithm of the concentration of protons gives a pH of 2.7

For 2 mM solution of NaOH and assuming complete dissociation:

$$NaOH \leftrightarrow OH^- + Na^+$$

Further, since water dissociates into protons and hydroxide ions and $K_w = 10^{-14}$

$$H_2O \leftrightarrow H^+ + OH^- \qquad pK_w = 14$$

Therefore, the concentration of OH^- is 2×10^{-3} M, and the concentration of H^+ is 5×10^{-12} M. The pH = 14 – 2.7 = 11.3.

PROBLEM 3.1.2

The pH of a 1 L distilled water solution is to be adjusted by adding 2 mM HCl or NaOH. Determine the amount of HCl or NaOH to be added to achieve a pH of (a) 4 and (b) 10.

Solution

Based on mass (or in this case, molar) balance for protons for a solution of HCl:

$M1V1 = M2V2$ where

$M1$ = molarity of the solution desired, mol/L = 10^{-4}

$V1$ = volume of the solution desired, L = 1 L

$M2$ = molarity of the solution used, mol/L = $2*10^{-3}$

$V2$ = volume of the solution used, L

Solving for V2, we get 50 mL of 2 mM HCl which is the amount of acid that is to be added to achieve a pH of 4. Similarly, 50 mL of 2 mM of NaOH are to be added to 1 L of water to achieve a pH of 10.

PROBLEM 3.1.3

The composition of a groundwater sample was determined and the results are shown in the table below. Determine the ionic strength of the water.

Ions	Conc., mg/L
F^-	0.747
Cl^-	3.140
Br^-	0.498
NO_2^-	0.037
NO_3^-	0.675
PO_4^{3-}	3.102
SO_4^{2-}	5.792
CO_3^{2-}	78.038
Na^+	19.701
K^+	2.802
Ca^{2+}	26.111
Mg^{2+}	8.138
NH_4^+	0.000
Mn^{2+}	1.475

Solution

Step 1. Note molecular weight and charge of each ion alongside the concentration data.

Step 2. Determine the concentration of each ion in mmoles/L (mM) by dividing the concentration in mg/L by their molecular weights.

Step 3. Calculate values of $C_iZ_i^2$ as shown in table below. Half of the sum of the last column is the ionic strength in moles/L (M).

Ions	Conc., mg/L	MW, mg/ mmoles	Conc, mM	Z	$CiZi^2$
F^-	0.747	19	3.93×10^{-2}	–1	3.93×10^{-2}
Cl^-	3.140	35.45	8.86×10^{-2}	–1	8.86×10^{-2}
Br^-	0.498	79.9	6.23×10^{-3}	–1	6.23×10^{-3}
NO_2^-	0.037	46	8.12×10^{-4}	–1	8.12×10^{-4}
NO_3^-	0.675	62	1.09×10^{-2}	–1	1.09×10^{-2}
PO_4^{3-}	3.102	95	3.27×10^{-2}	–3	2.94×10^{-1}
SO_4^{2-}	5.792	96	6.03×10^{-2}	–2	2.41×10^{-1}
CO_3^{2-}	78.038	60	1.30	–2	5.20
Na^+	19.701	23	$8.75 * 0^{-1}$	1	$8.75 * 10^{-1}$

(*Contd.*)

K^+	2.802	39	7.18×10^{-2}	1	7.18×10^{-2}	
Ca^{2+}	26.111	40	6.53×10^{-1}	2	2.61	
Mg^{2+}	8.138	24	3.39×10^{-1}	2	1.36	
NH_4^+	0.000	18	0.00	1	0.00	
Mn^{2+}	1.457	55	2.68×10^{-2}	2	1.07×10^{-7}	
				I =	5.44	mM
				I =	5.44×10^{-3}	M

PROBLEM 3.1.4

The electrical conductivity of a water sample was found to be 400 micro-S/cm. The conversion factor for determining total dissolved solids (TDS) for this water was found to be 0.65 (mg-cm/L-microS). Determine the TDS of this water sample.

Solution

TDS of water sample = 400*0.65 = 260 mg/L

PROBLEM 3.1.5

Wastewater analysis was done in the lab and following results were obtained:

a. 10 mL of the wastewater was taken in a crucible and allowed to dry at 105 °C in an hot air oven. Initial (empty) weight of crucible was 36.635 g and weight of crucible after drying was 36.638 g.

b. 50 mL of the wastewater was taken in a crucible and burned at 550 °C in a muffle furnace. The difference in weights of the crucible before and after burning was 0.006 g.

c. 50 mL of wastewater was passed through a 0.45 micron ashless filter and the difference in dry weights of the filter before and after passing the wastewater sample was 0.01 g.

d. The ashless filter paper with the solids was then burned in a muffle furnace at 550 °C. The difference in weight of solids before and after burning was 0.003 g.

Determine the concentrations of total solids (TS), total suspended solids (TSS), total fixed solids (TFS), total volatile solids (TVS), fixed suspended solids (FSS), and volatile suspended solids (VSS) in the wastewater sample.

Solution

Total solids (TS) = [Weight of crucible after drying (36.638 g) – empty weight of crucible (36.635 g)]*1000 (mL/L)/10 mL = 0.3 g/L

Total fixed solids (TFS) = [Weight of crucible after burning at 550 °C – empty weight of crucible]*1000 (mL/L)/50 mL = 0.006 g*1000 (mL/L)/50 mL = 0.12 g/L

Total volatile solids (TVS) = TS – TFS = 0.3 – 0.12 g/L = 0.18 g/L

Total suspended solids (TSS) = [Weight of filter after filtration and drying – weight of clean filter]*1000 (mL/L)/50 mL = 0.01*1000/50 = 0.2 g/L

Fixed suspended solids (FSS) = [Weight of filter after burning at 550 °C – weight of clean filter]*1000 (mL/L)/50 mL = 0.003 g*1000 (mL/L)/50 mL = 0.06 g/L

Volatile suspended solids (VSS) = TSS – FSS = 0.2 – 0.06 g/L = 0.14 g/L

TS	0.3	g/L
TSS	0.2	g/L
TFS	0.12	g/L
TVS	0.18	g/L
FSS	0.06	g/L
VSS	0.14	g/L

PROBLEM 3.1.6

Temperature of a water sample was measured and found to be 20 °C. Determine the dissolved oxygen saturation concentration for the water at this temperature ignoring salinity corrections.

Solution

DO saturation concentration for any water sample can be estimated using equation 3.1.7, ignoring salinity corrections.

The estimated concentration is 9.065 mg/L.

PROBLEM 3.1.7

Collect water samples from your neighborhood and analyze them for the water quality parameters described in this section.

Solution

The following water samples were collected from our laboratory and neighborhood and the results are summarized in the following table.

Water quality results for different water samples

Water sample	pH	Conductivity, microS/cm	Temperature, °C	Dissolved Oxygen, mg/L	Turbidity, NTU
Deionized water	6.51	-	-	-	0.0
Distilled water[a]	5.19	179.8	31.5	-	0.2
Untreated water (mixed river and GW)[a]	7.11	460	32.6	7.17	5.1
Tap water[c]	7.4	311	-	-	2.2
Ganga River water[b]	7.71	222	-	6.6	-
Groundwater (GW)[b]	6.8	198	-	5.27	-
Untreated water (mixed river and GW)[b]	7.17	218	-	7.17	-

a – Tripathi and Kumar 2014

b – Mohanta and Goel 2014

c – Apshanker and Kola 2014

PROBLEM 3.1.8

Cation–anion analysis was done for a groundwater sample using an ion chromatograph. The water sample pH was 7.0 and results are shown in the following table. Since the eluent for analysis is sodium bicarbonate, carbonate and bicarbonate ions could not be measured. Prepare a table showing concentrations of all cations and anions in meq/L; do a mass balance for cations and anions and estimate the carbonate and bicarbonate concentrations in this sample.

Solution

Cations	mg/L	mg/meq	meq/L	Anions	mg/L	mg/meq	meq/L
Fe(III)	6.1	18.67	0.327	**F^-**	0.32	19.00	0.017
NH_4^+	0.08	18	0.004	**Cr^-**	0	52.00	0.098
Na^+	14.3	23	0.622	**NO_3^-**	0.91	62.00	0.015
K^+	2.99	39.1	0.076	**NO_2^-**	0.04	46.00	0.001
Ca^{2+}	29.8	20	1.490	**Br^-**	0.79	80.00	0.010
Mg^{2+}	9	12.15	0.741	**PO_4^{3-}**	1.31	95	0.041
Mn(II)	2.69	27.45	0.098	**SO_4^{2-}**	56.33	48	1.174
		Total	3.358				**1.355**

Since the sum of all cations must equal the sum of all anions, the residual anions are assumed to be carbonate and bicarbonate in this case (eluent is sodium bicarbonate) = 2.003 meq/L = 2.003*61 = 122.2 mg/L.

In the above solution, all alkalinity was assumed to be due to bicarbonates alone since the pH was 7.0.

PROBLEM 3.1.9

For the same sample as shown in Problem 3.1.8, determine total hardness, carbonate hardness and non-carbonate hardness.

Total hardness (TH) = $[Ca^{2+}]$ + $[Mg^{2+}]$ = (1.49+ 0.74) meq/L * 50 mg of $CaCO_3$/meq = 2.23*50 = 111.5 mg/L as $CaCO_3$

Bicarbonate concentration $[HCO_3^-]$ = 122.22 mg/L or 2.003 (meq/L)* 50 mg of $CaCO_3$/meq = 100.15 mg/L as $CaCO_3$.

Carbonate hardness (CH) is equal to 100.15 mg/L as $CaCO_3$. Since TH is more than the total carbonate-bicarbonate concentration, non-carbonate hardness (NCH) = TH – CH = 11.35 mg/L as $CaCO_3$.

3.2 ORGANICS

Wastewater discharges contribute organic matter to aquatic systems. This organic matter is biodegraded in the natural environment simultaneously reducing the oxygen content of aquatic systems. Various methods like theoretical oxygen demand (ThOD), chemical oxygen demand (COD) and biochemical or biological oxygen demand (BOD) are used to quantify the potential for oxygen depletion due to the presence of organic matter. Total organic carbon (TOC) can also be used to assess the potential for oxygen depletion in aquatic systems and is generally associated with natural organic matter. Synthetic organic compounds (SOCs) are a much smaller but equally important source of organic matter in water bodies and a major cause of concern due to their toxicity and potential for bioaccumulation in ecosystems.

3.2.1 Theoretical Oxygen Demand (ThOD)

Theoretical oxygen demand (ThOD) is the amount of oxygen required to completely oxidize a particular organic substance and is calculated from simple stoichiometric considerations. In a comprehensive ThOD calculation, all organic and inorganic compounds are completely oxidized to their highest oxidation states. For example, organic compounds are converted to carbon dioxide and water, and other compounds like ammonia are converted to nitrate, P to phosphate, and S to sulphate. Therefore, ThOD is an estimate of the maximum oxygen demand of a sample or compound. Its use is limited to known compounds or mixtures where the elemental composition is known or the constituent compounds are known. It is important to note that in practice, oxygen needed for complete oxidation of a compound will always be in excess of ThOD to account for process inefficiencies, leaks, lack of mixing, and other factors.

Note: *Several textbooks account for conversion of organic-N to NH_3 but not NH_3 to nitrate in calculating ThOD. These ThOD calculations are done for comparison with COD values. However, this does not account for the additional oxygen demand from the conversion of NH_3 to nitrate.*

3.2.2 Chemical Oxygen Demand (COD)

Chemical oxygen demand (COD) is one of the most important water quality parameters and is defined as the amount of oxygen consumed to completely chemically oxidize the organic constituents present in water to inorganic end-products. It is expressed in milligrams per liter (mg/L or ppm) which indicates the mass of oxygen consumed per liter of solution. It is an indirect measure of the amount of organic matter in water. COD measurements are usually done for wastewaters (domestic or industrial) or for natural waters contaminated by domestic or industrial waste. It is an approximation of ThOD in the absence of knowledge about the constituent compounds or elemental composition of a sample. The COD test is a practical measure of the maximum oxygen demand of a water sample while ThOD is a theoretical measure of the oxygen demand of any sample.

The basis for the COD test is that nearly all organic compounds can be fully oxidized to carbon dioxide and water with a strong oxidising agent under acidic conditions at high temperature. It is a standardized laboratory test in which a water sample is digested with a strong chemical oxidant like potassium dichromate ($K_2Cr_2O_7$) in combination with boiling sulphuric acid (H_2SO_4) for 2 h at 150 °C (APHA, AWWA, WEF 2005).

Chemical oxygen demand is related to biochemical oxygen demand (BOD), another standard test for assaying the oxygen-demanding strength of waters. However, BOD is a measure of the amount of oxygen consumed by microbial oxidation and is most relevant to waters rich in organic matter. Further, the BOD test simulates natural conditions while the COD test allows determination of the maximum oxygen demand of sample. Another advantage of the COD test is its ability to measure the oxygen-consuming potential of slowly biodegrading material like cellulose which cannot be measured during a BOD_5 test. COD is always equal to or higher than BOD because there are many organic and inorganic compounds in water that cannot be oxidized by microorganisms (under ambient conditions) but are oxidized under COD test conditions (extreme conditions).

The COD test has its limitations:

1. It does not measure the oxygen-consuming potential associated with ammonia and its derivatives. If the concentration of ammonia is known, COD associated with it can be calculated separately. In practice, total Kjeldahl nitrogen (TKN) is measured in the laboratory, and the amount of oxygen required to convert this nitrogen to nitrate is estimated.

2. Other inorganic ions like nitrite, ferrous iron, sulphide, manganous (Mn) are oxidized quantitatively during the COD test. If the concentrations of these individual ions are known, then the total COD value obtained can be corrected to obtain the COD value associated with organic matter alone.
3. Interference in COD tests is mainly due to chloride ions which precipitate with Ag to form silver chloride and inhibit the catalytic role of Ag. Other halogens like bromine and iodine have similar effects and result in the production of the elemental forms of the halogen and chromic ion. This results in COD values that are higher than warranted.
4. A major problem associated with COD tests is the generation of hazardous waste due to the use of Hg, Cr(VI) and silver.

In a study with 565 chemicals, COD and ThOD values were determined and the overall average COD/ThOD ratio was found to be 0.85 with a standard deviation of 0.33 (Baker et al. 1999).

3.2.3 Biochemical Oxygen Demand (BOD)

Biochemical or biological oxygen demand (BOD) is a laboratory test to determine the oxygen demand, of wastes, due to biodegradation under standardized conditions that simulate natural conditions to the greatest extent.

Most water bodies receive varying amounts of wastes. These wastes may be surface runoff or wastewater discharges from industrial sources or municipalities leading to surface water contamination. The waste that is released to the water body can be biodegradable, in which case decomposition takes place. Microorganisms, especially bacteria, break down biodegradable organic matter into simple organic and inorganic substances which they use as substrate (food). When this decomposition takes place in the presence of sufficient oxygen, it is called aerobic decomposition. If oxygen is not present in sufficient amount, then it is called anaerobic decomposition. Aerobic decomposition produces harmless and stable end products like carbon dioxide, while anaerobic decomposition produces harmful and unstable end-products like ammonia and hydrogen sulphide.

Aerobic Decomposition: Organic matter + $O_2 \rightarrow CO_2 + H_2O$ + new cells + stable products (completely oxidized end-products like NO_3, PO_4, and SO_4)

Anaerobic Decomposition: Organic matter $\rightarrow CO_2 + H_2O$ + new cells + unstable products (reduced compounds like H_2S, NH_3, and CH_4)

Biochemical oxygen demand (BOD) is the amount of oxygen required by microorganisms to oxidize organic wastes aerobically and is generally expressed in units of mg/L. All biodegradable organic material in wastewater is converted to biomass, CO_2 and H_2O using O_2. BOD of a sample can be measured in a five-day BOD test and represents the oxygen demand of the biodegradable fraction of organic matter present in the sample.

The Standard Five-day BOD (BOD_5) test is the total amount of oxygen consumed by microorganisms during the first five days of biodegradation under standard test conditions. In this test, diluted wastewater is added to a BOD bottle and the concentration of dissolved oxygen (DO) in the sample before and after five days of the test is measured.

The steps followed in setting up the test and measuring BOD_5 are described here briefly (APHA, AWWA, WEF 2005):

1. Wastewater is diluted so that there is a measurable DO drop during five days of incubation at 20 °C. Based on an estimate of wastewater BOD, an appropriate dilution factor is chosen so as to achieve the necessary DO levels.
2. Nutrients (mostly inorganic) are added to the dilution water to ensure that biodegradable organic matter is the only limiting nutrient for microbial growth. The dilution water is aerated prior to being dispensed in the bottles to achieve an initial DO level equal to DO saturation. Addition of microbial seed is necessary especially if the sample is an industrial wastewater. Municipal wastewaters have more than adequate concentration of microbes in them to initiate growth, hence additional seed is not required.
3. The bottles are incubated in the dark to prevent algal growth in the bottle. Algae, if present, would generate oxygen during incubation and result in incorrect BOD results.
4. The BOD bottles are designed to maintain a water seal; this prevents re-aeration of the samples during incubation.
5. DO levels should be measurable before and after the test. The minimum DO drop should be 2 mg/L and the final DO should be at least 2 mg/L to ensure reasonably accurate results.
6. While standard test conditions are 5 days, it is common practice in India to do the BOD test for 3 days at 27 °C.

Five-day BOD is the ratio of drop in DO over 5 days and the dilution factor (Masters and Ela 2008).

$$BOD_5 = (DO_i - DO_f)/P \qquad 3.2.1$$

where DO_i = initial DO of the diluted wastewater, DO_f = final DO of the diluted wastewater, five days later; P = Dilution factor = Volume of wastewater/volume of wastewater plus dilution water (a standard BOD bottle is 300 mL).

In some cases, the dilution water has significant BOD of its own. In such cases, the dilution water BOD has to be subtracted from the BOD of the mixture to determine the BOD of the sample. To do this, two BOD bottles must be prepared, one containing seeded dilution water and the other containing a mixture of seeded dilution water and wastewater.

Change in the DO levels of both the bottles are noted and the BOD of wastewater is given by

$$BOD_5 = [(DO_i - DO_f) - (B_i - B_f)(1 - P)] / P \qquad 3.2.2$$

Where DO_i and DO_f are DO levels of the mixture before and after the test, and B_i and B_f are the DO levels of the seeded dilution water before and after the test, with P being the dilution factor.

BOD comprises of carbonaceous biochemical oxygen demand (CBOD) and nitrogenous biochemical oxygen demand (NBOD). During the initial growth phase, carbonaceous organic matter is degraded by microbes (mostly heterotrophic bacteria). In this phase, autotrophic nitrifying bacteria that convert ammonia to nitrite and nitrate are outcompeted since both groups are competing for O_2. As CBOD reduces to a significant level, the main substrate of the heterotrophic bacteria (organic carbon) is reduced while the nitrifiers become significantly large in number (they are slow growers and have poorer biomass yield compared to heterotrophs) and begin to exert NBOD. The effect of exerting NBOD is usually observable only after 7 days and is not measured in a standard BOD_5 test.

As the waste is being oxidized, the amount of organic matter remaining in the bottle decreases and eventually becomes zero or the amount of organic matter oxidized increases and eventually becomes constant.

L_0 is the total amount of oxygen required by microorganisms to oxidise the carbonaceous portion of the waste to CO_2 and H_2O and is called the ultimate CBOD (NBOD is neglected since it is generally exerted only after 5 days). BOD_t is the amount of BOD utilized at time t whereas L_t is the amount of BOD remaining at time t.

$$L_0 = BOD_t + L_t \text{ where } L_t = L_0 e^{-kt} \text{ and } BOD_t = L_0(1 - e^{-kt}) \qquad 3.2.3$$

The BOD reaction constant k is a factor that indicates the rate of biodegradation of wastes. As k increases, the rate at which oxygen is utilized increases. Temperature is one of the major factors that affects k. Standard procedure is to conduct the test at 20 °C. If the temperature is other than 20 °C then k is modified using the equation $k_T = k_{20}\theta^{(T-20)}$ where k_{20} is the value at 20 °C and k_T is the value at temperature T (expressed in °C). The most commonly used value for θ is 1.047 (Streeter and Phelps 1925).

NBOD is the amount of oxygen needed to convert ammonia or free nitrogen to nitrate. In aerobic environments, nitrite bacteria *Nitrosomonas* convert ammonia to nitrite and nitrate bacteria *Nitrobacter* convert nitrite to nitrate. This process is called nitrification. The total concentration of organic and ammonia nitrogen in wastewater is known as total Kjeldahl nitrogen (TKN). The oxygen demand associated with N-containing compounds (mainly amino acids) can be determined from TKN concentrations as follows:

$$NH_3 + (3/2)\, O_2 \rightarrow HNO_2 + H_2O \qquad \text{mediated by } \textit{Nitrosomonas}$$

$$HNO_2 + (1/2) O_2 \rightarrow HNO_3 \qquad \text{mediated by } \textit{Nitrobacter}$$

which implies that every mole of ammonia requires 2 moles of oxygen for its complete oxidation. In other words, (2 moles*32 g O_2/moles)/(14 g NH_3 – N/moles) = 4.57 g O_2/ g N. Therefore, the ultimate nitrogenous BOD is

$$\text{Ultimate NBOD} = 4.57 * \text{TKN} \qquad 3.2.4$$

3.2.4 Total Organic Carbon (TOC)

Natural organic matter (NOM) is a term used to describe all organic material present in natural waters. It is generally measured as total organic carbon (TOC). Most NOM is derived from two sources: decayed/decaying organic carbon (also called detrital organic carbon) and phytoplankton like algae. Decaying organic material may be of vegetal or animal origin. Algal material is fresh organic carbon which is generated by photosynthesis. There are significant differences in the character of these two types of NOM which lead to further differences during treatment by physico-chemical and biological processes. Major differences between these two types of NOM are summarized in Table 3.3.

Table 3.3 Differences in characteristics of NOM of algal origin versus humic material

Parameter	Algal material	Detrital organic carbon or humic material
Generation	Conversion of inorganic CO_2 to organic carbon (biomass)	Conversion of organic carbon to new biomass
Biogeochemical cycle of carbon	Fresh carbon	Carbon has been recycled several times
Biodegradability	High	Low, since the more biodegradable fractions have been removed
Chemical nature	Mostly aliphatic, no lignin is present	Mostly aromatic, significant amounts of lignin are present
Molecular weight	Low	High

NOM or TOC is generally not considered toxic but is responsible for many problems in water supply and water distribution systems. NOM is most apparent in water when it contributes to the color, taste, and odor of natural waters necessitating its removal for aesthetic reasons. It leads to the formation of disinfection by-products like chloroform when the water is chlorinated after treatment. It can reduce adsorber bed life and efficiency significantly making treatment by adsorption ineffective. NOM contributes to other problems such as increase in bacterial regrowth potential, sequestration of toxic heavy metals (by chelation) and adsorption of synthetic organic compounds like pesticides and polycyclic aromatic hydrocarbons.

TOC measured after filtration of a sample through 0.45 micron filter is termed dissolved organic carbon (DOC). TOC and DOC measurements are done using a TOC analyzer which is relatively expensive and has high maintenance and operating costs. A cheaper and more popular option is the use of a UV-visible spectrophotometer to measure ultraviolet absorbance at 254 nm which is a surrogate for DOC. Another parameter that is often reported for drinking water sources is specific ultraviolet absorbance (SUVA) which is determined by measuring ultraviolet absorbance at 254 nm and dividing it by the DOC concentration.

3.2.5 Synthetic Organic Compounds (SOC)

There has been an exponential increase in the manufacture and use of synthetic organic compounds (SOCs) in the last century. The best-known examples include pesticides, insecticides and herbicides which in some part have contributed to increased agricultural productivity and decline in various diseases. Other SOCs include pharmaceuticals, personal care products, antibiotics, prescription and non-prescription drugs, steroids and hormones, plasticizers, surfactants, and fire retardants

(Bhandari et al. 2009). These SOCs are often detected during water quality monitoring of rivers and streams in several parts of the world (Kolpin and Meyer 2002). Major sources of these SOCs are residential, agricultural and industrial activities (GRBEMP 2013; Bhandari et al. 2009).

Pesticides: SOCs like pesticides were designed to be persistent in the environment for their efficacy in killing pests and to make the compound cost-effective. Applications of pesticides to fields and homes resulted in these compounds being transported and transformed in every part of the environment. Reports of both acute and chronic toxicity effects in humans and animals due to exposure to these compounds became common. Publication of *The Silent Spring* by Rachel Carson in 1962 led to greater awareness and restraint including banning of many of these compounds. DDT is one of the best documented examples of the potential for bioaccumulation or biomagnification of an SOC and its subsequent consequences. Understanding the environmental fate and transport of many of these compounds remains an ongoing challenge.

Pharmaceuticals: Pharmaceuticals including prescription and non-prescription drugs are discharged into the environment by human use, industrial wastewaters, and animal husbandry activities. Many of these chemicals are endocrine disruptors. Endocrine disruption effects of hormonal pharmaceuticals, due to their high potency at extremely low concentrations, are of particular concern for humans and animals. Antibiotics are an especially important category of pharmaceuticals because of their potential to form and promote antibiotic resistance to human pathogens, and their potential to significantly impact natural microbial consortia (Goel 2015). Other classes of pharmaceuticals, such as analgesics and psycho-pharmacologicals, may also be important due to their strength and common use.

Personal Care Products: Personal care products (PCPs) include those compounds which are marketed for direct use primarily on the human body (mainly dermal contact). PCPs are generally not intended for injection or ingestion. There are thousands of chemicals that are constituents of PCPs. These are diverse and are used as active ingredients or preservatives in cosmetics, skin care, dental care, hair care products, soaps, cleansers, insect repellents, sunscreen agents, fragrances, and flame retardants. Many of these PCPs are used in large quantities, and often at dosages and frequencies higher than recommended.

The active ingredients in a number of PCPs are considered bioactive chemicals. This implies that they have the potential to affect the flora and fauna of soil and aquatic environments. In some cases, bioactive ingredients are first subject to metabolism by the consumer and the excreted metabolites and parent components are then subject to further transformation in the receiving environment. Personal care products differ from pharmaceuticals in that large quantities can be directly introduced into receiving environments (air, surface and groundwater, sewage, sludge and bio-solids, landfills, soils) through regular use such as showering, bathing, spraying, excretion or disposal of expired or used products. They can also bypass treatment systems. PCPs can be considerably persistent and can bio-accumulate in non-target aquatic organisms. Some PCPs have been found to exhibit negative hormonal and toxic effects on a number of aquatic organisms at concentrations in the µg/kg range.

Phthalate Plasticizers and Degradation Products: Plastic products have always been of major concern in terms of toxicity and persistence in the environment. They contain myriad additives including plasticizers, which can make up to 40 percent of plastic formulations. Plasticizers are low molecular weight organic compounds that are essential for effective processing and tailoring of plastic formulations. The production of flexible plastics with multiple applications ranging from automotive industry to medical and commodity products is due to plasticizers. They are manufactured in hundred millions of tons annually and represent an overwhelmingly large fraction of the plastic industry.

It has been established that plasticizers are toxic to some extent and some can exhibit endocrine-disrupting properties. In fact, these compounds may also leach out from the plastics as they are not chemically bound to the plastic polymers. Hence, leaching is a major process for contamination of the environment by the ubiquitous plasticizers. Meanwhile, it has been estimated that the average ingestion rate of plasticizers could be about 8 mg per person per day. Further, their biodegradation in the environment can release breakdown products which can be potentially more toxic than the parent compounds. Hence, there is a need to understand the fate and behavior of plasticizers and their degradation products in natural and engineered environments, including industrial and municipal effluents, sewage sludge, and landfill leachates.

Surfactants: Surfactants (surface active agents or wetting agents) are organic chemicals that reduce surface tension in water and other liquids. Surfactants represent major multipurpose groups of organic compounds. Nearly 3×10^{10} kg of surfactants are produced per year all over the world. The most familiar use of surfactants is in soaps, dishwashing liquids, laundry detergents, and shampoos. Other important uses are in industrial applications such as lubricants, emulsion polymerization, textile processing, mining flocculates, petroleum recovery, and a variety of other products and processes. Surfactants are also used as dispersants after oil spills.

There are hundreds of compounds that can be used as surfactants. These are usually classified by their ionic behavior in solutions: anionic, cationic, non-ionic or amphoteric (zwitter-ionic). The two major groups of surfactants are the anionics and non-ionics with a global production of around 2.5 and 0.5 million tons per year, respectively. Their main components are linear alkylbenzene sulphonates (LAS) for the anionics and alkylphenol polyethoxylates (APEOs) for the non-ionics.

Brominated Fire Retardants (BFRs): Fire retardants are substances that can delay or prevent combustion and some of these eventually enter the environment and are potentially toxic. BFRs are endocrine disrupting chemicals that are potentially toxic and bio-accumulative. There are more than 175 different types of compounds that are used as fire retardants. These compounds can be placed in four categories – halogenated organics, phosphorus-based, nitrogen based and other inorganic compounds. The halogenated compounds (chlorinated and brominated) are the most common fire retardants as they are less expensive and quite effective. In this group, brominated compounds are more common than the chlorinated types. The phosphorus and nitrogen-based fire retardants are mono-ammonium phosphate, di-ammonium phosphate, ammonium polyphosphate, and ammonium sulphate. These compounds individually or in combination are mixed with other chemicals such as corrosion inhibitors, alcohol, gum thickeners and surfactants in a fire retardant formulation. In the other inorganic compounds category, compounds like aluminium hydroxide, antimony oxides and chlorides may be included.

Brominated fire retardants (BFRs) are the most common fire retardants and have serious environmental effects. There are more than 75 different brominated fire retardants (Birnbaum and Staskal 2004). BFRs contribute to about 38 percent of the global bromine demand. The electronic industry is the major consumer of BFRs. Four main applications of BFRs in computers are: printed circuit boards, connectors, plastic covers and cables. BFRs are also used in many other products such as TV plastic covers, carpets, paints, upholstery and domestic kitchen appliances. Their effectiveness for fire prevention and hindrance to the spread of fire makes them very useful fire retardants. Since they are not bound to the polymers, they can enter the environment by leaching.

The bio-accumulation potential of SOCs and subsequent toxicity can be evaluated based on their water solubilities and their tendency to partition into organic versus aqueous phases (Sawyer et al. 2000). K_{ow} is a common parameter measured to determine this tendency of an SOC and is quantified as:

K_{ow} = Concentration of an SOC in octanol/ Concentration of SOC in water. Values of K_{ow} can vary from less than 1 for hydrophilic compounds like acrolein to >1 for hydrophobic compounds like DDT. Since K_{ow} values for SOCs can vary by several orders of magnitude, a more convenient expression is log K_{ow}.

Several studies have demonstrated clear linear relationships between log K_{ow} and bio-concentration factor (BCF). One such study defined the relationship between *Chlorella* (an algae) and 41 SOCs (Geyer et al. 1984). In general, the bio-concentration factor is defined with respect to fish tissue but it can be generally defined as the ratio of the concentration of an SOC in biota (any organism) to its concentration in water.

The degradation of SOCs is generally assumed to follow first-order decay and can be described as:

$dC/dt = -kC$ where k = first-order decay constant, 1/d

C = concentration of the compound at time t

Integrating the above differential equation results in

$\ln(C/C_0) = -kt$ where C_0 = concentration of the compound at time t = 0.

PROBLEM 3.2.1

Compare ThOD for various C6 compounds: glucose [$C_6H_{12}O_6$], hexane [C_6H_{14}], benzene [C_6H_6], and cyclohexane [C_6H_{12}].

Solution

$$C_6H_{12}O_6 + 6O_2 \rightarrow 6CO_2 + 6H_2O$$

ThOD of Glucose = (6*32) / (6*12 + 12 + 6*16) = 1.0667 g O_2 / g of glucose

= (6*32)/(6*12) = 2.67 g O_2 / g C

Similarly,

$$C_6H_{14} + 9.5O_2 \rightarrow 6CO_2 + 7H_2O$$

ThOD of Hexane = 3.5349 g O_2 / g of hexane = 4.22 g O_2 / g C

$$C_6H_6 + 7.5O_2 \rightarrow 6CO_2 + 3H_2O$$

ThOD of Benzene = 3.0769 g O_2 / g of benzene = 3.33 g O_2 / g C

$$C_6H_{12} + 9O_2 \rightarrow 6CO_2 + 6H_2O$$

ThOD of cyclohexane = 3.4286 g O_2 / g of cyclohexane = 4.0 g O_2 / g C.

Thus ThOD of the various C6 compounds follow the order: **Hexane >Cyclohexane >Benzene >Glucose**

PROBLEM 3.2.2

A series of BOD tests were conducted in a lab with different wastewater samples and the DO levels measured after 5 days were: Bottle A = 5 mg/L, Bottle B = 1 mg/L, Bottle C = 0 mg/L, Bottle D = 8 mg/L and Bottle E = 7 mg/L. All samples were diluted by a factor of 1/100. Initial DO in all bottles was 8 mg/L. Are these results useful? If yes, determine the BOD_5 in each bottle. If the results are not useful, state the reasons. Hint: What are the criteria for useful BOD_5 results?

Solution

	DO_f	DO_i	Comments
Bottle A	5.00	8.00	useful
Bottle B	1.00	8.00	useless, min should be 2.0
Bottle C	0.00	8.00	useless, min should be 2.0
Bottle D	8.00	8.00	useless, lack of drop shows toxicity/ non-biodegradable matter
Bottle E	7.00	8.00	useless, min drop should be 2.0
Bottle A, BOD_5	300	mg/L	

PROBLEM 3.2.3

Standard BOD_5 tests are conducted at 20 °C. Any change in temperature will result in a change in degradation rate. At what temperature will the degradation rate increase 3-fold compared to the rate at 20 °C?

Solution

Using the relation $k_t = k_{20}\theta^{(T-20)}$

Where k_t = biodegradation rate at a temperature T (in °C) and

k_{20} = biodegradation rate at 20 °C

θ = assumed to be 1.047

Degradation rate increases 3-fold implies $k_t = 3^*k_{20}$

$3^*k_{20} = k_{20}\theta^{(T-20)}$ can be simplified as $3 = 1.047^{(T-20)}$

Taking log of both sides of the equation and rearranging it results in

T = 20 + (log 3/log 1.047) which results in T = 43.92 ^{0}C

PROBLEM 3.2.4

A wastewater sample has ultimate BOD of 300 mg/L and BOD_5 of 200 mg/L measured at 20 °C. Determine the BOD_3 of this sample at 27 °C. Assume a temperature correction factor of 1.047.

Solution

$BOD_5 = L_0(1-\exp(-kt)) = 200$ mg/L and $L_0 = 300$ mg/L

Solving for $k_{20} = -0.21972$ 1/day

Using the equation for temperature correction: $k_{27} = -0.303$ 1/day

Therefore, BOD_3 for the same wastewater sample = 300(1–exp(–0.303*3)) = 179.1 mg/L.

PROBLEM 3.2.5

BOD tests were conducted on a single river water sample. 10 BOD bottles were prepared and sacrificed periodically for a total incubation time of 30 days. DO in the bottles was measured at the beginning of the test and after sacrificing each bottle. Data from this study is provided below. Draw a graph and explain each phase of the curve.

Time, days	DO consumed ($DO_0 - DO_t$), mg/L
0	0
1	1.09
2	2
3	2.71
4	3.3
5	3.8
7	4.5
9	6
11	7
15	8
20	8.5
30	8.7

Solution

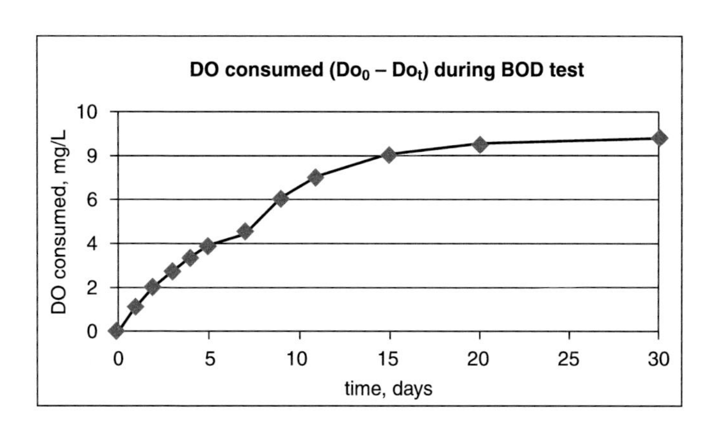

Generally, the BOD_5 test provides no information regarding the oxygen demand exerted by nitrifiers. In the initial phase of the experiment (or incubation period), oxygen and organic matter are available for heterotrophic bacteria to grow and dominate. This part of the oxygen demand is determined in a BOD_5 test and is termed carbonaceous oxygen demand or carbonaceous BOD (CBOD).

Ammonia is converted to nitrite and nitrate by autotrophic nitrifying bacteria. These bacteria are slow-growing bacteria and their impact on oxygen depletion is apparent only at a later stage when organic matter has been consumed substantially, organic-N has been converted to ammonia and heterotrophic bacterial growth is slowing down. Nitrifying bacteria exert some oxygen demand which is shown in Phase 2 of the curve and is termed nitrogenous BOD. Nitrogenous BOD can be evaluated using TKN values as illustrated in a later example.

Phase 1: Carbonaceous oxygen demand is exerted up to approximately day 7 by the growth of heterotrophic bacteria.

Phase 2: Nitrifiers become active and significant in numbers resulting in further depletion of oxygen in addition to that required by heterotrophic bacteria.

PROBLEM 3.2.6

The BOD_5 measured for a wastewater sample is 500 mg/L. The reaction rate constant for the sample was found to be 0.3 1/d. Determine the ultimate carbonaceous BOD of this sample.

Solution

Ultimate carbonaceous BOD, $L_0 = L_t/(1-e^{-kt})$

Ultimate carbonaceous BOD = 643.6 mg/L

PROBLEM 3.2.7

We need to estimate the ThOD of a wastewater (after primary settling) for doing biodegradation studies. We know its general composition in terms of carbohydrates (30 percent), proteins (20 percent), and lipids (50 percent). Average chemical formulas used in research/models for wastewater treatment are carbohydrates ($C_6H_{10}O_5$), proteins ($C_5H_7NO_2$), and lipids ($C_{57}H_{104}O_6$) (Sophonsiri and Morgenroth 2004). Determine the stoichiometric oxygen requirements for degrading this waste. Hint: Determine oxygen requirements in terms of grams O_2/gram C for each fraction and then determine the weighted requirement. What is the total organic carbon (TOC) content of this wastewater?

Solution

Calculation of ThOD of wastewater

Step 1: Calculate ThOD for various constituents of wastewater:

(i) Carbohydrates :

$$C_6H_{10}O_5 + 6O_2 \rightarrow 6CO_2 + 5H_2O$$

ThOD = 2.6667 g O_2 / g C or 6*16/(6*12+10+5*16) = 0.59 g O_2 / g carbohydrates

(ii) Proteins:

Assuming that the ammonia formed is oxidized to nitrate, the balanced reaction for the process is : $C_5H_7NO_2 + 7O_2 \rightarrow 5CO_2 + 2H_2O + NO_3^- + H_3O^+$

ThOD = 3.7333 g O_2 / g C or 0.99 g O_2/ g proteins

(iii) Lipids:

$$C_{57}H_{104}O_6 + 80O_2 \rightarrow 57CO_2 + 52H_2O$$

ThOD = 3.7427 g O_2 / g C or 1.448 g O_2/ g lipids

Step 2: ThOD of the wastewater is obtained by taking the weighted average of ThODs of carbohydrates, proteins and lipids. Thus, we have

ThOD = [(0.30*0.59) + (0.20 × 0.99) + (0.50 × 1.448)]/100 = 1.099 g O_2/g wastewater solids

Step 3: Calculation of total organic carbon (TOC)

To calculate TOC we first have to calculate grams of carbon per gram of respective constituents and then take the weighted average. Thus,

(i) Carbohydrates: (6*12) / (6*12+1*10+16*5) = 0.44 g C/ g Carbohydrate

Similarly,

(ii) Proteins: 0.53 g C / g Protein

(iii) Lipids: 0.77 g C / g Lipid

Thus TOC = (0.3*0.44) + (0.2*0.53) + (0.5*0.77) = 0.62 g C / g wastewater solids.

PROBLEM 3.2.8

The half-life of pesticide A is 100 days. Assuming its degradation follows first-order kinetics, determine the concentration in soil after 1 and 2 years of soil application. Answer can be reported as percent of original concentration, C_0.

Solution

For first-order decay of Pesticide A, ln $(C/C_0) = -kt$

Therefore, if the half–life = 100 days, it implies that C= 0.5 C_0 at t = 100 days.

Solving for k results in k = 6.9 x 10^{-3} 1/day

Therefore, after 1 year, $(C/C_0)*$ 100 = 8.06 percent

After 2 years, $(C/C_0)*$ 100 = 0.65 percent

PROBLEM 3.2.9

DDT has a bioaccumulation factor of 54,000. If the concentration in water is 0.01 micro-g/L, estimate the concentration in fish tissue.

Solution

BCF = $C_{(fish)}/ C_{(water)}$ = 54,000

Solve for $C_{(fish)}$ = 54000*0.01 micro-g/L = 540 micro-g/L.

3.3 INORGANICS

Various elements and inorganic compounds are found in all natural waters. The presence of different elements in water may be due to vegetal and animal detritus, surface runoff, dissolution of rocks and soil in contact with water, and entrainment of gases due to equilibrium with the atmosphere. Since most of these elements are associated with the growth or inhibition of growth of organisms, they are classified into three broad categories: macronutrients, micronutrients and toxic heavy metals.

3.3.1 Macronutrients

Various elements constitute the biomass of any living organism. Typical compositions of a bacterial cell (based on its dry weight) and the human body are shown in Table 3.4 along with the composition of the earth's crust for comparison. Major elements common to both living organisms are carbon, oxygen, nitrogen, hydrogen, phosphorus, sulphur, potassium, sodium and calcium. These elements constitute >1 percent of the total dry weight of the bacterial cell and are defined here as *macronutrients.* Macronutrient requirements for different organisms will vary depending on the chemical composition of the organism, as illustrated by the differences in requirements for sodium and calcium.

Law of the Minimum

Any element that is least available relative to the requirements of an organism is defined as the limiting nutrient since the growth of the organism is directly proportionate to the concentration of the limiting nutrient.

In general, any one of the macronutrients may be considered to be the limiting nutrient in an aquatic system. However, *it is important to note that any element (macro- or micro-nutrient) can be the limiting nutrient for the growth of an organism.*

Table 3.4 Elemental composition of a bacterial cell, human body and the earth's crust (all units are percent dry weight)

Element	Percent in bacteria (by dry weight)[27]	Percent in human body (by dry weight)[28]	Percent in earth's crust[29]
Carbon	50	61.7	0.022
Oxygen	20	9.3	46.4
Nitrogen	14	11.0	-
Hydrogen	8	5.7	0.14
Phosphorus	3	3.3	0.11
Sulphur	1	1.0	0.03
Potassium	1	1.3	2.1
Sodium	1	0.7	2.4
Calcium	0.5	5.0	4.1
Magnesium	0.5	0.3	2.3
Chlorine	0.5	0.7	0.013
Iron	0.2	-	5.4
Aluminum			8.1
Silicon			28.2
Titanium			0.5
Others....	0.3	-	-

27 Metcalf and Eddy 2003.

28 http://www.gly.uga.edu/railsback/1121CrustComposition.jpeg.

29 http://www.newton.dep.anl.gov/askasci/zoo00/zoo00432.htm.

3.3.2 Micronutrients

Micronutrients are defined as those elements that are essential for growth but can be toxic at higher than optimum levels. A typical growth curve is shown in Figure 3.2 where the organism's growth rate is plotted as a function of the magnitude of the growth factor. Growth factors can include concentration of the limiting nutrient, and environmental conditions like pH, moisture and temperature.

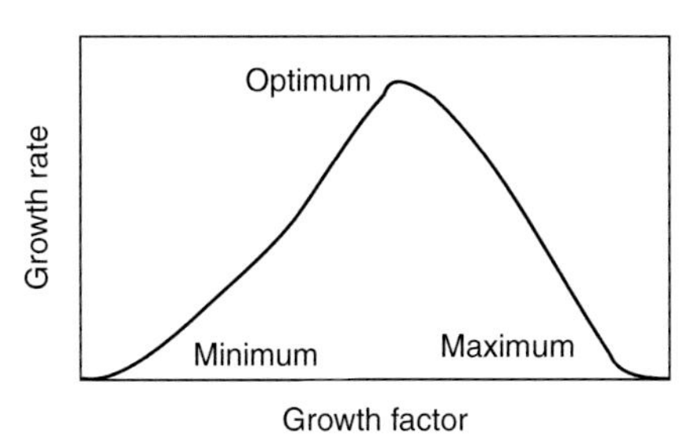

Figure 3.2 Generic graph showing relation between growth factors including nutrients versus growth rates.

Most micronutrients are metals and the most common ones are listed in Table 3.5 along with their sources and health effects on humans.

Table 3.5 List of micronutrients, their sources and health effects at sub-optimum and greater than optimum levels (GRBEMP 2013)

Micronutrients	Major sources	Effect on human health	
		Sub-optimum level	> Optimum levels
Iron	Natural deposits; mining[30]	Anaemia, poor central nervous system development[31] (Lozoff and Georgieff 2006)	Cause oxidative stress and cellular damage, conjunctivitis, choroiditis and retinitis (Higgins et al. 1992)
Copper	Natural/industrial deposits; wood preservatives, plumbing, mining, electroplating, smelting operations (USEPA 2007)	Anaemia (Williams 1983)	Gastrointestinal irritation (USEPA 2007)
Zinc	Smelting, electroplating (Agency for toxic substances and disease registry (ATSDR 2005))	Growth retardation, diarrhoea, alopecia, nail dystrophy, decreased immunity, and hypogonadism in males (Saper and Rash 2009)	Skin irritation, vomiting, nausea, anaemia, arteriosclerosis (ATSDR 2005)

(*Contd.*)

30 Iron. Water treatment solutions. http://www.lenntech.com/periodic/elements/fe.htm.

31 Iron deficiency. http://en.wikipedia.org/wiki/Iron_deficiency.

Micronutrients	Major sources	Effect on human health	
		Sub-optimum level	> Optimum levels
Calcium	Natural deposits; alloys (Martin and Griswold 2009)	Osteopenia, osteoporosis (Nordin 1997)	Kidney stones and sclerosis of kidneys and blood vessels[32]
Magnesium	Natural deposits, glass, cement manufacture (Martin and Griswold 2009)	Gastrointestinal disorders (Rude 1988)	Nausea, vomiting, impaired breathing[33]
Nickel	Metal alloys, electroplating, batteries, chemical production (Das et al. 2008)	Anaemia (Anke et al. 1984)	Heart, liver damage, contact dermatitis, lung and nasal cancer (Duda-chodak and Blaszcyk 2008)
Cobalt	Natural deposits, metal alloys, ceramics and paints (Payne 1977)	Pernicious anaemia	Polycythaemia, cardiomegaly or diffuse interstitial pulmonary fibrosis
Manganese	Natural deposits, municipal wastewater discharges, mining, combustion of fossil fuels[34]	Skeletal deformation	Central nervous system abnormalities (Santamaria 2008)

3.3.3 Toxic Heavy Metals

Toxic heavy metals find their way into aquatic systems through natural as well as anthropogenic sources. Surface runoff from polluted or unpolluted areas and dissolution of soil and rocks are major sources of toxic heavy metals. Anthropogenic sources of heavy metals in the environment comprise of discharges of municipal and industrial wastewaters and leachates from open dumps of solid waste materials or contaminated sites. These metals do not serve any known physiological function and can be toxic at very low levels (in the parts per billion range). A summary of some toxic heavy metals, their sources and their effects on public health is provided in Table 3.6. Arsenic is a well-known example of a geogenic pollutant that is toxic at 'ppb' levels and is found in groundwater in many regions of the world; more information regarding arsenic is provided in Chapter 7. Mercury is a well-documented example of an anthropogenic pollutant in aquatic systems as well as in the rest of the environment.

***Mercury poisoning*:** The most significant case of mercury poisoning was the Minamata disaster which led to the term Minamata disease. From 1951 to 1968, a fertilizer manufacturing company–Chisso Corporation released wastewater containing methyl mercury along with various other contaminants into Japan's Minamata Bay.[35] A disease affecting the central nervous system was reported in 40 patients during the period April to October 1956, out of which 14 patients had

32 Kidney stone. http://en.wikipedia.org/wiki/Kidney_stone.

33 Hypermagnesemia. http://en.wikipedia.org/wiki/Hypermagnesemia.

34 Manganese and Its Compounds: Environmental Aspects. https://www.who.int/ipcs/publications/cicad/cicad63_rev_1.pdf. 2004.

35 http://en.wikipedia.org/wiki/Minamata_disease.

died. Those affected included children, women, cats, and birds and it was concluded that the disease was due to food poisoning probably associated with shellfish and fish contaminated with heavy metals. It was only in 1958 that suspicion turned to methyl mercury. Subsequent studies found maximum Hg levels in the hair of patients suffering from Minamata disease to be 705 parts per million(ppm) and in non-symptomatic residents of Minamata, it was 191 ppm. In comparison, Hg levels in the hair of residents outside the Minamata Bay area were only 4 ppm.

Today, Hg contamination is a global phenomenon and though the levels of Hg releases are not as dramatic as the Minamata Bay incident, chronic low-level exposures are now unavoidable due to release of Hg along with flue gases from thermal power plants and industrial facilities. Mercury in flue gases is in volatile form; its boiling point is 357 °C. As these gases cool, Hg solidifies on the particles in flue gases or on any other available surfaces, and is deposited on land or water by processes of wet or dry deposition. The particles with Hg deposits can be transported over long distances on a global scale.

Table 3.6 List of toxic heavy metals, their sources and health effects (GRBEMP 2013)

Pollutants	Major source	Effect on human health
Arsenic	Natural deposits; smelters, glass, electronics wastes; orchards (USEPA 2001)	Skin, nervous system toxicity
Selenium	Natural deposits; mining, smelting, coal/oil combustion (Martin and Griswold 2009)	Liver damage
Mercury	Natural deposits; crop runoff; batteries, electrical switches, chlor-alkali plants, thermal power plants, fluorescent lamps, hospital waste (damaged thermometers, barometers, sphygmomanometers), electrical appliances, etc. (Martin and Griswold 2009)	Gastrointestinal ulceration, central nervous system damage
Cadmium	Natural deposits; galvanized pipe corrosion; batteries, paints (Martin and Griswold 2009)	Kidney effects
Lead	Natural/industrial deposits; plumbing, solder, brass alloy faucets (Martin and Griswold 2009)	Kidney, nervous system damage
Chromium	Natural deposits; mining, electroplating, pigments (Martin and Griswold 2009)	Liver, kidney, circulatory disorders

3.4 MICROBES

Various living organisms are found in natural waters. From virus particles to bacteria, worms and other higher organisms, the list of organisms that may be present in aquatic systems is seemingly infinite! The classification of organisms has been a somewhat controversial issue and major milestones in this area are noted in Table 3.7. Prior to the currently extant 3-Domain classification first proposed by Woese in 1978, Whittaker's five-kingdom classification was the most popular. In Whittaker's five-kingdom classification, all living organisms are either prokaryotes (bacteria or monera) or eukaryotes (Whittaker 1969). All prokaryotes are unicellular organisms, while eukaryotes can be unicellular (Protista) or multicellular (Plantae or Animalia); Fungi can be unicellular or multicellular. The current 3-Domain classification is based on rRNA sequencing and evolutionary relationships. The three Domains are Archaea, Bacteria and Eukarya (Tortora et al. 2004). Organisms belonging to Archaea and Bacteria are prokaryotes while those belonging to Eukarya are eukaryotes.

Table 3.7 Major milestones in the classification of organisms (Tortora et al. 2004)

Year	Person	Milestone
Before 1735		All organisms were classified as plants or animals
1735	Carolus Linnaeus	Formal 2-Kingdom classification as Plantae or Animalia
1866	Ernst Haeckel	Proposed another Kingdom called Protista that included bacteria, protozoa, algae and fungi
1937	Edward Chatton	Coined the term prokaryote to differentiate bacterial cells from cells of plants and animals
1961	Roger Stanier	Provided current definition of prokaryotes: cells with nuclear material but without a nuclear membrane
1969	Robert H. Whittaker	Proposed Five kingdom classification: Monera for all prokaryotes, Protista (algae and protozoa), Fungi, Plantae and Animalia
1978	Carl R. Woese	Proposed elevating the 3 cell types to a level above kingdom naming them Domains: Archaea, Bacteria and Eukarya based on similarities in ribosomal RNA

The current system of nomenclature of all organisms was established in 1735 by Carolus Linnaeus in which each species is identified by two names: *Genus species.* It is often abbreviated after first mention to *G. species,* for example, *Escherichia coli* is *E. coli,* or *Staphylococcus aureus* is *S. aureus.* A list of microorganisms that are likely to be found in aquatic systems is provided in Table 3.8 along with methods for their identification and quantification.

Microbes or microorganisms are defined as free-living cells (cells or cell clusters) that are not dependent on each other for their primary functions of growth, energy generation and reproduction. This is in contrast to cells of higher organisms that cannot live freely and can only survive as part of the larger organism. While the majority of microbes are either beneficial or benign, some species are *pathogenic* to humans and other animals.

Table 3.8 Microbes and methods for their identification and quantification

Organisms	Size range	Examples	Identification methods	Quantification methods
Virus	10 to 400 nm	Poliomyelitis virus	SEM (Scanning Electron Microscope), TEM (Transmission Electron Microscope)	Plaque assays, particle counting
Bacteria	0.1 to 5 microns	*E. coli*	SEM, TEM, optical microscopy, staining, biochemical assays	Plate counts, membrane filtration, epifluorescence microscopy
Protozoa	10 to 100 microns	*Amoeba, Giardia* and *Cryptosporidia*	Optical microscopy, SEM, TEM	Membrane filtration and epifluorescence microscopy
Algae	1 micron to 100 meters	*Anabaena*	Optical microscopy	Chlorophyll concentration, optical microscopy
Fungi	Hyphae from 2 to 10 microns in diameter and several centimeters in length	Unicellular: Molds, yeasts Multicellular: mushrooms	Optical microscopy	Plate counts and optical microscopy

Pathogens are defined as any disease or infection causing organism. A list of pathogenic organisms associated with aquatic systems and diseases related to their presence is provided in Table 3.9. The most common routes of transmission of these diseases are ingestion of water contaminated by human or animal feces, and inadequate personal hygiene. There are four categories of water-related diseases (Tate and Arnold 1990; Masters and Ela 2008):

1. Waterborne diseases: Ingestion of feces-contaminated water can lead to various diseases like cholera, typhoid, hepatitis, gastroenteritis, and dysentery. Diarrheal diseases are the most important category of waterborne diseases and are responsible for most of the morbidity and mortality in infants in developing countries. Globally, more than 2 billion people are estimated to be at risk of diarrheal diseases.

   A common phenomenon in many cities in India and other developing countries is the increase in incidence of these diseases during the monsoon season. This is due to contamination of water supply lines by sewage pipes due to leaks and water-logging.

2. Water-washed diseases: Diseases that occur due to improper hygiene and sanitation are included in this category. Inadequate amounts of water for washing and bathing can result in diseases of the skin and eyes. Examples include scabies, dysentery, leprosy, conjunctivitis, and skin infections.

3. Water-based diseases: If water is the habitat for the pathogenic species or if the pathogen is dependent on or spends an essential part of its life cycle in water, and transmission occurs through direct contact with humans, those diseases are considered to be water-based diseases. An important example of a water-based helminthic disease is schistosomiasis in which contact with water contaminated by the fecal matter of an infected person can result in infection in those in contact with the same water.[36] Schistosomiasis is caused by blood flukes which are trematode worms. An estimated 206.4 million infected people required treatment in 2016 and 91.4 percent of these people live in Africa.
4. Water-vectored diseases: Insect vectors like mosquitoes which require water for their habitat can infect humans with the pathogenic species. No direct contact of water with humans is necessary for the transmission of this disease. Mosquitoes can transmit bacterial, viral or protozoan pathogens. It is estimated that 700 million people are infected and 1 million deaths occur every year due to mosquito-borne diseases. A list of the most common mosquito-borne diseases, pathogenic microbes that depend on mosquitoes for their transmission and the mosquito species is provided in Table 3.10.

Table 3.9 Some pathogenic microbes and diseases caused by them

Pathogenic organism	Microbe	Disease caused
Hepatitis A virus	Virus	Infectious hepatitis
Poliovirus	Virus	Poliomyelitis
Salmonella typhi	Bacterium	Typhoid
Vibrio cholera	Bacterium	Cholera
Giardia lamblia	Protozoan	Giardiasis (gastroenteritis)
Cryptosporidium	Protozoan	Cryptosporidiosis

Table 3.10 Mosquito-borne diseases and the pathogenic species that cause the infections[37]

Pathogenic species	Microbe type	Mosquito species	Disease
Plasmodium sp.	Protozoan	*Anopheles*	Malaria
Arbovirus	Virus	*Culex*	Encephalitis
Roundworm	Worm	*Culex*	Filaria, Helminthiasis
Dengue virus	Virus	*Aedes aegypti*	Dengue fever
Zika virus	Virus	*Aedes aegypti*	Zika fever
Chikungunya virus	Virus	*Aedes aegypti and Aedes albopictus*	Chikungunya

36 http://www.who.int/news-room/fact-sheets/detail/schistosomiasis.

37 https://en.wikipedia.org/wiki/Mosquito-borne_disease.

For purposes of water quality assessment, microbes or unicellular organisms are the most important category, and both prokaryotic and eukaryotic unicellular microorganisms are discussed here.

3.4.1 Virus

Virus particles are not considered to be living organisms since they lack metabolic or reproductive abilities which are necessary characteristics of all living organisms (Madigan et al. 2003). They are static, stable structures (nucleic acid with a protein coat) in the extracellular state and are non-living. The viral nucleic acid can be DNA or RNA and it can single or double-stranded. Once a virus infects a host cell by injecting its *genome* (all the genetic material of an organism) into it, it can be defined as a living organism. After infection, copies of the virus particles are formed causing the host cell lysis and viruses are released into their environment. Viruses are responsible for causing genetic mutations and various diseases in humans and other organisms.

Many viral infections occur due to exposure to feces-contaminated water and include poliomyelitis virus and hepatitis A virus.

3.4.2 Bacteria

All bacteria are prokaryotes in contrast to all other living organisms that are eukaryotes. Prokaryotes are considered to be the most primitive organisms on an evolutionary basis. Fossil evidence suggests that prokaryotes were present 3.5 billion years ago while eukaryotes have been found only around 2 billion years ago (Madigan et al. 2003). Table 3.11 provides a summary of major differences between prokaryotes and eukaryotes.

Prokaryotes are further categorized as Archaea and Eubacteria, or simply Bacteria. From an environmental or water quality perspective, Archaea are important since they inhabit 'extreme' environments such as high temperature locations like hydrothermal vents (deep sea vents), hot water springs and produce methane. These prokaryotes are not known to be pathogenic. Examples include *Methanosarcina barkeri*, and *Thermoplasma acidophilum*. Bacteria, on the other hand, are found in most 'normal' environments including soil, water, sediment, and air. Examples of Bacteria commonly found in water are *Pseudomonas aeruginosa*, *Klebsiella pneumoniae*, *Escherichia coli*, and *Vibrio cholera*. All known prokaryotic pathogens belong to the domain Bacteria.

Genome sequencing of Archaea and Bacteria

The first Bacterial genome was sequenced in 1995 and the first Archaeal genome was sequenced in 1996. Genome sequencing of 659 Bacteria and 52 Archaea was available for analysis (Koonin and Wolf 2008). Bacterial genomes ranged in size from 180 kilo bases (Kb) to 13 million bases (Mb) with a bimodal distribution; the highest peak was at 2 Mb and the second peak was at 5 Mb. Archaeal genomes have a smaller size range from 0.5 Mb to 5.5 Mb with a distinct singular maximum at 2 Mb.

Table 3.11 Major differences between Prokaryotes and Eukaryotes

	Prokaryotes	Eukaryotes
Nucleus or nuclear material	No nuclear membrane or real nucleus, only nuclear region	True nucleus with nuclear membrane
Membrane-enclosed organelles	Absent	Present
Cell wall	Usually present, and complex in structure	When present, it is simple in structure
Chromosome (DNA)	Single, circular strand of DNA	Multiple linear chromosomes
Ribosomes	70S	80S; mitochondria and chloroplasts are 70S
Reproduction	Binary fission: two daughter cells are produced by division of a single parent cell	No binary fission; Protists (unicellular eukaryotes) can reproduce asexually; sexual reproduction occurs by mitosis and meiosis

Bacteria are ubiquitous in the environment, and water supplies are no exception. Most bacteria are beneficial to human health; in fact, our ability to digest complex foods is, in part, due to the flora of the gastrointestinal tract. However, some species are always pathogenic while others are opportunistic pathogens. An example of a water-borne bacterial pathogen is *Salmonella typhi* which causes typhoid. *Escherichia coli* is a common example of an opportunistic pathogen that is also a regular inhabitant of the gastro-intestinal tract of humans and other warm-blooded animals.

Microbe-host relationships

Microbial interactions with other organisms are termed microbe-host relationships. The host may be human or any other organism. There are three main types of microbe-host relationships (Madigan et al. 2015):

1. **Neutral** where there is no effect on the host; the normal microbial flora of the human body is either neutral or symbiotic.
2. **Symbiotic** or mutually beneficial where both, microbe and host are benefitted.
3. **Pathogenic** where the microbe is antagonistic to the health of the host.

The outcome of the relationship or the occurrence of disease in the host depends on microbial virulence, i.e., the ability to damage the host's body and the host's immunity or resistance to attack. The outcome of any microbe-host pair is not constant since external or environmental factors will influence it. Further, many microbes are opportunistic pathogens that cause disease due to environmental factors, strain specificity and lower immune response of the host.

Direct Methods for Detecting and Enumerating Bacteria in Water

The objective of monitoring bacteria in water is to detect fecal contamination of water supplies. Isolation and identification of individual bacterial species is time-consuming, expensive and is not required under normal conditions. Monitoring of a group of bacteria, i.e., coliforms, for water quality is essential according to WHO guidelines and standards in most countries and has been in practice since 1893 (Winslow et al. 1916). Coliforms are indicator organisms that are used as surrogate parameters for pathogenic bacteria and may or may not be pathogenic themselves. They are defined as aerobic or facultatively aerobic, Gram-negative, non-spore forming, rod-shaped bacteria that ferment lactose with gas formation within 48 hours at 35 °C (Madigan, Martinko and Parker 2003). Most, but not all coliforms, are regular inhabitants of the gastro-intestinal tract of humans, birds and other animals and are also found in soil. *E. coli* and *Klebsiella pneumoniae* are examples of enteric bacteria. Two tests are usually done for coliforms: total and fecal coliforms. The presence of total coliforms in water supplies indicates the possibility of fecal contamination which is then confirmed by testing for fecal coliforms.

Various standard methods for monitoring coliforms and other bacteria are available and some common methods are described here:

1. **Plate Counts**

   Plate counts are used to determine the number of viable cells or colony forming units (cfu) in a sample of water or wastewater. Standard glass petri dishes of 9 cm diameter are used for this method. Nutrient media along with agar (to provide a gel-like substratum for bacterial growth) is added to these plates. Water or wastewater samples are added to the plates and the plates are incubated in the upside-down position for 24 to 48 hours at 35 °C. After incubation, single colonies are visible in the media or on the surface of the media. Each colony is assumed to represent a single cell. It is important to note that in reality, cells are often found in aggregate or cluster form and therefore, the number of colonies is an underestimate of actual cell concentrations. The cell counts are reported as heterotrophic plate counts (HPC) or viable cell (VC) counts.

   There are two methods for doing plate counts: spread plate and pour plate. In both cases, the sample is diluted to a level that will result in a quantifiable number of colonies. The statistically significant range for detection of colonies on standard petri plates/dishes is 30 to 300 colonies (APHA et al. 2005).

   Spread plating: Standard petri plates are prepared with autoclaved media containing agar. The media is allowed to come to room temperature. The sample to be plated is diluted appropriately so that quantifiable or countable numbers of colonies can be obtained. An aliquot of 0.1 to 1 mL of the diluted sample is spread (with a sterile glass or plastic L-shaped rod) to cover the entire surface of the media in the plate. It is important to plate at least 3 or more dilutions of the sample to ensure at least one quantifiable result. The plates are incubated for a specified period in upside-down position and counted for colonies at the end of the incubation period. Some examples of spread plate results are shown in Figure 3.3.

***Pour plating*:** For pour plating, 0.1 to 1 mL of sample is poured into the plate after appropriate dilution. Autoclaved media is added to the pour plate and gently spread to cover the entire surface of the plate. The plates are incubated and counted as mentioned above. The pour plate method is easier in terms of sample preparation but can lead to underestimation of bacterial counts due to cell death from heat shock (the agar at the time of pouring is about 45 to 50 °C). This can be a matter of concern especially for drinking water samples. The pour plate method has been described in Standard Methods since 1915.[38]

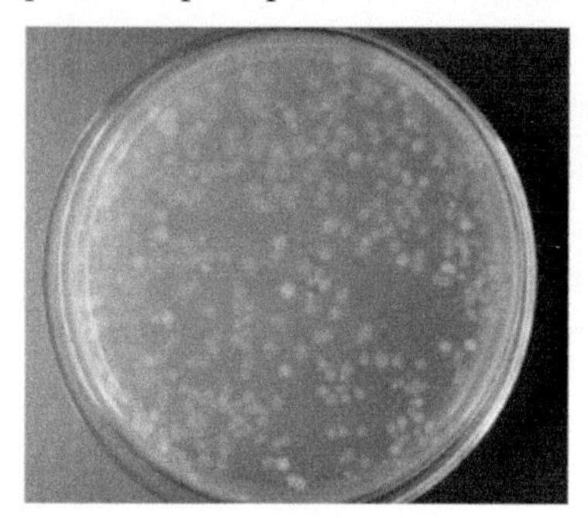

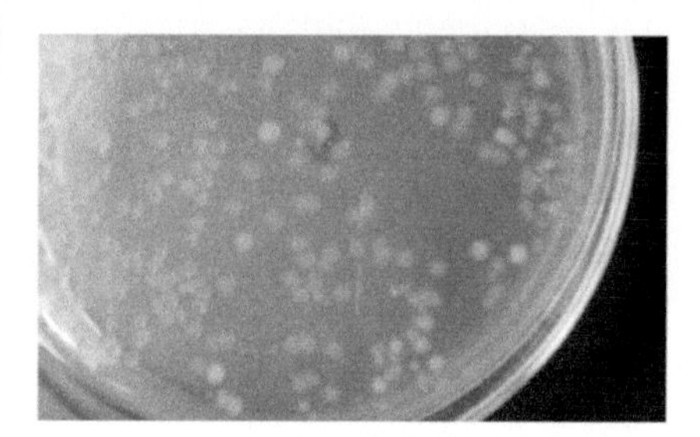

Figure 3.3 Bacterial colonies seen on spread plates after incubation.

2. Membrane Filtration

Membrane filtration is a method for enumerating bacteria in drinking water samples. One of the earliest publications describing the use of membrane filtration for enumerating bacteria in drinking water samples was in 1951 (Goetz and Tsuneishi 1951). The main advantage of membrane filtration is that large volumes of sample can be filtered and cells captured on the filter can then be incubated on media. The result is greater sensitivity and lower detection limits for bacterial cells compared to plating methods. For example, if the objective is to detect 1 or 2 fecal coliforms in 100 mL of sample, plating may or may not give positive results with a sample size of 0.1 to 1 mL. However, there is no limit to sample size with membrane filtration and the probability of capturing and detecting a single coliform increases in proportion to the sample volume filtered.

As shown in Figure 3.4, different volumes of water sample were passed through the membrane filters by vacuum filtration. After filtration, the membrane was placed in a petri dish containing a moist absorbent pad with nutrients for the growth of heterotrophic bacteria. The petri dish was then incubated for 24 hours at 35 °C. The red color of the colonies is due to a chromogenic compound added to the media for making identification easy and the printed grid helps in counting colonies. Differences in colony size are indicative of the presence of cell clusters or aggregates in the water sample. Different nutrient media can be used for a specific species or a group of organisms. Fecal coliforms can be detected with M-FC media while total coliforms can be detected with m-Endo media. Sterile membrane packs with the specific nutrient media required for detection of different groups of bacteria are available commercially.

38 https://catalog.hathitrust.org/Record/000494277.

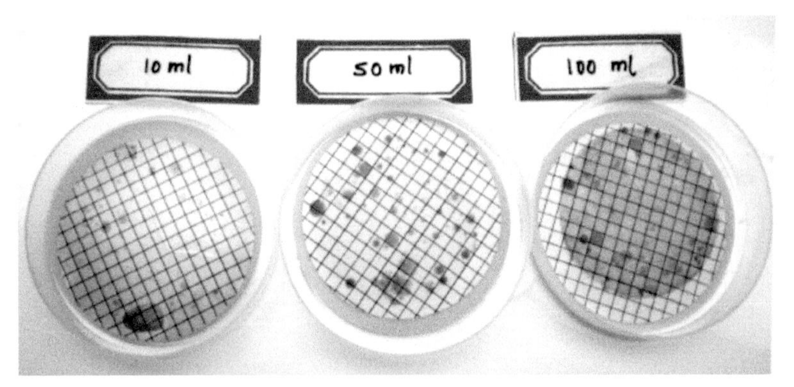

Figure 3.4 Photograph of bacterial colonies detected in different volumes of the same water sample and grown on membrane filters resting on absorbent nutrient pads (Source: B. Mahto) (see color plate).

3. Microscopic Methods

The unaided human eye is capable of resolving objects as small as 0.2 mm depending on various factors such as light intensity and wavelength, contrast with background, size and shape of object and the shadows it casts. Optical microscopes can extend this limit to about 0.2 microns so that microbes like bacteria and protozoa are distinguishable. However, bacteria with a specific gravity and composition that is very close to water are not easily distinguished unless they are stained with a dye. Gram stains are the most common procedure for staining and identifying bacteria under bright-field optical microscopes. Two examples of bacteria: Gram negative coliforms and Gram positive staphylococci are shown in Figure 3.5.

Another method is acridine orange direct cell (AODC) count method where a fluorescent dye like acridine orange is used to stain bacteria and other microbes (APHA et al. 2005). These organisms can then be observed using phase-contrast microscopes or epifluorescence microscopes as shown in Figure 3.6. This method is used in conjunction with membrane filtration to observe and enumerate bacteria.

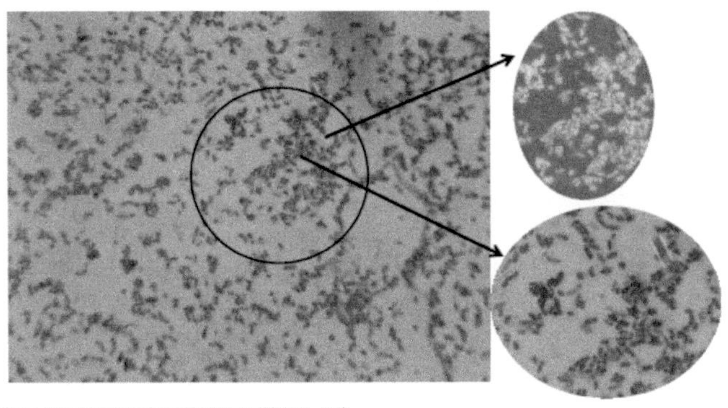

Type of microorganism: Coli from (Gram –ve)
Method of isolation: fermentation tube method
Source: Wastewater (1:100 dilution)
Magnification : 100X (Motic Microscope)

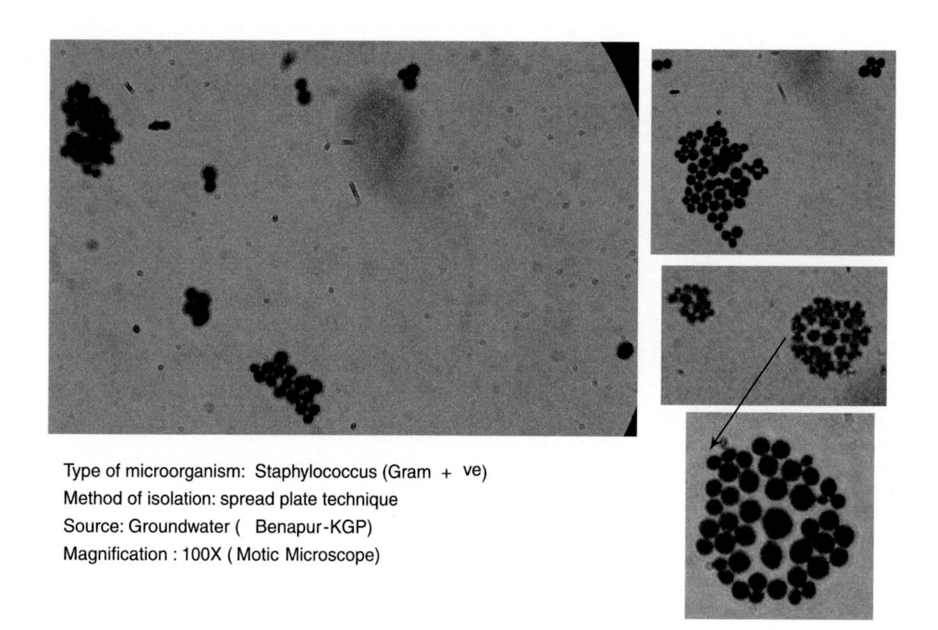

Figure 3.5 Gram-stained bacteria under an optical microscope with 100x objective lens (Source: Iti Sharma) (see color plate).

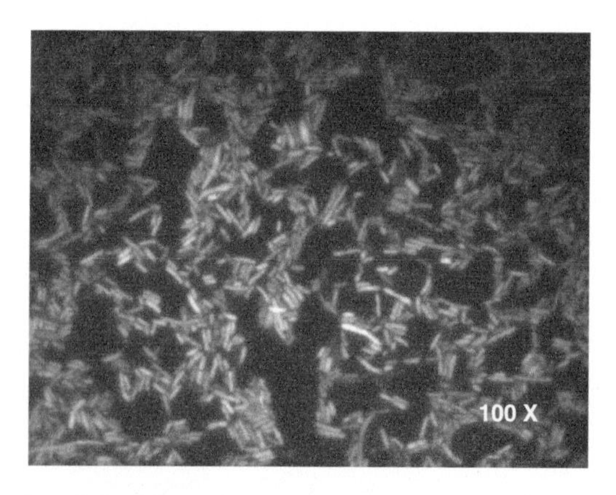

Type of microorganism: Coli form
Date: 3-4-12
Method of isolation: Fermentation tube method
Source: wastewater (1:100 dilution)
Image:Fluorescence microscope (Dye: Acridine orange), Filter: TRITC (Motic Image Advanced 3.2)

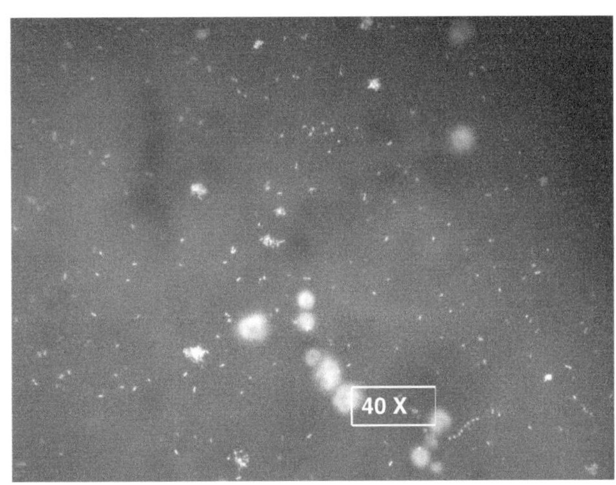

Type of microorganism: unknown
Date: 29 -3-12
Method of isolation: Filtration (0.2μm)
Source: tap water
Image: fluorescence microscope (Dye:Acridine orange), Filter: FITC (Motic Image Advanced 3.2)

Figure 3.6 Bacteria observed under epifluorescence microscope based on the AODC method and different filters (Source: Iti Sharma) (see color plate).

Electron microscopes are used to further extend the limits of the human eye and can magnify objects as small as 20 nm with scanning electron microscopes (SEMs) and 2 nm with transmission electron microscopes (TEMs). SEM allows visualization of the whole cell and its morphology while TEM allows cell organelles to be detected. Higher magnification allows identification of bacterial species to some extent based on their morphology and can be confirmed using standard biochemical identification techniques (Madigan, Martinko and Parker 2003). Examples of images generated using SEM are shown in Figure 3.7 for *E. coli.*

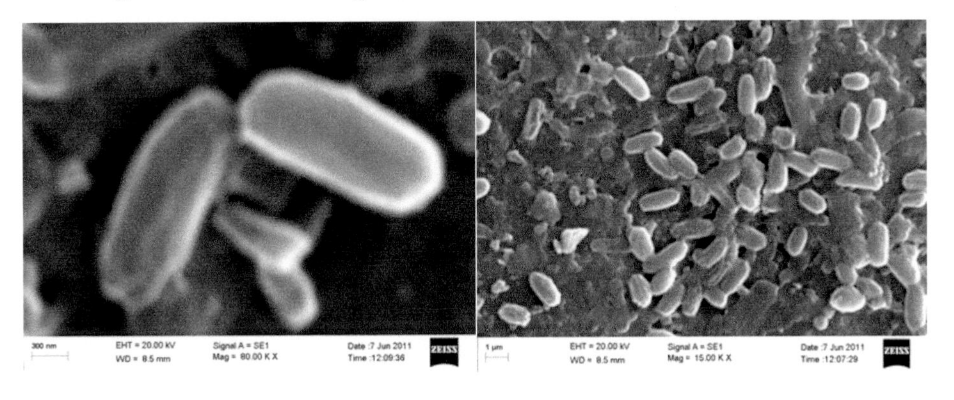

Figure 3.7 Scanning electron micrographs of E. coli (Source: S Mandal, B Mahto).

3.4.3 Protozoa

Protozoa are unicellular, eukaryotic microbes that lack cell walls, are usually colorless and motile. Examples of genera include *Amoeba* and *Paramecium* (Madigan et al. 2003). They can be distinguished from algae by their lack of chlorophyll, and from yeasts by their motility and lack of cell wall. Protozoa such as *Amoeba, Giardia* and *Cryptosporidium* are common pathogens in water supplies and are responsible for causing amoebiasis, gastroenteritis or giardiasis and cryptosporidiosis, respectively.

3.4.4 Algae

Algae are found in soil and aquatic environments and are primary producers in the food chain. Their nutrient requirements are minimal and they convert CO_2 to biomass in the presence of light. Their presence in water supplies can be a major concern due to 'blooms' which are large surficial growths of algae found on surface water bodies. Various extracellular compounds or 'exudates' are excreted by the algal cells contributing to color, taste, and odor problems, and NOM concentrations. Some of these 'exudates' are responsible for toxicity to humans and other organisms.

Algae are eukaryotes that are found in unicellular, colonial (e.g., Volvox), or multicellular forms (e.g., Kelp). Unicellular algae are termed Microalgae while the multicellular forms are termed Macroalgae. Some algae are motile, usually with flagella. They are phototrophic organisms that have chlorophyll-containing organelles called chloroplasts which give them their characteristic green color (Madigan et al. 2003). However, brown and red algae are also found and *Gonyaulax*, a dinoflagellate is associated with the classic 'red tides' that occur frequently on European shores. These red algae are associated with fish kills and poisoning in humans due to ingestion of water or shellfish. Toxicity is due to a neurotoxin that is produced by the algae and can kill at nanogram levels. *Pfisteria* is another toxic dinoflagellate that was responsible for causing massive fish kills and various human health problems in North Carolina in 1991.

Green algae constitute the algal group *Chlorophyta*. Other major groups include Diatoms, *Euglenophyta, Dinoflagellata, Chrysophyta, Phaeophyta,* and *Rhodophyta*. Algae in all groups except *Euglenophyta* have cell walls. The cell walls are made of cellulose and in the case of diatoms, they are composed of silica. Diatoms are often used along with sand in filter beds in water treatment plants. They range in size from 2 to 200 microns and are found in freshwater, seawater and soil. Living diatoms are estimated to be producing 20 percent of total oxygen produced on the planet each year while dead diatoms contribute to an enormous amount of sediment accumulation on the ocean floor.[39]

3.4.5 Fungi

Fungi along with bacteria are the major decomposers in any ecosystem and are responsible for the recycling of nutrients trapped in biomass in various bio-geochemical cycles. They are present in all

39 https://en.wikipedia.org/wiki/Diatom.

parts of the environment: air, water, soil, and food and can grow on any surface including rocks. Fungi can be unicellular or multicellular, and are heterotrophic eukaryotes. The presence of chitin in the fungal cell wall differentiates it from all other organisms. There are three major groups of fungi: yeasts and molds which are unicellular, and mushrooms which are multicellular.

Lichen is a special case of a symbiotic relationship between algae and fungi that allows both species to survive in an inhospitable environment. The fungi attach themselves to a rock surface and secrete acids that dissolve the substratum releasing nutrients while the algae utilize these nutrients and atmospheric carbon dioxide for their growth. The organic matter produced by the algae serves as food for the fungi to grow on.

Fungi, especially basidiomycetes, are responsible for the decomposition of complex organic materials like paper, cloth and wood. Lignocellulosic material which gives strength and rigidity to woody materials can be decomposed by white rot and brown rot fungi.

3.5 WATER QUALITY IN FLOWING WATER BODIES

Flowing water bodies where water is renewed rapidly over time are known as *lotic* systems in contrast to standing water bodies which are called *lentic* systems where water does not get renewed at all or is renewed very slowly. Examples of lotic systems are rivers and streams while lentic systems include lakes, ponds, and reservoirs. There are significant differences in water quality in the two types of water bodies.[40]

3.5.1 Dissolved Oxygen Sag Curve

Oxygen is one of the most important parameters that determine the health of an aquatic system. When waste is disposed into a surface water body, dissolved oxygen is depleted due to the microbial decomposition of waste in a process termed deoxygenation. In any surface water body, the water is oxygenated by equilibrium with the atmosphere. This process is of greater significance in lotic (flowing) systems since the processes of diffusion and dispersion occur simultaneously. This is in contrast to standing water bodies where reaeration or oxygenation is due to diffusion only, and there is no or little dispersion in the depths of the water body. The two processes of oxygenation and deoxygenation can be modeled for a lotic system by the Streeter–Phelps model or DO Sag Curve. The model is based on the assumption that a stream or river behaves like an ideal plug-flow reactor which means that flow lines do not intersect, there is no mixing in the longitudinal direction and complete mixing in the radial direction.

Saturated dissolved oxygen concentration (DO_{sat}) can be defined for a given water body depending on the salinity and temperature of the water as described in Section 3.1.6. If the actual dissolved oxygen concentration is DO, then DO deficit is defined as the difference between actual and

40 http://www.preservearticles.com/2012021623375/complete-information-on-lentic-and-lotic-aquatic-systems.html.

saturation DO levels, i.e., D = DO_{sat} – DO. The point of discharge of waste can be defined on a time scale or distance scale as $t = 0$ or $x = 0$, respectively as shown in Figure 3.8. Distance and time are related by the velocity of the river, $v = x/t$. As water flows downstream from the point of discharge, the two processes of deoxygenation and oxygenation occur simultaneously resulting in variations in DO deficit.

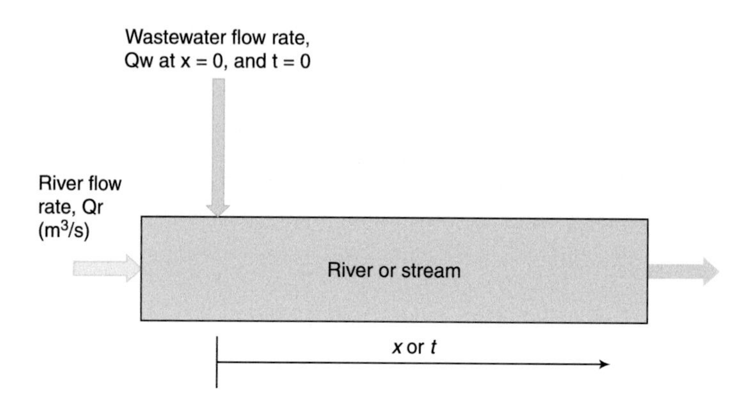

Figure 3.8 Schematic for the Streeter–Phelps model for DO concentration in a flowing water body.

Rate of change of deficit = rate of deoxygenation – rate of oxygenation

$$\frac{dD}{dt} = k_d L_0 e^{-k_d t} - k_o D \qquad 3.5.1$$

Integrating from t = 0 to t results in

$$D = \frac{k_d L_0}{(k_o - k_d)}(e^{-k_d t} - e^{-k_o t}) + D_0 e^{-k_o} \qquad 3.5.2$$

where D is the DO deficit which is the difference between saturated dissolved oxygen DO_s and actual dissolved oxygen DO, L_0 = ultimate carbonaceous BOD, k_d = deoxygenation rate constant and k_o = oxygenation or reaeration rate constant. Also,

$$DO = DO_s - D \qquad 3.5.3$$

Substituting for D from the above equation:

$$DO = DO_s - \frac{k_d L_0}{(k_o - k_d)}(e^{-k_d t} - e^{-k_o t}) - D_0 e^{-k_o t} \qquad 3.5.4$$

results in the ***Streeter–Phelps oxygen sag equation*** (Streeter and Phelps 1925).

Assuming the stream to be of constant cross-sectional area,

$$x = v * t \tag{3.5.5}$$

where v is the velocity of water in the river or stream, and x is the distance traveled from the disposal point of sewage in time t.

Now the above equation reduces to:

$$D = \frac{k_d L_0}{(k_o - k_d)}\left(e^{-\frac{k_d x}{v}} - e^{-\frac{k_o x}{v}}\right) + D_0 e^{-\frac{k_o x}{v}} \tag{3.5.6}$$

Time at which DO level reaches minimum, that is, the stream conditions are worst is called **critical time.** Critical time is calculated by setting the derivative of oxygen deficit (D) as zero. Thus:

$$t_c = \frac{1}{(k_o - k_d)} \ln\left\{\frac{k_o}{k_d}\left[1 - \frac{D_o (k_o - k_d)}{k_d L_o}\right]\right\} \tag{3.5.7}$$

At the critical point corresponding to critical time, rates of deoxygenation and oxygenation are equal. Upstream of the critical point, deoxygenation rate is greater than oxygenation rate as organic matter concentrations are high. Downstream of the critical point, the oxygen starts getting replenished as the rate of deoxygenation decreases with depletion of organic matter, and the rate of oxygenation starts dominating. This phenomenon is known as 'self-purification' of a river or stream.

When wastewater is discharged without treatment, DO levels are reduced to zero resulting in anaerobic conditions in some stretches of the river. Anaerobic conditions are evident when the water has high turbidity, dark color and is malodorous. This model has served as a useful tool for monitoring the health of a river or stream in terms of DO.

Factors affecting the DO sag curve

- ***Seasonal changes*:** During summer, due to decrease in water level, waste concentration increases, resulting in higher BOD concentration in the river or stream, thereby bringing down DO levels.
- ***Temperature effect*:** With increase in temperature, waste decomposition is faster. Thus, rate of deoxygenation increases and DO level decreases. The critical time is achieved sooner in summer compared to winter due to difference in water temperature.
- ***Effect of photosynthesis*:** Photosynthetic organisms like algae and aquatic plants add DO during the daytime, and consume DO during night time due to respiration.
- ***Nitrification*:** Another secondary dip in DO sag curve is observed downstream of the discharge point due to conversion of organic nitrogen and ammonia to nitrites and nitrates.
- ***Waste concentration*:** As waste concentration increases, the amount of DO consumed also increases leading to greater depletion in DO levels.

The above model is based on a single point source of wastewater, i.e., BOD to the river. However, in most river systems there are multiple discharges of wastewater (BOD) to the river which can contribute to increasing depletion of DO to the extent of making some stretches of the river completely anaerobic. Several rivers in India and elsewhere receive both point (wastewater discharges, industrial and municipal) and non-point sources (surface runoff from urban areas and agricultural fields) of pollution which contribute to DO depletion. In many cases, the river's self-purification capacity is destroyed due to these multiple discharges of wastewater and runoff.

In one study, measured DO and BOD data for River Yamuna were used along with reported wastewater discharges in the Streeter–Phelps equation to determine the applicability of the model to Indian river conditions. Large errors in estimated values were found, especially for the stretches where the DO levels were <6 mg/L. These errors were attributed to incomplete reporting of all sources of wastewaters to the river (Kumar and Goel 2018). Low flow volumes in the river, especially during the dry seasons, also lead to severe pollution in some stretches of the river. In two other studies of River Ganga water, measured DO levels did not drop below 4 mg/L despite large-scale organic inputs from point and non-point sources suggesting that photosynthetic activity and high volumes of river water contribute to the self-purification capacity of this river (Tare et al. 2003; Mohanta and Goel 2016).

3.5.2 Nitrogen Species in Rivers

As described in Section 3.2.3, nitrogenous BOD is exerted in a BOD test at a much later time as compared to carbonaceous BOD. Nitrogenous BOD is attributed to the sequential hydrolysis and oxidation of organic N, ammonia, and nitrites to their most oxidized form, nitrate. At any point of discharge of municipal wastewater, whether it is treated or not, the dominant form of nitrogen is organic-N (mainly from urea and uric acid). Immediately after discharge, uric acid and urea are hydrolysed to ammonia. Ammonia is then converted to nitrites by *Nitrosomonas* and nitrites are converted to nitrates by *Nitrobacter*.

Another major source of ammonia or nitrates in a river is surface runoff from agricultural fields where fertilizers have been applied.

3.5.3 Environmental Flows (E-flows)

A major issue in river management is maintenance of environmental flows (e-flows) for sustaining ecosystems. **Environmental flows** can be described as 'the quality, quantity, and timing of water flows required to maintain the components, functions, processes, and resilience of aquatic ecosystems which provide goods and services to people'. The flows of the world's rivers are increasingly being modified when water is withdrawn for agriculture, urban use, and hydropower and when it is not returned to the river through drainage or groundwater flow. Thus the flow of many rivers has been reduced or seasonally altered changing the size and frequency of floods, the length and severity of droughts, and adversely affecting ecosystems. Several rivers run dry before they reach their destination, including, among others, the Aral Sea and the China Hai River.[41]

41 http://water.worldbank.org/topics/environmental-services/environmental-flows.

PROBLEM 3.5.1

Draw DO sag curves for a river with a single point source discharge of wastewater and a temperature of 20 °C. The DO deficit immediately downstream of the discharge is 5 mg/L, L_0 = ultimate BOD at the same point is 10 mg/L, DO saturation is 10 mg/L and k_o = 0.2 1/d and k_d = 0.4 1/d. Determine the minimum DO concentration and the distance from discharge at which DO concentration is minimum, assuming the velocity of the river is 24 km/h?

Solution

The solution to this problem can be derived either graphically or analytically. In either case, equation 3.5.6 is solved to determine DO concentrations along the river length and equation 3.5.7 is solved to determine the time at which DO concentrations are at a minimum, i.e., the critical time – t_c. Critical time can be used to solve for critical distance, x_c, using the river velocity value.

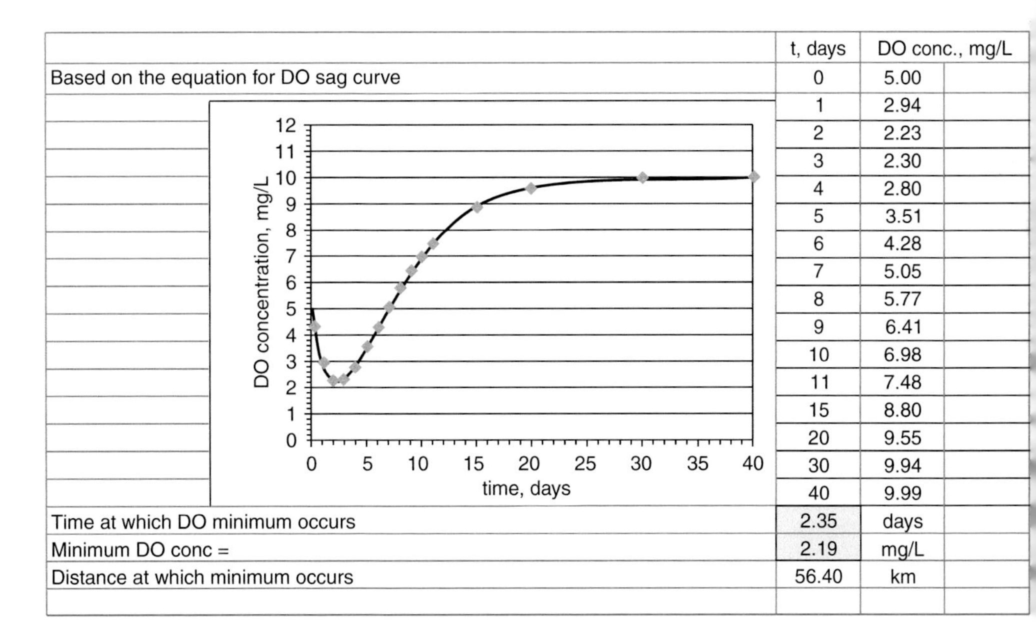

Based on the equation for DO sag curve

t, days	DO conc., mg/L
0	5.00
1	2.94
2	2.23
3	2.30
4	2.80
5	3.51
6	4.28
7	5.05
8	5.77
9	6.41
10	6.98
11	7.48
15	8.80
20	9.55
30	9.94
40	9.99

Time at which DO minimum occurs	2.35	days
Minimum DO conc =	2.19	mg/L
Distance at which minimum occurs	56.40	km

PROBLEM 3.5.2

The oxygenation and deoxygenation rate constants are a function of temperature. Use the temperature correction equation: $k_T = k_{20*}\Theta^{(T-20)}$ where $\Theta = 1.047$ to draw DO sag curves for the previous problem for 2 different water temperatures: 10 and 30 °C. Comment on the difference.

Solution

DO sag curves were drawn for two different temperatures after correcting the rate constants. There is a small difference in t_c values for the two curves; due to rapid reaction kinetics, critical time is achieved earlier at a temperature of 30 °C rather than at 10 °C. However, at a low temperature of 10 °C there is a much bigger difference in the time it takes for the river to reach DO saturation levels. This implies that rivers and streams in temperate regions or at low temperatures take a lot longer to recover from the burden of pollution, i.e., they are more sensitive, as compared to those at higher temperatures.

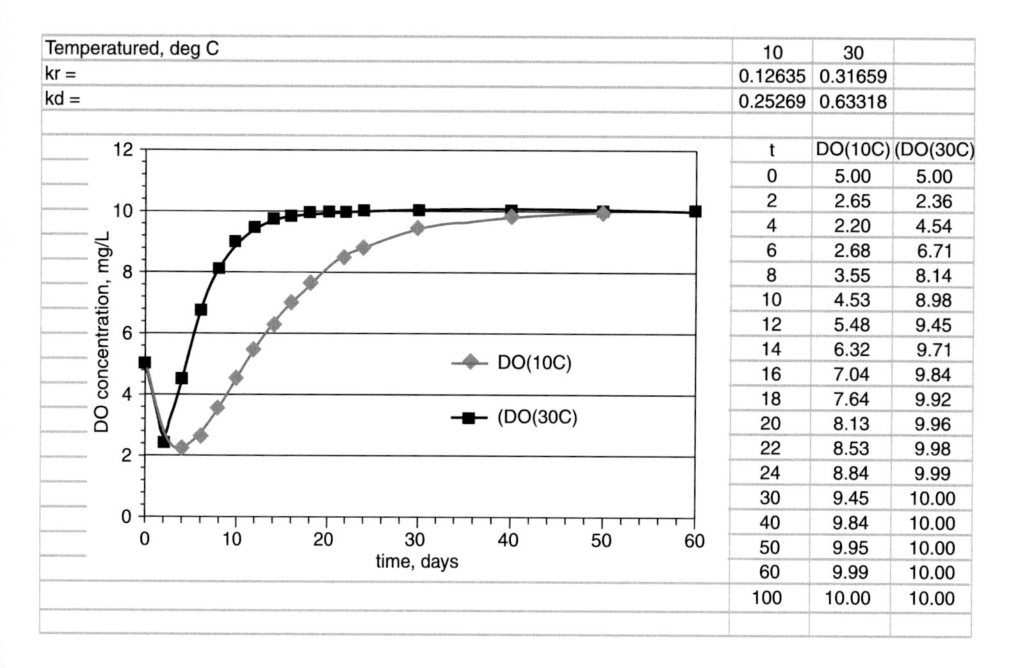

Temperatured, deg C	10	30
kr =	0.12635	0.31659
kd =	0.25269	0.63318

t	DO(10C)	(DO(30C)
0	5.00	5.00
2	2.65	2.36
4	2.20	4.54
6	2.68	6.71
8	3.55	8.14
10	4.53	8.98
12	5.48	9.45
14	6.32	9.71
16	7.04	9.84
18	7.64	9.92
20	8.13	9.96
22	8.53	9.98
24	8.84	9.99
30	9.45	10.00
40	9.84	10.00
50	9.95	10.00
60	9.99	10.00
100	10.00	10.00

PROBLEM 3.5.3

Effluent discharge standard for BOD_3 to an inland river is 30 mg/L. Rework Problem 3.5.1 for a small stream with 1 m^3/s flow. If L_0 = 10 mg/L (downstream of discharge point 1), DO saturation = 8 mg/L and there is an initial DO deficit of 2 mg/L, calculate the minimum DO and the distance at which it occurs.

Solution

Assuming: $k_d = 0.4$ 1/d, $k_o = 0.2$ 1/d

Initial deficit = 2 mg/L

Ultimate BOD (L_0) = 10 mg/L

DO saturation = 8 mg/L

Velocity = 24 km/d

Use equation 3.5.7 to determine Critical time (t_c)= 2.99 d

Distance at which minimum DO occurs = 71.74 km

Use equation 3.5.4 to determine DO_{min} = 1.95 mg/L

PROBLEM 3.5.4

***Sensitivity analysis*:** It is possible to determine which parameter in the Streeter–Phelps equation has greater impact on the predicted DO concentrations by using the following baseline conditions and varying each parameter as follows.

***Baseline conditions*:** A river with a single point source discharge of wastewater and a temperature of 20 °C has a DO deficit of 3 mg/L, immediately downstream of the discharge. L_0 which is the ultimate BOD at the same point is 10 mg/L, DO saturation is 10 mg/L and k_o = 0.2 1/d and k_d= 0.4 1/d. Determine the minimum DO concentration and the time at which the minimum DO concentration occurs.

Vary the following parameters one at a time, keeping all other parameters at the values stated above, and draw DO sag curves for each set of conditions.

DO deficit: 5 and 1 mg/L

Ultimate BOD concentration, L_0 = 20 and 7 mg/L

Deoxygenation constant, k_d = 0.3 and 0.5 1/d

Oxygenation constant, k_o = 0.1 and 0.3 1/d

Solution

DO sag curves for each set of conditions are shown in the following figure. The times (critical times) at which minimum DO concentrations occur for each of these conditions are shown in a table. In the range of values tested, the model was most sensitive to changes in k_o values followed by D, while changes in k_d and L_0 had very similar impact on predicted DO concentrations. Increase in the L_0 value resulted in anaerobic conditions. A decrease in the oxygenation rate constant meant that the river DO did not come back to DO saturation concentration within 30 days as it did in all other cases.

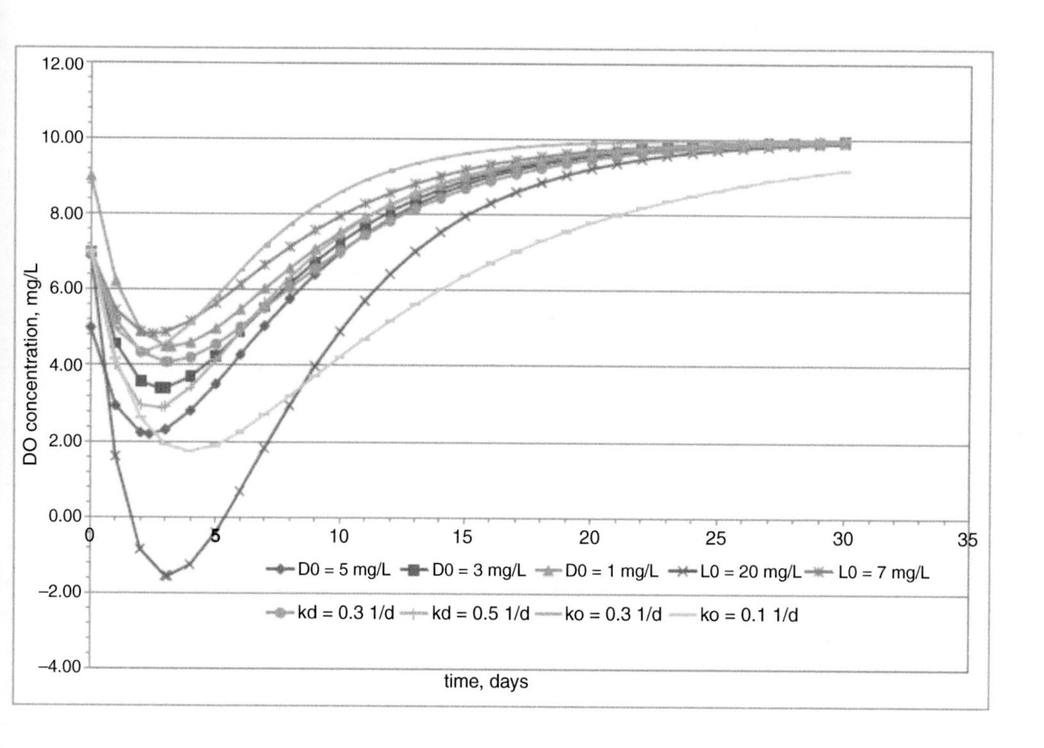

Parameter varied	Parameter value	t_c, days	Change in t_c, days
DO	5 mg/L	2.35	−0.42
DO	**3 mg/L**	**2.77**	**0**
DO	1 mg/L	3.22	0.45
L_O	20 mg/L	3.10	0.33
L_O	7 mg/L	2.49	−0.28
k_d	0.3 1/d	3.10	0.33
k_d	0.5 1/d	2.50	−0.27
k_o	0.3 1/d	2.15	−0.62
k_o	0.1 1/d	3.94	1.17

3.6 WATER QUALITY IN STANDING WATER BODIES

Standing water bodies (lentic systems) like lakes and reservoirs are affected by DO depletion and other environmental problems that may or may not affect lotic water bodies. While all surface water bodies are affected by nutrient discharges, accumulation of nutrients in water and sediments affects standing water bodies to a greater extent than flowing water bodies, resulting in their eutrophication. Further, due to thermal stratification in temperate and colder regions, nutrients accumulate below ice during winter exacerbating water quality problems often with serious consequences. Issues of eutrophication and thermal stratification are dealt with in detail here.

3.6.1 Eutrophication

Many communities rely on ponds, tanks, lakes, and large and small reservoirs for their water supply needs. Natural sources of water to these water bodies are rainfall and surface runoff from the surrounding area, whereas water losses are due only to evaporation and infiltration. Surface runoff, especially in areas with settlements, brings with it a heavy burden of silt, organic matter and nutrients like nitrogen, phosphorus and trace metals essential for microbial growth (both algal and bacterial) and pollutants resulting in reduced water quality over time. These materials accumulate in the standing water body leading to an increase in primary productivity. The increase in primary productivity, i.e., increase in phytoplankton levels, leads to concomitant increase in turbidity. This process of water quality degradation is natural and is associated with the aging of all standing water bodies and is termed *eutrophication.* Eutrophication is accelerated when nutrient inputs increase excessively due to increased urbanization and is sometimes called 'cultural eutrophication' (Masters and Ela 2008). Frequently seen layers of green algae (green scum) in and at the surface of almost all standing water bodies, especially in the summer, are the clearest examples of eutrophication.

Lakes are generally classified based on their nutrient levels as oligotrophic, mesotrophic or eutrophic. Oligotrophic lakes have low nutrient levels resulting in low primary productivity and high water quality. Many Alpine and Himalayan lakes are oligotrophic lakes. Mesotrophic lakes have moderate levels of nutrients and visible proof of primary productivity like algal and weed growth. Most lakes that are used for recreational purposes or as drinking water sources after treatment are examples of mesotrophic lakes. Eutrophic lakes are marked by high levels of primary productivity leading to high chlorophyll concentrations, and highly degraded water quality.

Water quality impacts of eutrophication

As mentioned above, increased inputs of nutrients lead to excessive primary productivity in the form of phytoplankton like algae and weeds. Due to excessive algal growth, layers of scum are visible on the surface of standing water bodies (and sometimes on the fringes of flowing water bodies). These phytoplankton contribute to the presence of exudates (extracellular by-products of phytoplankton growth) that may be toxic, and lead to color, taste and odor problems, high turbidity, and high NOM levels in the water body, resulting in general degradation of water quality. NOM in a drinking water source leads to disinfection by-product (DBP) formation during disinfection with halogenated compounds like chlorine. These disinfection by-products may be cytotoxic or genotoxic, and potentially carcinogenic.

The scum layer at the surface of the water body can lead to the formation of an anaerobic zone below the algal layer since it prevents penetration of light into the water column and the dead planktonic cells contribute organic matter which is utilized by facultative and anaerobic bacteria for their growth. These bacteria will deplete DO in the lower layers of the water column leading to loss of habitat for fish and other higher organisms. Eutrophication can also result in excessive macrophyte (weeds) growth leading to loss of open water.

A simple model is described in Problem 3.6.1 to illustrate the eutrophication process in a lake or reservoir. It is a snapshot of the processes occurring in a standing water body, i.e., time has not been considered as a factor here. A continuous input of nutrients (organic carbon, essential and trace nutrients and pollutants) to a standing water body with no outputs is assumed, leading to net accumulation of these nutrients and acceleration of water quality degradation over time. The end effects are depletion of oxygen, accumulation of pollutants, greater turbidity, and net decrease in water quality.

Many solutions are possible for reducing the rate at which water bodies can eutrophy. Algal harvesting is a simple and inexpensive solution with the added advantage that the harvested algae can be used as a fertilizer supplement. Dredging (and skimming of floating scum) of large ponds and reservoirs is a more difficult and expensive solution and has been tried with limited success in developed countries. Watershed protection for large water bodies is another long-term measure that is essential for slowing down the eutrophication processes that standing water bodies are prone to. Watershed protection must necessarily include prevention of all wastewater discharges to a water supply body. Other essential measures include creating a natural 'buffer zone' of vegetation around the water body to filter the surface runoff to some extent and retain nutrients in soil instead of allowing them to flow to the water body.

3.6.2 Thermal Stratification

In general, the density of all solvents including water increases with decrease in temperature. When water is placed in any large container, a marked stratification is evident with cold water at the bottom and warmer water at the top. However, water has many anomalous properties compared to any other solvent and one of its most important properties is the lack of linear relationship between its density and temperature below 4 °C. Water is most dense at 4 °C, approximately 1000 kg/m^3, while ice at 0 °C is lighter than water between 0 to 4 °C as noted in Table 3.12.

Table 3.12 Relation between temperature and density of water

Temperature, °C	Density, kg/m^3
40	992.2
30	995.7
25	997.0
20	998.2
15	999.1
10	999.7
4	999.999
0	999.87
-10	998.16

The above phenomenon leads to *thermal stratification* in all water bodies and *turnover* of water in temperate and colder regions of the world. Thermal stratification in lakes leads to three distinct strata or layers as shown in Figure 3.9:

a. **Epilimnion:** This is the top layer in the water column and is marked by a negligible decline in temperature with depth. Mixing at the top due to wind and wave action is responsible for the almost uniform temperature profile in the top layer. This layer can extend from 1 m for small lakes to 50 m for large reservoirs or other water bodies. The epilimnion is a layer of warm water floating on top of a layer of cold water.

b. **Thermocline or metalimnion:** This layer is defined by the steepest change in temperature with increase in depth. The depth of this layer ranges from 5 to 10 m depending on the nature and size of the water body. It is also possible that the thermocline is associated with the transition from photic to aphotic zone. As light penetrates into the water column there is some heating of water. Below the photic zone, there is no energy input into the water column leading to the formation of the hypolimnion.

c. **Hypolimnion:** This layer is formed under the thermocline and is marked by a very slow change or insignificant change in temperature. It may also be associated with the aphotic zone.

These strata are stable and constant over time unless the temperature of water at the surface drops to 4 °C or DO levels in water are a function of temperature, depletion or deoxygenation due to biodegradation and reoxygenation due to equilibrium with the atmosphere and photosynthesis. As shown in Figure 3.9, the profiles for DO and temperature vary with season due to changes in water temperature at the surface for a relatively unpolluted water column. Reviews of thermal stratification of lakes can be referred to for more details (Boehrer and Schultze 2008; Kirillin and Shatwell 2016).

Summer: During summer, surface water temperatures are highest and reoxygenation occurs at the surface due to wind and waves. Due to the relatively uniform temperature profile in the epilimnion, DO levels will be inversely proportional to temperature. At the top of the thermocline, DO levels will start increasing in proportion to the decrease in temperature and due to inputs of DO from photosynthesis. This is likely to lead to peak DO levels at the bottom of the thermocline since the temperature is lowest here. Below the thermocline, temperature becomes low and uniform again, and DO levels are expected to rise if temperature were the only factor. However, the hypolimnion is a deep, cold, aphotic zone where no inputs of DO can occur due to photosynthesis or reoxygenation from the atmosphere and DO is depleted due to biodegradation of organic material present in the entire water column. This leads to low DO levels in the hypolimnion.

Autumn: As summer turns to autumn and surface water temperature starts dropping, the temperature and DO profiles start changing. If the surface water temperature is ≤ 4 °C, the density-based stratification of summer breaks up as the epilimnion is no longer lighter than the layers below it and is now at its densest at 4 °C. The entire water column is now mixed by the breaking up of the density gradient leading to the phenomenon of 'turnover'.

Winter: As surface water temperatures continue to dip and eventually go down to 0 °C, a layer of ice is formed at the top that is lighter than the layers of water below it. This layer acts as an insulating blanket and is responsible for the health of most temperate and cold ecosystems as water below the layer of ice remains at a relatively warm and dense 4 °C. Fish and higher organisms can thrive and survive in these ecosystems despite intensely cold conditions at the surface. However, these ecosystems are extremely sensitive to the effects of pollution since there is no input of DO by equilibrium with the atmosphere during winter, or by photosynthesis and DO is depleted by the biodegradation of organic pollutants in the relatively warm meta- and hypolimnion.

Spring: With the onset of spring and rising surface water temperatures, the ice layer thaws and surface water temperatures climb to 4 °C. At this temperature, the density-stratification of winter is broken up and the contents of the entire water column 'turnover' again. In polluted lakes and reservoirs, this often leads to visible fish kills and panic regarding specific contaminants. However, DO depletion in winter below the ice layer due to biodegradation of organic matter is generally responsible for these massive fish kills which are visible only after the ice thaws.

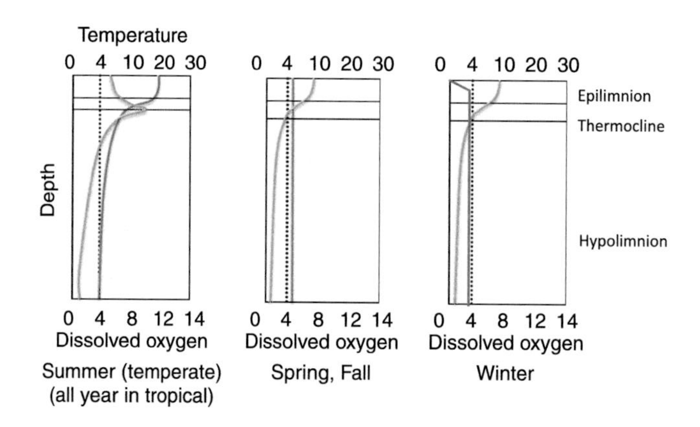

Figure 3.9 Change in Temperature and DO profiles in water with season (see color plate).

PROBLEM 3.6.1

Rainwater has been stored in a pond of 10,00,000 L volume. Concentrations of nitrogen (N) and phosphorus (P) in pond water were found to be 0.1 and 0.04 mg/L, respectively. If we assume that all other nutrients are available in adequate amounts, and growth of algae is limited by the availability of either N or P, we can determine the maximum (based on stoichiometry) amount of algal growth and subsequent bacterial decomposition of dead algae that can occur in this pond.

a. How much algae and oxygen will be produced in this pond? Algal cell composition is assumed to be $C_{106}H_{263}O_{110}N_{16}P$.

b. If bacteria represented by $C_{60}H_{87}O_{23}N_{12}P$ degrade dead algae, a balanced chemical reaction can be used to determine the amount of oxygen consumed in the decomposition process. The question then is, is there sufficient oxygen for decomposing dead algae?

c. Will this pond eutrophy and why?

Algal Production

A simple balanced chemical reaction for algal growth can be shown as follows:

$$106CO_2 + 122H_2O + 16HNO_3 + H_3PO_4 \rightarrow C_{106}H_{263}O_{110}N_{16}P + 138O_2$$

The first issue is to determine which nutrient is limiting growth, N or P. In this case, the growth of 1 mole (3550 g of algae) of algae requires 16 moles N (or 16*14 = 224 g N) and 1 mole P (31 g P). That is the ratio of N:P that is required is 16:1 (molar) or 7.22:1 (weight). The ratio that is present in the pond is 2.5:1 by weight, proving that N is the limiting nutrient for algal growth. In other words, the amount of N that is available for algal growth should be 7.22 times the amount of P, but is actually only 2.5 times the amount of P present.

Based on the N available, we can calculate the maximum possible algal production.

Concentration of N = 0.1 mg/L implies 7.14×10^{-3} mM, obtained by dividing 0.1 mg/(14 mg/milli-mole N)

Concentration of algae produced is dependent on the limiting nutrient, i.e., N. If the N concentration is 7.14×10^{-3} mM and 1 mole of algae require 16 moles of N, then the moles of algae produced will be

$$= 7.14 \times 10^{-3} \text{ mM of N } /(16 \text{ moles of N/mole of algae}) = 4.46 \times 10^{-4} \text{ mM of algae.}$$

MW of algae = 3550 mg/mmoles can be used to convert mM to mg/L resulting in an algal concentration of 1.58 mg/L.

Similarly, the concentration of oxygen produced is 138 moles oxygen per 16 moles of N or 1 mole of algae. Algae: O_2 molar ratio is 1: 138, therefore, for 4.46×10^{-4} mM of algae, 0.06155 mM of oxygen will be produced. This amounts to 1.97 mg/L of oxygen.

Bacterial decomposition of dead algae

All algae that are produced are eventually going to die and sink within the pond resulting in a net input of organic carbon to the pond in excess of what is brought by surface runoff alone. If bacterial cell composition is assumed to be $C_{60}H_{87}O_{23}N_{12}P$ then a balanced chemical reaction for decomposition of algal cells can be used:

$$C_{106}H_{263}O_{110}N_{16}P + 51.5O_2 \rightarrow C_{60}H_{87}O_{23}N_{12}P + 46CO_2 + 86H_2O + 4HNO_3$$

This shows that for every mole (3550 g) of algae that is decomposed by bacteria, 51.5 moles (1648 g) of oxygen are consumed and 1 mole (1374 g) of bacteria are produced.

Therefore, for decomposing 1.58 mg/L of algae, 0.73 mg/L of oxygen will be consumed.

Based on the stoichiometry of the process, yes, there is sufficient amount of oxygen for bacterial decomposition of all dead algae.

Despite the fact that there remains a surplus of oxygen after algal decomposition, the pond is **likely** to eutrophy for the following reasons:

i. Eutrophication is the gradual and natural accumulation of silt and organic matter in lakes and ponds due to the presence of excess nutrients. As long as there are nutrient inputs to the water body, algal and bacterial growth will occur, resulting in net accumulation of organic matter and sediments in the water column.

ii. Significant production of algae will be stimulated by the availability of N and P. New biomass will continue to be produced as long as the two processes of algal photosynthesis and bacterial decomposition continue. Bacterial decomposition ensures that N and P from dead algae is recycled and made available for continued algal and bacterial growth.

iii. Stoichiometric predictions show that no oxygen deficit will occur due to the dual processes of algal photosynthesis and bacterial decomposition of dead alga. This would be true in reality only if the lake or pond is completely mixed. However, we know that algal production occurs only in the surface euphotic layers while bacterial decomposition can occur throughout the water column. Oxygen produced in the surface layers is likely to be lost to the gaseous phase (atmosphere) given its low solubility in water, and only a limited amount will diffuse downwards through the water column. Availability of oxygen for decomposition will depend on mass transfer processes like diffusion where depth of the water column and extent of light attenuation (depth of euphotic zone) are major factors.

iv. For a small, shallow pond, turbidity not depth is likely to limit the depth of the euphotic zone. Formation of algal mats is a likely scenario and is commonly observed. Algal mat formation will limit the availability of oxygen to the bottom layers for decomposition to a greater extent than algal growth in clearer, deeper lakes and ponds.

3.7 WATER QUALITY INDEX (WQI)

Water quality indices (WQI) provide a uniform scale for comparison of water quality in any part of the world with a single value since it is difficult to compare several water quality parameters for different water sources at the same time. WQI can be used to compare water quality spatially as well as temporally. Several water quality indices are available in the literature and one of the earliest ones was formulated by the USA-National Sanitation Foundation (NSF) in 1970 and is described here. The NSF–WQI was based on the Delphi approach where a panel of 142 experts from all over the United States with expertise in various aspects of water quality management provided a list of the most important water quality parameters and their relationship with water quality (Canter 1999). Charts correlating the value of each water quality parameter with a quality index were generated by averaging the responses of experts and are available in the same text or on various websites. A summary of these water quality parameters and their quality indices is provided in Tables C1 and C2.

This method was simple, but qualitative (due to the Delphi approach and chart-based values) and data for only nine water quality parameters are required. The nine water quality parameters, their weights and a table for interpreting the value of WQI are shown in Tables 3.13 and 3.14. The equation used for calculating WQI is as follows:

$$WQI = \sum WiQi / \sum Wi \qquad 3.8$$

If data for all nine water quality parameters are available, then the denominator is 1. If data for less than the nine recommended parameters are available, then the denominator is the total of the weights of the measured parameters. WQI has been measured in several studies from around the world and only two are cited here: Mohanta and Goel (2014), Kumar and Goel (2018).

Table 3.13 Water quality parameters included in NSF–WQI and their weights

Water quality parameter (Qi)	Weight (Wi)
Dissolved oxygen (DO), percent of DO saturation	0.17
Fecal coliform (colonies or cfu/100 mL)	0.16
pH	0.11
Biochemical oxygen demand, mg/L	0.11
Temperature change, °C	0.1
Total phosphate, mg/L	0.1
Nitrates, mg/L	0.1
Turbidity, NTU	0.08
Total solids (TS), mg/L	0.07
Overall WQI	**1**

Table 3.14 Water Quality Index Legend

Range	Quality
90–100	Excellent
70–90	Good
50–70	Medium
25–50	Bad
0–25	Very Bad

PROBLEM 3.7.1

Water quality data were collected for one location and the following data are available:

Fecal coliforms = 100 cfu/100 mL, DO 7.0 mg/L, Temperature = 25 °C, pH = 7.0, Turbidity = 25 NTU, Total solids = 50 mg/L, BOD = 4 mg/L, nitrates = 5 mg/L, phosphates = 40 mg/L. Determine the water quality index based on these data.

Solution

Of the nine recommended water quality parameters, temperature change was not measured, only a single temperature reading was available. Therefore, ignoring this parameter, the total weightage used in calculating WQI was 0.9. For determining DO as a percentage of DO-saturation, temperature versus DO_s equation has to be used as noted in Section 3.1.6. The DO_s value for a temperature of 25 °C was 8.265 mg/L. Therefore, percent of DO_s = 84.7 percent and the corresponding Q-value from the relevant chart is 91.[42] The overall WQI for this location is 63 which implies that the water quality is medium or moderate. High concentrations of phosphates and fecal coliforms resulted in low WiQi values implying that the two parameters are major contributors to the medium water quality as is evident from their Qi values.

Water quality parameter	Weight, Wi	Value	Qi, %	WiQi
Dissolved oxygen (DO), % of DO saturation	0.17	7 mg/L; 84.7%	91	0.155
Fecal coliform (colonies or cfu/100 mL)	0.16	100	44	0.070
pH	0.11	7	88	0.097
Biochemical oxygen demand, mg/L	0.11	4	61	0.067
Temperature change, °C	0.1	–		0.000
Total phosphate, mg/L	0.1	40	2	0.002
Nitrates, mg/L	0.1	5	65	0.065
Turbidity, NTU	0.08	25	57	0.046
Total solids (TS), mg/L	0.07	50	87	0.061
Total weightage	0.9			56.250
WQI				**63**

42 http://home.eng.iastate.edu/~dslutz/dmrwqn/water_quality_index_calc.htm.

Study Outline

General water quality parameters

Evaluation of basic water quality parameters is essential for determining the overall water quality of a water resource. Often, these parameters are the basis of regulations for drinking water standards and other uses of water.

General water quality parameters include pH, ionic strength, conductivity, solids of different types, temperature, dissolved oxygen, turbidity, alkalinity, hardness, and color.

Organics in water

Organic compounds are present in all natural waters: surface and groundwaters. These organic compounds can be of natural or anthropogenic origin, while the bulk of the organic matter is of natural origin. Organic compounds are responsible for the depletion of oxygen in rivers and streams which lead to several environmental problems such as damage to aquatic ecosystems, degradation of water quality leading to nuisance conditions in terms of poor color, taste and odor of water, and high treatment costs.

Synthetic organic compounds like pesticides, insecticides, herbicides, pharmaceutical compounds, surfactants, fire retardants, and many others have been detected with increasing frequency in surface and groundwater bodies. Their presence in drinking water sources is a direct and serious health concern. These compounds also bio-accumulate in the food chain and can lead to adverse health effects for humans and other organisms.

Inorganics in water

Various inorganic compounds are found in all natural water bodies and are broadly classified as macronutrients, micronutrients and toxic heavy metals.

Macronutrients are those elements that constitute >1 percent of the dry weight of the organism. These include carbon, hydrogen, oxygen, nitrogen, phosphorus, sulphur, potassium, sodium, and calcium.

Micronutrients are elements or compounds that are required in small amounts (they constitute <1 percent of the dry weight of the organism) and can be toxic at higher concentrations. Examples include Mg, Fe, Zn, Ni, Cu, Co, Mn, and several others.

Other elements like mercury, cadmium, arsenic, selenium, chromium, and lead are not known to have any beneficial effects at any concentration and are toxic heavy metals.

Microbes in water

Microbes are defined as unicellular organisms that can carry out their life processes of growth, energy generation and reproduction independently, i.e., they are free-living cells. Microbial cells are unlike cells within higher organisms like plants and animals that cannot exist independently but only as part of the larger organism.

Several microbial species of virus, bacteria, protozoa, algae and fungi are found in aquatic systems and many of these species are pathogenic (disease-causing).

Water-related diseases are a major concern and motivation for water and wastewater treatment. Contamination of water supplies by sewage is responsible for the spread of water-borne diseases like hepatitis, typhoid, and cholera.

Water-vectored diseases like malaria and dengue are spread by mosquitoes whose habitat and reproductive phases are dependent on water.

Bacteria are the most important microbes in water and wastewater treatment. Coliforms are indicator organisms that are used as surrogate parameters for pathogenic bacteria and may or may not be pathogenic themselves.

Coliforms are defined as aerobic or facultatively aerobic, Gram-negative, non-spore forming, rod-shaped bacteria that ferment lactose with gas formation within 48 hours at 35 °C. *Escherichia coli* is the best known example of coliforms and only one of its strains is considered to be an opportunistic pathogen.

Water quality in flowing water bodies

Dissolved oxygen (DO) is a primary water quality parameter for flowing water bodies since the nature and health of aquatic ecosystems depends on DO.

The Streeter–Phelps model (1925) can be used for estimating DO concentrations in a flowing stream over time or distance from the point of discharge. It accounts for depletion of oxygen due to biodegradation of organic matter present in wastewater discharges and oxygenation of water due to equilibrium with the atmosphere. As long as DO concentration is below the saturation level, there is a net driving force for oxygenation of the river or stream water. The model can be used to determine the critical time at which DO is minimum in the stream and is useful for management of river health.

Maintenance of e-flows has become an essential part of any water resources management program. The health of any aquatic ecosystem depends on maintaining e-flows throughout the year.

Water quality in standing water bodies

There are two major issues associated with standing water bodies: eutrophication and thermal stratification.

Eutrophication is defined as the degradation of water quality in standing water bodies due to an excessive influx of nutrients. The influx of nutrients is from surface runoff and wastewater discharges due to urbanization and other changes in land use and land cover.

A model based on the stoichiometry of algal and bacterial growth along with an understanding of the law of the minimum can be used to predict algal growth and oxygen consumption in water bodies due to eutrophication.

Changes in water temperature lead to changes in its density thereby resulting in three clear strata in all standing water bodies. The topmost layer is the epilimnion, the bottom layer is the hypolimnion and the transition region between the two layers is the thermocline. Seasonal changes in ambient temperature lead to changes in water temperature. Since DO concentration is a function of temperature and salinity, the DO profile of the water column in any lake or pond changes with seasons as well.

In temperate and cold climates, there is a 'turnover' and breaking up of these layers resulting in complete mixing of the lake's contents in spring and fall. In winter, if a layer of ice forms on top, it serves as a blanket for the aquatic ecosystem to survive even though ambient temperatures may be sub-zero. The thawing of ice in spring and 'turnover' of the layers can result in massive fish kills due to depletion of DO in the water column during winter.

Water quality index

Water quality indices provide a single value that can be used to compare water quality in different locations and over time. The NSF–WQI is one of the most popular and oldest WQI and is based on the Delphic method; it comprises of nine water quality parameters.

Quality rating charts for these nine water quality parameters are used along with pre-defined weights to determine WQI at a particular location and over time.

Study Questions

1. Define strong and weak acids with one example of each along with their pK_as.
2. Should the activity of an ion or compound or its concentration be used in equilibrium calculations? Explain in detail.
3. When and why is electrical conductivity a good approximation of TDS? Explain.
4. Define total solids, total suspended solids, and total dissolved solids.
5. What is the importance of temperature as a water quality parameter?
6. What is the relationship between temperature and dissolved oxygen? Why is it important?
7. What makes water turbid? How is turbidity measured?
8. What is alkalinity and how is it measured?
9. What is hardness and why should hardness be removed from water?
10. Why do deep groundwaters tend to have lower dissolved oxygen levels and lower microbial concentrations in comparison to surface waters?
11. Why do organics in water exert an oxygen demand? What methods can be used to estimate oxygen demand and what are the appropriate conditions for each of these methods? Is total organic carbon related to oxygen demand, explain your answer.
12. Why are synthetic organic compounds (SOCs) becoming a major concern in water and wastewater treatment?
13. What are macronutrient and micronutrients? Can they be toxic under certain conditions, explain your answer.

14. Name some toxic heavy metals and list their health effects.
15. What are the microbes of interest in water and wastewater treatment? Name the five major groups of microbes.
16. What are coliforms and why are coliforms used as indicators of fecal contamination?
17. Draw concentration profiles of all four nitrogen species in a river starting from the point of discharge. Explain each profile in detail.
18. Why is it important to maintain e-flow in a river?
19. Define eutrophication and explain the role of algae in it.
20. Does thermal stratification happen in all standing water bodies? Why are a large number of dead fish often found when the ice thaws in spring in cold climates?
21. What is the use of the NSF–WQI? Are there other water quality parameters that should be included besides the nine that are recommended?
22. What water quality parameters are regulated for drinking water use and why?
23. Why are some macronutrients and micronutrients considered to be pollutants? Explain with examples.
25. Is turbidity related to total solids or to total suspended solids concentrations?

Self-Study Questions

1. What instruments are used for determining each water quality parameter mentioned in this chapter?
2. Make solutions of various salts with known concentrations and measure the electrical conductivities of each solution. Develop a correlation between the measured electrical conductivity and calculated ionic strength.
3. Apply the Streeter–Phelps model to a river with multiple wastewater discharges and validate with real data if possible.
4. Collect water quality data and determine the water quality index (WQI) for different locations and at different times. Compare WQI values and comment on differences.
5. Use potassium hydrogen phthalate (KHP) to make total organic carbon standards ranging from 1 to 10 ppm. Measure the TOC (or DOC) and UV absorbance at 254 nm for these standards and determine the correlation between the two parameters. Determine UV absorbance at 254 nm for various water samples and use the above correlation to estimate the DOC concentration.
6. Conduct a BOD test in the laboratory for 20 to 30 days and draw graphs showing CBOD and NBOD.

CHAPTER 4

Mass Transfer and Transformation

Learning Objectives

- *Analyze and quantify natural or engineered systems based on a mass balance approach*
- *Describe Brownian diffusion and the Stokes–Einstein equation for particle diffusion*
- *State Fick's first and second laws of diffusion for dissolved compounds*
- *Derive equations for different reaction orders including unknown reaction order*
- *Model or analyze the effect of environmental variables like temperature and pH on reactions and their kinetics*
- *Compare and contrast the different types of ideal reactors*
- *Describe the salient features of a batch reactor and derive a mass balance equation for it*
- *Describe the salient features of a CSTR and derive a mass balance equation for it*
- *Describe the salient features of a PFR and derive a mass balance equation for it*
- *Analyze pulse and step inputs of tracers to different types of reactors*
- *Determine average hydraulic residence times and dispersion numbers for the different types of reactors*
- *Use tracer data to model engineered and natural systems and predict their behavior/efficiency*
- *List and explain factors affecting growth of organisms including bacteria in nature or in the lab*
- *Determine bacterial growth rates for a specific set of conditions*
- *Describe different biochemical or metabolic pathways by which organisms derive energy for maintenance and creation of new biomass*
- *Derive equations for bacterial growth rates*

A mass balance approach can be utilized for analyzing and quantifying the behavior of contaminants in any system: natural or engineered. Basic concepts of mass transfer and transformation can be applied to all processes including those in water and wastewater treatment, no matter how simple or complex they may be. Mass transfer and transformation are based on the laws of conservation of mass. The concept of applying mass balances to any system and using the steady-state simplification to predict system behavior is introduced in the first section. The concept of reactions or mass transformation and their application to chemical or biological reactions or transformations is dealt with in the second section of this chapter. Zero-, first-, second-, mixed-order reactions are defined in this unit. The third section includes definitions of the three basic types of ideal reactors and how their behaviors differ. Application of mass balances along with reaction kinetics to the analysis of each reactor and prediction of outflow characteristics are covered in this section. Since all 'real' systems or reactors deviate to some extent from ideal reactor behavior, tracer curves are used to evaluate and quantify the extent of reactor deviation from ideality. The use of tracer curve data to define the behavior of 'real' systems is covered in the fourth section. The last (5th) section introduces bacterial growth and Monod kinetics which are pseudo 1st-order kinetics that are used to model bacterial growth rates in the lab or the environment.

4.1 MASS TRANSFER: USING MASS BALANCES

The law of conservation of mass states that 'mass can neither be created nor destroyed but can be only be transformed'. In practice, we can use this concept to track the movement (transport or transfer) of mass from one point or phase to another or its transformation from one form to another (this can include a phase change or a change in chemical identity in the same phase). All natural and engineered systems can be analyzed using this approach.

For a given reactant, the mass balance equation can be written as:

Rate of ACCUMULATION of reactant within the system boundary = Rate of INFLOW of reactant into the system boundary – Rate of OUTFLOW of reactant out of the system boundary ± Rate of CHANGE (production [positive sign] or loss [negative sign]) of reactant within the system boundary

The above equation is termed the equation of continuity. The first three terms in the above equation refer to mass transport while the last term refers to mass transformation processes. Often, one or more terms in the above equation may be zero. For example, in a batch reactor where there is no inflow or outflow, the second and third terms are equal to zero. The rate of generation, i.e., the fourth term will be zero in the analysis of reactor hydraulics since no mass transformation has to be accounted for.

Mass can be transferred or transported by three different processes: advection (or convection), diffusion, and dispersion.

Advective or convective flow: Transport of particles or compounds of interest due to movement of the bulk fluid in which they are present.

***Diffusion*:** Transport of the compound of interest within the fluid due to either temperature (Brownian diffusion) or concentration gradients.

***Dispersion*:** Transport of the compound of interest within the medium due to velocity gradients only.

In general, mass balances for flow reactors are based on convective (advective) flow only. However, diffusion and dispersion are major issues in resolving many environmental problems and in reactor design, and are dealt with in Sections 4.1.2 to 4.1.4.

Notation regarding derivatives

ΔC or Δt (delta C or delta t) are discrete increments in concentration or time, respectively. In practice, these are measured quantities that can be used for modeling and other calculations.

∂C/∂t (del C or del t) denotes the partial derivative of concentration of a compound over time. It implies that the concentration is measured at the same fixed position, i.e., the x, y and z coordinates are held constant. In other words, there is no change in position (Bird, Stewart, and Lightfoot 1960).

dC/dt denotes the total derivative of the concentration with respect to time. If measurements of C are taken over varying time and space intervals, the change in concentration over time is the sum of the partial derivatives over time and space.

$$\frac{dC}{dt} = \frac{\partial C}{\partial t} + \frac{\partial C}{\partial x}\frac{dx}{dt} + \frac{\partial C}{\partial y}\frac{dy}{dt} + \frac{\partial C}{\partial z}\frac{dz}{dt}$$

4.1.1 Transport by Advection

Advection is defined as the transport of mass by flow of bulk fluid. Generally, horizontal flows are termed advection as against vertical flows which are termed convection.[43] To illustrate the concept involved in mass balance analysis, consider the bulk flow of a fluid such as water. The mass balance for water can be written as:

Rate of accumulation in the reactor (dM/dt) = rate of mass inflow into the reactor – rate of mass outflow from the reactor.

M = mass of water flowing into the reactor volume (M), t = time (T)

The fourth term in the general equation of continuity is dropped in the above case since it is assumed that water is not being generated or consumed during the transport process.

Mass (M) in grams is related to volume (V) in m^3 by density $\rho = M/V$ (kg/m^3) and $Q = V/T$ (m^3/s). In practice, water is measured in terms of volume not mass and the above mass balance is modified as below; assuming no change in temperature, density of water is assumed to be constant.

$$dM/dT = \rho^* dV/dt \quad (4.1.1)$$

43 https://en.wikipedia.org/wiki/Advection.

Two other terms need to be defined: Q_0 = volumetric rate of inflow of water (0 for $t = 0$, inflow is also called influent) and Q_e = volumetric rate of outflow of water (e = effluent from wastewater literature).

The mass balance is now simply:

$$\rho^*dV/dt = \rho^*Q_0 - \rho^*Q_e \qquad 4.1.2$$

Since density of water is assumed to be constant, the equation of continuity is:

$$dV/dt = Q_0 - Q_e \qquad 4.1.3$$

Mass balance for contaminant A through a reactor volume (V) is shown in Figure 4.1.

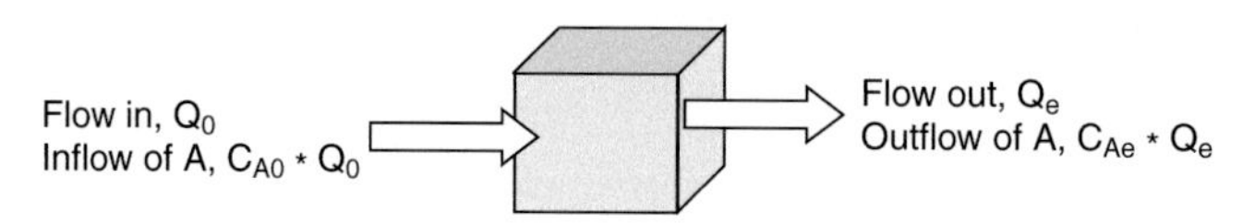

Figure 4.1 Mass balance analysis for contaminant A flowing through volume V.

4.1.2 Transport by Diffusion

Transport of particles and dissolved compounds in any quiescent (no flow) fluid medium can occur only due to random motion of the fluid molecules. The transport of *particles* under quiescent conditions is described as *Brownian diffusion* while the transport of *dissolved compounds* is described as *molecular diffusion* (Nazaroff and Alvarez-Cohen 2004). Diffusion is a microscopic (or molecular level) phenomenon as against dispersion which is a macroscopic phenomenon.

Brownian diffusion

Brownian diffusion was first described by Robert Brown, a botanist, in 1827 when he observed the motion of pollen grains in water. This phenomenon is a function of fluid temperature; the driving force for the diffusion of particles or molecules in a fluid is the thermal energy of the fluid. Stokes–Einstein equation relates the Brownian diffusion coefficient for particles, D_m, in water as (Crittenden et al. 2012):

$$D_m = kT/3\pi\mu d_p \qquad 4.1.4$$

where k = Boltzmann's constant, 1.38 x 10^{-16} erg/K or g/(K-cm^2-s^2), T is absolute temperature (degree Kelvin, 273.15 + degree Celsius), μ = dynamic viscosity of the fluid, d_p = diameter of the particle.

Dynamic viscosity of water for any given temperature in degree Celsius can be determined using the following regression equation from *Weast's Handbook of Chemistry and Physics*, 1978 (Hendricks 2011).

$$10^*\mu\ (\text{g/cm-s}) = 0.0017802356694 - 5.6132434302^*10^{-5}\ (T) + 1.0031470384^*10^{-6*}(T)^2 - 7.5406393887^*10^{-9}\ (T)^3 \qquad 4.1.5$$

Molecular diffusion

Transport of a dissolved compound due to a concentration gradient is described by Fick's laws of diffusion which were first postulated by Adolf Fick in 1855. When the compound is transported due to a concentration gradient under quiescent conditions, i.e., no flow conditions, it is termed molecular diffusion. A simple example of molecular diffusion is when a drop of ink is added to a glass of water. With no mixing, the ink will spread in all directions and eventually, it will be uniformly distributed within the entire volume of water. It is important to note that molecular diffusion can occur under flowing conditions as well, unlike dispersion which occurs by definition under flowing conditions only.

Fick's first law of diffusion

$$J = -D_m \frac{\partial C}{\partial x} \qquad 4.1.6$$

Where J = diffusive flux in one dimension, moles/m^2-s

$\frac{\partial C}{\partial x}$ = concentration gradient in x-direction (one-dimension only), moles/m^4

D_m = molecular diffusion coefficient, m^2/s

Fick's second law of diffusion

Incorporating Fick's first law of diffusion into the equation of continuity for a conservative compound (thus ignoring the rate of change term), results in Fick's second law of diffusion (Nazaroff and Alvarez-Cohen 2004):

$$\frac{\partial C}{\partial t} - D_e \frac{\partial^2 C}{\partial x^2} \qquad 4.1.7$$

4.1.3 Transport by Eddy or Turbulent Diffusion

Turbulence is defined as a flow condition that is irregular and chaotic and flow lines intersect randomly, as against laminar flow where flow is regular and the flow lines do not intersect. In other words, the velocity vectors in laminar flow do not change magnitude or direction while those in turbulent flow can change direction or magnitude randomly.

When the compound is transported by mixing or turbulence leading to 'microscale turbulence' and eddy formation, the diffusion coefficient is termed D_e = eddy or turbulent diffusion coefficient (Metcalf and Eddy 2003). Eddy diffusion can occur under both, no flow or flowing conditions. For example, if a drop of ink is added to a glass of water and mixed, the ink is transported throughout the volume of water by eddy or turbulent diffusion. Similarly, if a dye is added to a flowing stream, transport of the dye under laminar conditions would be due to molecular diffusion while transport of the dye under turbulent conditions would be due to eddy or turbulent diffusion.

Analogous to Fick's first law of diffusion, a contaminant moves from a region of high concentration to low concentration in a turbulent flow field and the proportionality constant is the eddy diffusion coefficient, D_e as shown below:

$$J = -D_e \frac{\partial C}{\partial x} \tag{4.1.8}$$

4.1.4 Transport by Dispersion

Dispersion is the process by which a compound mixes in the fluid due to turbulence generated by flow conditions, i.e., due to deviations from ideal laminar flow conditions, and it is a macroscopic phenomenon. Dispersion leads to the spreading of a plume or smoke from a smoke stack. Another example of dispersion is to add dye at one end in the center of a glass cylinder that is being filled with water continuously at a very low flow rate. The dye will move to the center of the tube without getting dispersed. If the flow rate is steadily increased, the dye will start dispersing into the entire volume of the cylinder in proportion to the flow rate and the degree of turbulence due to higher flow rates. In groundwater systems, the tortuosity of the flow path can lead to dispersion of contaminants.

The range of values for the different diffusion and dispersion coefficients are shown in Table 4.1.

Table 4.1 Diffusion and dispersion coefficients and their values in water (Metcalf and Eddy 2003)

	Values, cm^2/s
Molecular diffusion, D_m	10^{-8} to 10^{-4}
Eddy diffusion, D_e	10^{-4} to 10^{-2}
Dispersion, D	10^{2} to 10^{6}

When diffusion and dispersion are clubbed together, they are termed hydrodynamic dispersion (Masters 1998).

PROBLEM 4.1.1

A river with a flow rate of 1000 cumecs (m^3/sec) has a BOD of 10 mg/L and receives multiple discharges of wastewater from both sides. One wastewater discharge has a flow rate of 10 cumecs and a BOD of 300 mg/L (it is untreated sewage) while the other discharge from the opposite bank is 50 cumecs and 200 mg/L of BOD (wastewater after primary settling). Calculate the BOD downstream of the two discharges. The DO levels in the river upstream of the wastewater discharges is 5 mg/L and wastewaters have zero DO. Calculate the DO levels downstream of the two discharges.

Solution

The solution is derived by doing a volumetric balance for the three streams: river and 2 wastewater discharges downstream of the discharges, i.e., total flow rate, $Q_t = Q_r + Q_1 + Q_2$

For determining concentration of *BOD* and *DO* downstream of the discharges, a mass balance is done:

Total mass = ΣC_i (mg/L) $*Q_i$ (L/s) where C_i = concentration of species i

River flow and concentration data	Wastewater flow and concentration data	
Initial flow of the river, $Q_r = 1000\ m^3/s$	$Q_1 = 10\ m^3/s$	$Q_2 = 50\ m^3/s$
Initial *BOD* of river, $BOD_r = 10$ mg/L	$BOD_1 = 300$ mg/L;	$BOD_2 = 200$ mg/L
Initial *DO* of river, $DO_r = 5$ mg/L	$DO_1 = 0$ mg/L	$DO_2 = 0$ mg/L

Flow rate of the river downstream of the discharges, $Q_t = Q_r + Q_1 + Q_2 = 1060\ m^3/s$

BOD of the river downstream of the discharges, $BOD_t = [\Sigma BOD_i \,.\, Q_i]/Q_t = 21.70$ mg/L

DO of the river downstream of the discharges, $DO_t = [\Sigma DO_i \,.\, Q_i]/Q_t = 4.72$ mg/L

PROBLEM 4.1.2

A city has a population of 2.9 lakhs (2011 Census) while a residential academic campus in this city has an additional resident population of 25,000.

a. If the average water consumption in the city is 300 Lpcd while that in the campus is 500 Lpcd, determine the amount of wastewater being discharged by the campus into the city's municipal sewers. Also, determine the total amount of wastewater that needs to be treated at a proposed wastewater treatment plant. Assume 90 percent of the water consumed is discharged as wastewater.

b. An industry is proposing to discharge its wastewater to the city's sewers also. The amount of wastewater is likely to be 1 MLD with a BOD of 6000 mg/L. The municipal wastewater has an average BOD concentration of 300 mg/L. Determine the total flow to be treated and BOD loading to the proposed wastewater treatment plant.

Solution

a. Wastewater generated in academic campus = 0.9 x 25,000 persons x 500 L/person-day = 11.25 MLD

   Wastewater generated in city = 0.9 x 2.9 x 10^5 persons x 300 L/person-day = 78.3 MLD

   Total wastewater generated = (11.25 + 78.3) MLD = 89.55 MLD

b. Additional discharge of industrial wastewater = 1 MLD

   Total wastewater discharge = 90.55 MLD

   BOD concentration in combined industrial and municipal flows = 89.55*10^6 (L/d)*300 (mg/L) + 10^6 (L/d)*6000 (mg/L)) = 370.72 mg/L

PROBLEM 4.1.3

Determine the diffusivity of particles of size 0.001, 0.01, 0.1 and 1 micron using the Stokes–Einstein equation at a temperature of 25 °C.

Solution

Create a table for the four different particle sizes and use the following equation to predict the dynamic viscosity of water at 25 °C (Hendricks 2011):

$$10^*\mu \text{ (g/cm-s)} = 0.0017802356694 - 5.6132434302^*10^{-5} \text{ (T)}$$
$$+ 1.0031470384^*10^{-6*}(\text{T})^2 - 7.5406393887^*10^{-9} \text{ (T)}^3$$

For a temperature of 25 °C, μ (g/cm-s) = 8.861 x 10^{-3}

$$k = 1.38 \times 10^{-16} \text{ g/(cm}^2\text{-s}^2\text{-°K)}$$

The following values for molecular diffusivity of particles are obtained by solving the Stokes–Einstein equation for different particles sizes.

d_p, micron	d_p, cm	D_m, cm^2/s
0.001	1.00×10^{-7}	4.92×10^{-6}
0.01	1.00×10^{-6}	4.92×10^{-7}
0.1	1.00×10^{-5}	4.92×10^{-8}
1	1.00×10^{-4}	4.92×10^{-9}

PROBLEM 4.1.4

A village pond is being used for long-term storage of water. The source of water is a neighboring stream and the pond has no outlets. The pond volume is 1000 m^3 with an average water depth of 5 m.

a. If the mean annual evaporation rate is 1 cm/d, and the infiltration rate is 0.5 cm/d, what is the mean annual flow rate into the pond, assuming that the pond volume remains constant throughout the year?
b. The influent water contains nutrients like carbon, nitrogen and phosphorus. Total Kjeldahl nitrogen (TKN) concentration in the influent is 50 mg/L. Assume that all TKN will be converted to nitrate, and nitrate does not evaporate but can infiltrate into the subsurface. Can you predict the concentration of nitrate-N over time in the pond? Assume that there are no reactions with nitrate.

Solution

a If volume of pond = 1000 m^3

Average annual depth of water = 5 m

Pond surface area = 200 m^2

A water balance around the pond can be done:

Mass accumulated = Mass in – Mass out (reactions are ignored)

Mass out = Evaporation + Infiltration = (0.01 m/d + 0.005 m/d)*200 = 3 m^3/d

Assuming pond volume remains constant throughout the year:

Mass in = Mass out assuming temperature, density and volume remain constant during this period.

Therefore, volumetric flow rate of water into the pond = 3 m^3/d of water flowing into the pond = 3 m^3/d

b Rate of Mass in [nitrate] = 50 g/m^3 *3 m^3/d = 150 g/d

Rate of Mass out [nitrate] is due to infiltration only = 0.005 m/d*200 m^2/d*50 g/m^3 = 50 g/d

Therefore, rate of mass of nitrate accumulation in pond = 100 g/d

Concentration of nitrate over time can be predicted based on the accumulation rate of nitrate in the pond at the rate of 100 g/d.

PROBLEM 4.1.5

River water samples were taken for water quality analysis. Total Kjeldahl nitrogen (TKN) was measured in the water before and after filtration through a 0.45 micron membrane. The concentration of TKN in water after filtration was 40 percent less than the concentration without filtration. Can you explain these results?

Solution

Nitrate can be taken up by biomass (can be algal, bacterial or higher organisms in the ecosystem of the pond or it may be adsorbed to particles). Filtration will remove all suspended solids including particles and organisms (both dead and alive) and therefore, nitrate present in suspended solids is removed on the filter. Filtered water will contain N in dissolved form only. Since the filtered water has 40 percent less N in it compared to the raw water, this 40 percent is in the suspended solids captured on the filter.

4.2 MASS TRANSFORMATION

A chemical species is said to have reacted when its chemical identity changes, i.e., mass transformation occurs. *Conservative* compounds are defined as those that do not change their chemical identity, and are used as tracers, e.g., chloride, and dyes. *Non-conservative* compounds can react or be transformed, i.e., change their chemical identity. The identity of a chemical species is determined by the *kind*, *number*, and *configuration* of atoms of that species (Levenspiel 1998).

4.2.1 Types of Reactions

Reactions can be classified based on many different criteria and several examples are provided here.

Irreversible reactions: The reaction proceeds in one direction only until reactants are completely utilized. All chemical reactions are strictly speaking reversible reactions. The only exceptions are nuclear reactions. Practically, reactors are often designed so that equilibrium is NOT reached and reactants are utilized to the maximum extent, i.e., products are removed as fast as they are produced, thus driving the reaction in the forward direction.

Reversible reactions: The reaction can proceed in either direction tending towards equilibrium. Equilibrium concentrations of reactants and products are defined by thermodynamic considerations and apply to all chemical and biological reactions.

Homogeneous and heterogeneous reactions are classified based on the number of phases involved in the reaction.

Homogeneous reactions: are those reactions in which the reactants are in the same phase, i.e., single phase. For example, reactions between a gas and another gas, or between two liquids, are homogeneous. Homogeneous reactions may be either reversible or irreversible.

Heterogeneous reactions: are those reactions in which the reactants are in two or more phases. For example, reactions between a liquid and a solid or a reaction that requires the presence of a solid-phase catalyst are heterogeneous. Examples include burning of liquid materials like petrol or oil, or solid materials like solid waste or coal in air or with oxygen, or adsorption of dissolved compounds on solid substrate.

Catalytic and non-catalytic reactions are categorized based on the presence or absence of a catalyst in the reaction.

Catalytic reactions: are those where a catalyst is used to increase the reaction rate but the catalyst is itself unchanged. All biological reactions can be termed catalytic reactions since they are mediated by enzymes which serve as catalysts.

Non-catalytic reactions: are those where no catalyst is used.

All chemical reactions can be classified based on their end-products into the following three categories:

Synthesis: A synthesis reaction occurs when reactants combine to form a new product. An example of a synthesis reaction is when reactants X and Y combine to form a new product *XY*. Polymerization is a synthesis reaction where a polymer is created by combining several monomers.

Decomposition: A decomposition reaction occurs when a reactant is broken down into two or more new products; a reversal of the previous reaction. Decomposition of *XY* would result in the formation of X and Y products. Reversal of the polymerization reactions are decomposition reactions, where a polymer decomposes into individual monomers or fragments of the polymer.

Exchange: Exchange reactions include synthesis and decomposition. An example of an exchange reaction is when the two reactants *XY* and *AZ* result in *AX* and *YZ*. Complete oxidation of an organic compound like glucose in the presence of oxygen with end products like carbon dioxide and water is a very common example of an exchange reaction.

Single and multiple reactions are defined on the basis of the number of reactions involved.

Single reactions: When a single stoichiometric equation and single rate equation can represent the progress of the reaction, it is a single reaction.

Multiple reactions: Multiple reactions require more than one kinetic expression to follow the changing composition of all reaction components. They can be series, parallel or mixed reactions as shown below:

- Series: $A \rightarrow R \rightarrow S$
- Parallel: Parallel reactions are of two types –
  - Competitive (using the same reactant): $A \rightarrow R$; $A \rightarrow S$
  - Side-by-side (simultaneous): $A \rightarrow R$; $B \rightarrow S$
- Complex: reaction schemes that have both above types (series and parallel). For example, for the reaction $A + B \rightarrow R$; $R + B \rightarrow S$, the reaction is parallel with respect to B, and in series with respect to A and R.

  Elementary and non-elementary reactions are defined based on their stoichiometry.

Elementary reactions: These are reactions where the rate equation corresponds to the stoichiometric equation: $A + B \rightarrow R$; $-r_A = kC_AC_B$

Non-elementary reactions: When the rate equation DOES NOT correspond to the stoichiometric equation. This is explained based on the assumption that what is observed as (or represented by) a single reaction is actually the net effect of many elementary reactions. The best and most common examples of non-elementary reactions are the biological reactions.

4.2.2 Reaction Kinetics

The stoichiometry and the rate of reaction are the principle concerns from the view point of process selection and design.

Rate of Reaction

The rate of reaction is defined as the rate of change (decrease or increase) of moles of reactive substance (reactant or product) per unit volume (for homogeneous reactions) or per unit surface area (for heterogeneous reactions). For heterogeneous reactions, it can also be defined as rate of change of mass per unit time.

For the following chemical reaction: $A \rightarrow B$

r_A = rate of reaction of species A per unit volume (moles/L-min) or rate at which A is reacting or $-r_A$ = the rate of disappearance of species A per unit volume;

r_B = the rate of formation of species B per unit volume.

Reaction rate or rate law or reaction kinetics is an intensive quantity and is a function of the state of the system. The state of a system is defined by other intrinsic parameters such as temperature, concentration, and composition. It is important to note that the rate law is an algebraic equation not a differential equation. For example, for the above reaction: $r_A = k\ C_A$ is an algebraic equation not a differential equation.

$$r_A = [k_A(T)]\ [f(C_A, C_B, ...]$$

Generally, rate of reaction r_A is the rate law where k_A is the specific rate of reaction referred to a particular species. For the following reaction:

$$NaOH + HCl \rightarrow NaCl + H_2O$$

$$k = k_{NaOH} = k_{HCl} = k_{NaCl} = k_{H_2O}$$

In general, the dependence of rate of reaction r_A on concentrations of species is determined experimentally, using best-fit models of concentration versus time data. These experiments are done in batch reactors.

Order of Reaction

The order of reaction refers to the power to which concentration is raised. If power is based on stoichiometry, the reaction is elementary. Often, determination of power (or reaction order) is based on experimental data, i.e., power is determined by the best-fit curve for data. General form of the expression is:

$$-r_A = k_A C_A^{\alpha} C_B^{\beta} \qquad 4.2.1$$

Overall order of reaction in the above equation is $n = \alpha + \beta$ where α = order of reaction with respect to A and β = order of reaction with respect to B.

For example: $2NO + O_2 \rightarrow 2NO_2$ where the reaction is 1st-order with respect to O_2, 2nd-order with respect to NO, and it is 3rd-order – overall. The rate equation is then:

$$-r_{NO} = k_{NO}\ C_{NO}^{2} C_{O}^{2} \qquad 4.2.2$$

The order of a reaction with respect to a certain reactant is defined as the power to which its concentration term in the rate equation is raised.

For example, given a chemical reaction $2A + B \rightarrow C$ with a rate equation

$$r = k[A]^2[B]^1$$

The reaction is second-order with respect to A, first-order with respect to B and the overall order of the reaction is 2 + 1 = 3. The reaction rate constant is represented by k. The order of a reaction may be zero or a fraction depending upon the mechanism of the reaction. Depending upon the order of the reaction, some common types of rate expressions are listed below (the equations are derived for decay or degradation by putting a negative sign for *k* which is typical for many environmental problems) (Metcalf and Eddy 2003; Sengupta 2014):

Zero-order since it is not dependent on C:

$$r = -k$$

$$dC/dt = -k \quad 4.2.3$$

$$\int_{C_0}^{C} dC = -k \int_{C_0}^{C} dt$$

Integrating the above equation from $C = C_0$ at $t = 0$ to $C = C$ at $t = t$ results in

$$C - C_0 = -kt, \text{ therefore } C = C_0 - kt \quad 4.2.4$$

First-order where the reaction is directly proportionate to C

$$r = -kC$$

$$dC/dt = -kC \quad 4.2.5$$

Integrating the above equation from $C = C_0$ at $t = 0$ to $C = C$ at $t = t$ results in

$$\int_{C_0}^{C} \frac{dC}{C} = -k \int_{0}^{T} dt$$

$$\ln C - \ln C_0 = -kt \text{ implies } \ln (C/C_0) = -kt \text{ or in linear equation form: } \ln C = -kt + \ln C_0$$

$$\text{or in another way } C = C_0 \exp(-kt) \quad 4.2.6$$

Second-order where the reaction is directly proportionate to C^2 or $C_A{}^*C_B$, where C_A and C_B are the concentrations of two different reactants.

$$r = -kC^2$$

$$dC/dt = -kC^2 \quad 4.2.7$$

Integrating the above equation from $C = C_0$ at $t = 0$ to $C = C$ at $t = t$ results in

$\int_{C_0}^{C} \frac{dC}{C^2} = -k \int_{0}^{t} dt$ can be integrated based on the general integral for $\int x^a = \frac{x^{a+1}}{a+1}$

$$\int_{C_0}^{C} \frac{C^{-2+1}}{-2+1} = -k\int_{0}^{t} dt$$

$(-1/C) - (-1/C_0) = -kt$ can be simplified further to $(1/C_0) = (1/C) - kt$ 4.2.8

Mixed-order or saturation kinetics

$$r = -kC / (K + C)$$

where K is the saturation constant or carrying capacity depending on the application.

Dividing the right hand side of the equation by C results in

$$dC/dt = -k /(K/C + 1) \quad 4.2.9$$

which can be rearranged as

$$(K/C + 1)dC = -k\,dt \quad 4.2.10$$

Integrating the above, results in

$$\int_{C_0}^{C} \frac{K}{C} dC + \int_{C_0}^{C} dC = -k\int_{0}^{t} dt$$

Since K is a constant, the solution to the above integrals is

$$K^*\ln(C/C_0) + (C - C_0) = -kt \quad 4.2.11$$

Converting to the form of a linear equation by dividing all terms by Kt

$$(1/t)^*\ln(C/C_0) = -(1/Kt)^*(C - C_0) - k/K \quad 4.2.12$$

In the above linear form of the equation, the slope is –1/K and the intercept is –k/K.

Differential method for unknown order

$$r = dC/dt = -kC^n \quad 4.2.13$$

Taking logarithms for all terms

$$\text{Ln}(-dC/dt) = \ln k + n \ln C \quad 4.2.14$$

Concentration versus time data can be used to determine reaction order n and reaction rate k using this equation.

Effect of temperature on the rate constants

All rate constants need to be corrected for temperature. The temperature dependence of the reaction rate constant given by the van't Hoff-Arrhenius relationship is presented below:

$$k = k_0 \exp(-E_\alpha / RT) \qquad 4.2.15$$

For two different temperatures, the equation can be written as:

$\ln (k_2/k_1) = E_\alpha{*}(T_2 - T_1) / (R{*}T_1{*}T_2))$ and further simplified as

$$\ln \frac{k_2}{k_1} = \frac{E}{R}\left[\frac{1}{T_1} - \frac{1}{T_2}\right] \qquad 4.2.16$$

Where k_2 = reaction rate constant at temperature T_2, (1/s, 1/min, etc.)

k_1 = reaction rate constant at temperature T_1, (1/s, 1/min, etc.)

T = temperature in K, where K = 273.15 + T °C

E_α = a constant characteristic of the reaction, i.e., its activation energy, kJ/mol.

R = ideal gas constant, 8.314 J/mol–K.

Activation energy (E_α) is a function of the degree of conversion or completion of the reaction α. Reactions occurring in water and wastewater processes are non-specific in nature due to the large variety of compounds and other substances present in solution. Further, these processes are generally carried out over a relatively narrow temperature range approximating environmental conditions. The following empirical equation is commonly used for water and wastewater processes:

$$k_2 / k_1 = \theta^{(T_2 - T_1)}$$

Typical values of θ vary from 1.020 to 1.10.

PROBLEM 4.2.1

Determine the order and reaction rate constant for the reaction $X \rightarrow Y + Z$ using the following data obtained in a lab study. Test by curve-fitting and determining equation of line and R^2 for the data for 0-, 1st- and 2nd-order reactions.

time, min	C, mg/L
0	50
10	42
20	35
30	30
40	25
50	21
60	17

Solution

	0-order	1st-order	2nd-order
time, min	C, mg/L	$\ln(C/C_0)$	$1/C_0$-$1/C$
0	50	0	0
10	42	−0.17435	−0.00381
20	35	−0.35667	−0.00857
30	30	−0.51083	−0.01333
40	25	−0.69315	−0.02
50	21	−0.8675	−0.02762
60	17	−1.07881	−0.03882

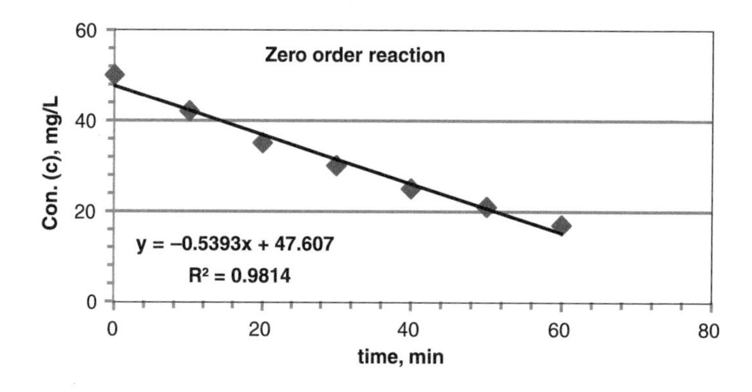

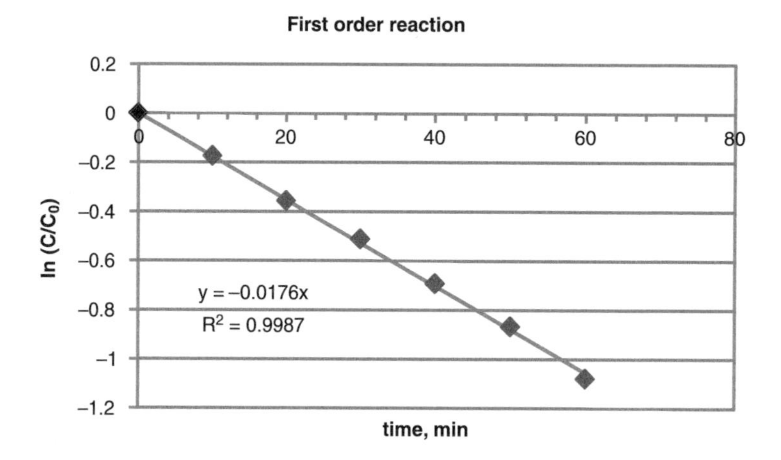

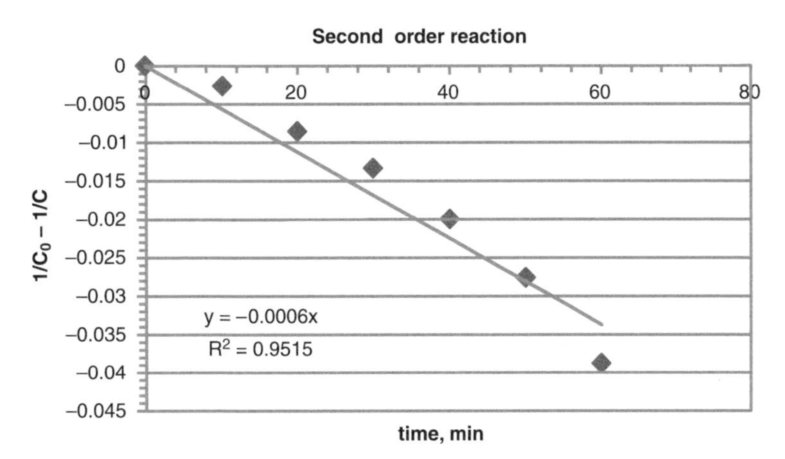

Since best-fit based on R^2 values was for 1st-order, therefore, the reaction is assumed to be 1st-order and k = –0.01761/min.

PROBLEM 4.2.2

The initial concentration of a contaminant is 100 mg/L. The reaction rate constant (k) is 0.2 but since the rate law is unknown, the units are unknown as well. Assume zero-order, 1st-order and 2nd-order kinetics and determine the time taken to achieve 90 percent removal of the contaminant in each case. Hint: Ensure that units are consistent in each case.

Solution

For zero-order kinetics, using equation 4.2.4: $C = C_0 - kt$ and solving for t, time for 90 percent removal = 450 min

For 1st-order kinetics, using the equation $\ln(C/C_0) = -kt$, time for 90 percent removal is 11.5 min

For 2nd-order kinetics, using the equation $1/C_0 - 1/C = -kt$ and solving for t, time for 90 percent removal is 0.45 min

PROBLEM 4.2.3

If the 1st-order reaction rate constant at 20 °C is 0.2 1/d, what will the reaction constant be at 40 °C? Assume the coefficient for temperature activity is 1.03.

Solution

Using the formula: $k_2 / k_1 = \theta^{(T_2 - T_1)}$ and assuming k_2 = reaction rate constant at T_2 = 40 °C, k_1 = reaction rate constant at T_1 = 20 °C, θ = 1.03, value for k_2 = 0.36 1/d.

PROBLEM 4.2.4

Concentrations of volatile suspended solids were monitored over time in batch biodegradation experiments in the laboratory and the data are shown in the following table. Determine the order of the reaction and reaction rate constant.

Time, days	Concentration, C (mg/L)
1	0.498
3	0.43
5	0.3763
8	0.322
13	0.2073
22	0.1826
35	0.1696
65	0.148

Solution

The solution to this problem is based on the use of equation 4.2.13 and its linear form 4.2.14.

Step 1: Set up a table with the data and calculations as shown below. The first two columns are the given data. It is easier to deal with dimensionless concentrations and this is done by normalizing the concentrations by dividing them by C_0 (column 3).

Step 2: Column 4 is $(-\Delta C/\Delta t)$ which is calculated by taking the difference in two consecutive concentration points and dividing by the difference in the corresponding time points. The calculations begin with the second point: $-(C_2 - C_1)/(t_2 - t_1)$. Column 5 is the logarithm (natural or base 10) of the values in column 3 and column 5 is the logarithm of the concentration values.

Step 3: Plot a graph of ln C (x-axis) versus $\ln(\Delta C/\Delta t)$. Results for this data are shown in the following graph. A linear equation of the form can be found: $\mathrm{Ln}(-dC/dt) = \ln k + n \ln C$. The slope of the line is n and the intercept is ln k.

From the graph, the slope of the curve is the reaction order, n = 3.6733 and the reaction rate constant, k is determined from the intercept, log k = –2.2737 which results in a k value of 0.1 with units of $(1/(\mathrm{mg/L})^n)$

Note: In this data set, a good fit was obtained with this model with an R^2 value of 0.9355. However, all data sets may not fit this generic model and other individual reaction orders should be tried to get a best-fit for the data in such cases.

Time, days	Concentration, C (mg/L)	Normalized concentration, C/C_0	$\Delta C/\Delta t$	$\ln(\Delta C/\Delta t)$	lnC
1	0.498	1.000			
3	0.43	0.863	0.06827309	–2.684	0.000
5	0.3763	0.756	0.05391566	–2.920	–0.147
8	0.322	0.647	0.03634538	–3.315	–0.280
13	0.2073	0.416	0.04606426	–3.078	–0.436
22	0.1826	0.367	0.00551093	–5.201	–0.876
35	0.1696	0.341	0.00200803	–6.211	–1.003
65	0.148	0.297	0.00144578	–6.539	–1.077

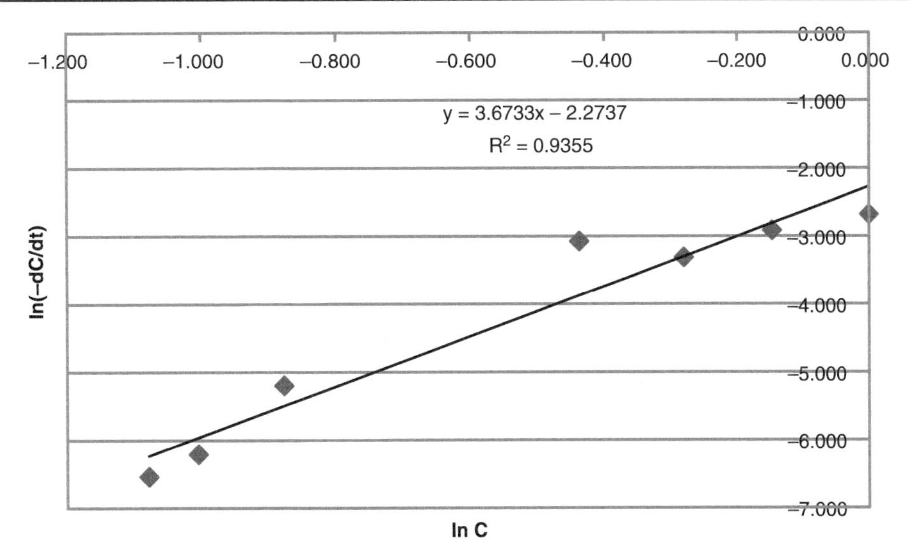

4.3 IDEAL REACTORS

Ideal reactors are well-defined in terms of mass and energy inputs and outputs and degree of mixing. All reactors can be categorized as either batch or flow reactors. Batch reactors are defined as reactors which have no inputs or outputs of mass or energy during the period of interest. Flow reactors are defined as those which have flow of mass or energy through the reactor; this flow may or may not be continuous. There are two types of ideal flow reactors: complete mix reactor (more commonly known as the continuously stirred tank reactor [CSTR]), and plug flow reactor (PFR). All real reactors are some combination of the 3 ideal reactor types and in general, engineered reactors are designed as one of these types. The 3 ideal reactors are described in this section: batch reactor, PFR and CSTR. An understanding of the behavior of these reactors can be applied to natural systems to model and predict their behavior and to design efficient engineered reactors. A general mass balance for a dissolved compound with concentration C around the reactor would be:

or $$\frac{dC}{dt}V = QC_0 - QC + rV \qquad 4.3.1$$

4.3.1 Batch Reactors

Ideal Batch Reactor

In a batch reactor, there is no flow into or out of the reactor. Since there are no inputs or outputs of mass or energy during operation, batch reactors are most popular for determining rate laws. Mass is added to the reactor, it is treated, and then discharged. In a sequential batch reactor, the cycle is repeated. In general, the liquid contents of the reactor are mixed completely, i.e., there is no spatial variation in the concentration of compound A as shown in Figure 4.2.

Figure 4.2 An Ideal Batch Reactor.

For a batch reactor with no inputs or outputs, i.e., $Q = 0$, equation 4.3.1 becomes results in $r = dC/dt$.

$$\frac{dC}{dt} = r_c$$

If the rate of reaction is defined as first-order (i.e. $r_c = -kC = dC/dt$), integrating between the limits $C = C_0$ at $t = 0$ and $C = C$ at $t = t$, we get

$$\int_{C=C_0}^{C=C} \frac{dc}{C} = -k\int_{t=0}^{t=t} dt = kt$$

$$= C/C_o = e^{-kt} \qquad 4.3.2$$

4.3.2 Continuously Stirred Tank Reactors (CSTR)

In a complete mix reactor or CSTR, it is assumed that complete mixing occurs instantaneously and uniformly throughout the reactor, as fluid particles enter the reactor as shown in Figure 4.3. The concentration of compound A leaving the reactor is equal to the concentration of the compound within the reactor, i.e., there is no spatial variation in the concentration of compound A in the reactor.

Fluid particles leave the reactor in proportion to their statistical population.

The mass balance equation can be written as

$$\frac{dC}{dt}V = QC_0 - QC + r_c V \qquad 4.3.3$$

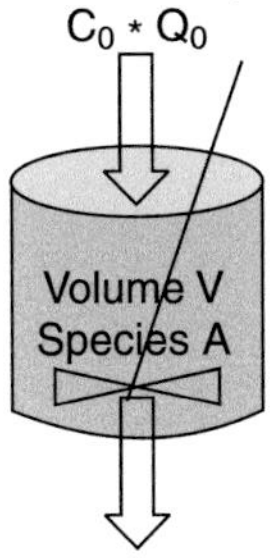

Flow in, Q_0
Inflow of A,
$C_0 * Q_0$
Volume V
Species A
Flow out, Q_e
Outflow of A,
$C * Q_e$

Figure 4.3 Ideal Continuously Stirred Tank Reactor (CSTR).

Assuming first-order kinetics ($r_c = -kC$), and dividing the entire equation by V

$$\frac{dC}{dt} = (Q/V)C_0 - (Q/V)C - kC = (Q/V)C_0 - C[(Q/V) + k] \qquad 4.3.4$$

The above equation can be rearranged as:

$$C' + \beta C = (Q/V)\, C_0$$

Where $\quad C' = dC/dt$ and $\beta = k + Q/V$

This is a first-order linear differential equation, whose solution is given by

$$C = \frac{Q}{V}\frac{C_0}{\beta} + Ke^{-\beta t} \text{ where } K \text{ is a constant} \qquad 4.3.5$$

When $\quad t = 0, C = C_0$ and K is equal to

$$K = C_0 - \frac{Q}{V}\frac{C_0}{\beta}$$

Substituting for K and simplifying, we get

$$C = \frac{Q}{V}\frac{C_0}{\beta}(1 - e^{-\beta t}) + C_0 e^{-\beta t} \qquad 4.3.6$$

which is the non-steady-state solution.

For steady-state $dC/dt = 0$, the solution becomes

$$C = \frac{C_0}{[1 + k(V/Q)]} = \frac{C_0}{[1 + kr]} \qquad 4.3.7$$

which can also be obtained from non-steady-state solution by assuming that t tends towards infinity, i.e., $t \to \infty$ at steady-state.

4.3.3 Plug Flow Reactors (PFR)

In a plug flow reactor (PFR), the reactor is much longer than it is wider; the length to width ratio is ≥ 10. Flow is assumed to be laminar, i.e., the streamlines or flow lines do not intersect and each differential element or volume is assumed to be 'isolated' from its neighbors. Since there is no mixing of fluid elements, each fluid element behaves as a batch reactor moving at the same uniform speed as its neighbors. The concentration of compound A at any point in space within this reactor is a function of the reaction time.

Fluid going through a PFR may be modeled as flowing through the reactor as a series of infinitely thin 'plugs', each with a uniform composition, traveling in the axial (longitudinal) direction of the reactor, with each plug having a different composition from the ones before and after it as shown in Figure 4.4. The key assumption is that as a plug flows through a PFR, the fluid is perfectly mixed in the radial direction (perpendicular to the direction of flow or sideways) but not in the axial direction (forwards or backwards). Each plug of differential volume is considered to be a separate entity, effectively an infinitesimally small batch reactor, tending to zero volume. Each differential volume in an ideal PFR remains in the reactor for exactly the theoretical detention time (unlike a differential volume in a CSTR which may come out in a few seconds or never at all).

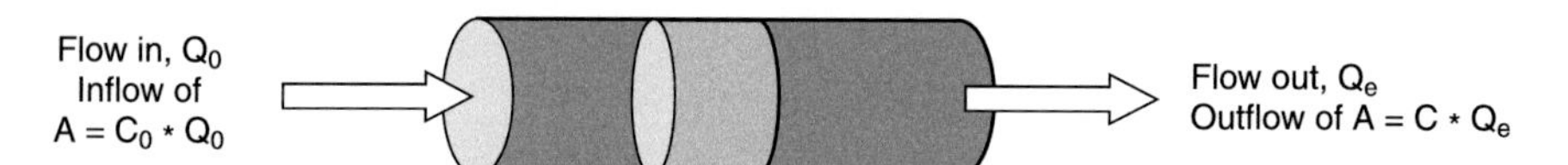

Figure 4.4 Ideal Plug Flow Reactor.

Writing a mass-balance equation for a differential volume ΔV, we have

$$\frac{\partial C}{\partial t}\Delta V = QC|_x - QC|_{x+\Delta x} + r_c \Delta V$$

Substituting the differential form for the term $C|_{x+\Delta x}$, we get

$$\frac{\partial C}{\partial t}\Delta V = QC - Q\left(C + \frac{\Delta C}{\Delta x}\Delta x\right)C + r_c \Delta V$$

Substituting AΔx for ΔV where A is the area perpendicular to the direction of flow (cross-sectional area) and Δx is the thickness of the plug.

Dividing by $A\Delta x$ gives,

$$\frac{\partial C}{\partial t} = -\frac{Q}{A}\frac{\Delta C}{\Delta x}\Delta x + r_c$$

Taking the limit as $\Delta x \to 0$ yields

$$\frac{\partial C}{\partial t} = -\frac{Q}{A}\frac{\partial C}{\partial x} + r_c$$

For steady-state condition $(\partial C/\partial t) = 0$, and if the rate of reaction is defined as $r_c = -kC$, integrating between the limits $C = C_o$ to C and $x = 0$ to L (L = length of the reactor), we get

$$\int_{Co}^{C} \frac{dC}{C} = -k\frac{A}{Q}\int_{0}^{L} dx = -k\frac{AL}{Q} = -k\frac{V}{Q} - k\tau \equiv C / C_o = e^{-kt} \qquad 4.3.8$$

Equation 4.3.8 is equivalent to the equation derived for the batch reactor, τ being the theoretical detention time.

4.3.4 Tracer Curves

Reactors often deviate from 'ideal' conditions due to various reasons leading to inefficient use of reactor volume and reduced performance. Some of the major factors that result in non-ideal flow conditions are (Metcalf and Eddy 2003; Henry and Heinke 1996):

Channelling: Non-uniform distribution of fluid in the reactor is termed channelling. This can be a problem in packed bed towers with counter-current flows.

Short-circuiting: Incomplete mixing or short-circuiting of flow can happen for various reasons. Temperature differences can result in density gradients leading to incomplete mixing, wind-driven circulation patterns may be limited to the top layers of the reactor leading again to incomplete mixing; mechanical or other types of mixing that are provided may be inadequate.

Stagnant regions: Dead zones where no mixing occurs often exist in reactors depending on the design of the inlet and outlet zones.

Dispersion: Movement of tracer is due to advection (convection) and dispersion. Hydrodynamic dispersion includes transport due to velocity gradients, turbulent eddies, and molecular diffusion.

Tracer tests are conducted to determine the extent to which a reactor deviates from 'ideal' flow conditions and allows for a more accurate assessment of the hydraulic performance of any reactor.

Some of the important applications of tracer studies are assessment of short circuiting in sedimentation tanks and biological reactors, assessment of the contact time in chlorine contact basins, assessment of hydraulic approach conditions in UV reactors, assessment of flow patterns in constructed wetlands and other natural treatment systems.

Important characteristics of a tracer

- It should not affect the flow, i.e., it should have the same density as water when diluted
- It must be conservative (non-reactive) so that a mass balance can be performed
- Tracer should be easy to detect and monitor
- Tracer should have low molecular diffusivity
- It should not sorb or react with exposed surfaces of reactors and particles in wastewater.

Examples of tracers include Congo red, fluorescein, fluorosilicic acid (H_2SiF_6), hexafluoride gas (HF_6), lithium chloride (LiCl), pontacyl brilliant pink B, potassium permanganate, rhodamine WT and sodium chloride or sodium nitrate.

Conduct of tracer tests

Generally, a tracer is introduced into the influent end of the reactor or basin to be studied. Depending on influent and effluent characteristics, two types of tracer inputs can be used.

Pulse or slug of dye is injected over a very short period of time (instantaneously) and mixed. Since a measured mass of compound is injected instantaneously, the concentration C_0 at $t = 0$ is defined as M/V where M = mass of tracer injected and V = volume of reactor.

Continuous input of dye is introduced along with the influent for a period that is 3 to 4 times the design hydraulic time of the reactor.

In both methods, the concentration of tracer in the effluent is measured throughout the test and different curves are defined based on these data.

E-curve (RTD – residence time distribution or exit age distribution): The age distribution of the fluid elements leaving the reactor. Applies to a pulse input, and defines the output for the reactor.

$$\int_0^\infty Edt = 1$$

I-curve: the internal age distribution of the fluid elements in the reactor. Area under both, E and I curves, is always equal to 1.

F-curve (step input): Cumulative residence time distribution for a step input, or C/C_o versus θ (t/τ) or t is known as the F-curve. For an ideal CSTR, $F = 1 - e^{-\theta}$ and is derived below (Henry and Heinke 1996):

For a step input to a CSTR, the mass balance equation can be written as

Accumulation = Input – output (there is no generation or degradation term for a tracer)

$$V\ dC/dt = QC_0 - QC$$

Dividing the equation by QC_0 and noting that $\tau = V/Q$ = design hydraulic residence time results in

$$(\tau/C_0)*dC/dt = 1 - C/C_0$$

Using normalized time, $\theta = t/\tau$, and normalized concentration, $F = C/C_0$ and noting that

$$d\theta = dt/\ \tau \text{ and } dF = dC/C_0$$

Substituting the normalized variables,

$$1 - F = dF/d\theta$$

Integrating from $\theta = 0$ to θ

$$\int_0^\theta d\theta = \int_0^F \frac{dF}{1-F}$$

$$\theta = (-1)^*\ln(1 - F) + \ln(1 - 0)$$

$$-\theta = \ln(1 - F)$$

Therefore, $\exp(-\theta) = 1 - F$ or

$$F = 1 - e^{-\theta} \quad 4.3.9$$

Therefore for a step input to an ideal CSTR, at $t = \tau$, $F = 0.632$ which means that the concentration leaving the reactor is 63.2 percent of C_0.

C-curve (pulse input): Similarly for a pulse (slug) input to an ideal CSTR, the equation for the normalized concentration C can be derived.

$$V\, dC/dt = QC_0 - QC$$

For a pulse input, $C_0 = 0$ at $t = 0$. Dividing the above equation by V results in:

$$dC/dt = -(Q/V)^*C$$

Rearranging the terms and integrating both sides,

$$\int dC/C = -(Q/V) \int dt$$

Since $Q/V = 1/\tau$, integrating from $t = 0$ to t and $C = C_0$ to C

$$\text{Ln } C/C_0 = -1/\tau\,[t] = -\theta$$

$$C/C_0 = \exp(-\theta) \quad 4.3.10$$

which is the equation of the C-curve.

At $t = \tau$, $C/C_0 = 0.368$, i.e., approx 37 percent of the pulse input to the reactor is still in the reactor after τ_{CSTR}.

PROBLEM 4.3.1

Compare and contrast the three ideal types of reactors.

Solution

Characteristic	Batch	PFR	CSTR
Flow through system	No	Yes	Yes
Degree of mixing	100%	0%	100%
Removal efficiency	High	High	Low
Vulnerability to shock loads	High	High	Low
Volume required for same efficiency	Low	Low	High
Process control	High	Medium	Low

PROBLEM 4.3.2

Determine the volume of a reactor required to achieve 90 percent conversion of a reactant R under two different flow conditions: PFR and CSTR. Assume flow rate, Q = 500 m^3/d; reactant R is converted to product P in a 1st-order reaction with a rate constant of 0.8 1/d; and the reactors are to be analyzed at steady-state.

Solution

Flow rate, Q = 500 m^3/d

1st-order rate constant, k = 0.8 1/d

(i) For a CSTR

Under steady-state, $C/C_0 = 1/[1+k(V/Q)]$

Here C = 0.1 C_o (Since 90 percent has been converted), substituting values yields

Solving for theta (or V/Q) = 11.25 days

Volume of reactor, V = 5625 m^3

(ii) For PFR

Under steady-state, $C/C_0 = e^{-k\tau}$

Solving for theta (or V/Q) = 2.88 days

Volume of reactor, V = 1439.116 m^3

We see that the required volume for PFR is far lesser than the required volume for CSTR, for the same conditions.

PROBLEM 4.3.3

If reactant R is to be converted in a sequential batch reactor and has an initial concentration of 300 mg/L, determine the residence time necessary to achieve 95 percent conversion assuming that the reaction is (i) first-order; (ii) second-order. Assume a rate constant of 0.8 day-1 for the first order reaction and 0.8 L mg-1 day-1 for the second order reaction.

Solution

Initial concentration, C_0 = 300 mg/L

(i) for first order conversion:

We have $C/C_0 = e^{-kt}$

Solving this equation for 95% conversion, $C/C_0 = 0.05$ results in t = 3.74 days.

(ii) for second order conversion:

The reaction rate equation is the only term, since there are no inputs and outputs

Therefore, $C = C_0/[1 + kC_0T]$

or $\mathbf{T = [(C_0/C) - 1]/C_0k = 0.08}$ **days**

PROBLEM 4.3.4

Compare the volume of a CSTR to the volume of a PFR for achieving 90 percent conversion in each of the following cases: 0-order reaction rate, 1st-order reaction rate, and 2nd-order reaction rate

Solution

$C_0/C_e = 10$ for 90 percent conversion. Derive expressions for T (hydraulic detention time) for CSTR and PFR for each of the reaction orders.

Using the relationship between detention time and volume of reactor (V), find the ratio of $V_{CSTR}:V_{PFR}$ for the same Q (flow rate)

For zero-order reaction: ratio = 1

For 1st-order reaction: ratio = 3.9

For 2nd-order reaction: ratio = 10

PROBLEM 4.3.5

A tracer test (step input) was run on an experimental lab reactor. Nitrate served as the tracer and the influent concentration was 100 mg/L. The reactor volume was 4 L and the influent flow rate was 2 L/h. Nitrate concentration in the effluent was monitored continuously and the data are provided below. Draw a tracer response curve and determine the mean residence time (T) for these reactor conditions. Explain why T is not equal to the design hydraulic residence time?

Note: For the tracer test, no reactions occurred in the reactor.

Time	Conc.
0	0
15	13.5
30	25.2
45	35.7
60	44.2
90	58
120	78.5
150	83.4
180	87.45
240	93.6
300	96.65
360	96.65

Solution

Since the reactor volume is 4 L and the flow rate is 2 L/h, the design hydraulic residence time = $V/Q = \tau = 2$ h = 120 min. There is a step input of tracer to the reactor, and values for the F-curve can be calculated using equation 4.3.9 for an ideal CSTR. They can then be compared to the real F values and used to determine the real hydraulic residence time and, therefore, deviation from ideal behavior as is apparent in the following figure.

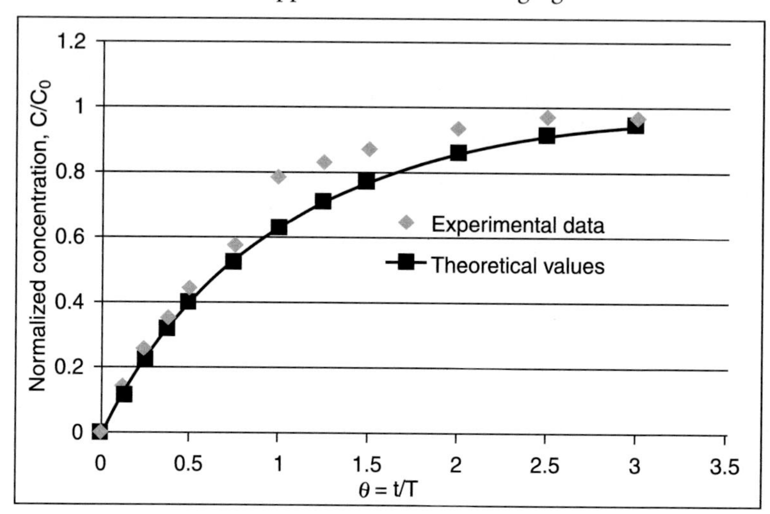

Step 1: Columns 1 and 3 in the following table are the data provided in the problem. Both time and concentration values have to be normalized for solving this problem. Determine values for $\theta = t/\tau$ (Column 2 in table). Determine the normalized concentrations C/C_0 – real (Column 4) and F-curve values (for an ideal reactor) using equation 4.3.9 (Column 5) for an ideal CSTR.

Step 3: Determine 1-F values using Column 4 data since the real value of θ can be obtained using these values, see column 6.

Step 4: Plot the values in column 6 against different values of theta as shown in the following graph and determine the exponential equation for this data. Curve-fitting the experimental data resulted in 1-F = 0.95358*exp(-1.261*x).

For an ideal reactor, $1\text{-}F = \exp(-\theta_i)$ and can be written as $\ln(1\text{-}F) = -\theta_i$. At $\theta_i = 1$, 1-F = 0.368.

To determine the actual or real hydraulic detention time, θ_r, we take the natural logarithm of the best-fit equation and equate it to the one for the ideal reactor. Therefore, for the real data, $\ln(0.368) = -\theta_r = -1.261x = -0.952$. $\theta_r = t/T$ and we can solve for x = 0.755*T, i.e., the actual or real hydraulic residence time is 90.6 min since T = 120 min.

This can be verified based on the concentration data. For an ideal reactor, 63.2 percent of C_0 should be leaving the reactor at T. In our case, 78.5 percent of C_0 is leaving at T which means that some of the tracer is leaving earlier than expected, i.e., there is some short-circuiting.

1	2	3	4	5	6
Time, min	$\theta = t/T$	Conc. C, mg/L	C/C_0 (real)	C/C_0 (ideal)	1 – F
0	0	0	0	0.000	1
15	0.125	13.5	0.135	0.118	0.865
30	0.25	25.2	0.252	0.221	0.748
45	0.375	35.7	0.357	0.313	0.643
60	0.5	44.2	0.442	0.393	0.558
90	0.75	58	0.58	0.528	0.42
120	1	78.5	0.785	0.632	0.215
150	1.25	83.4	0.834	0.713	0.166
180	1.5	87.45	0.8745	0.777	0.1255
240	2	93.6	0.936	0.865	0.064
300	2.5	96.65	0.9665	0.918	0.0335
360	3	96.65	0.9665	0.950	0.0335

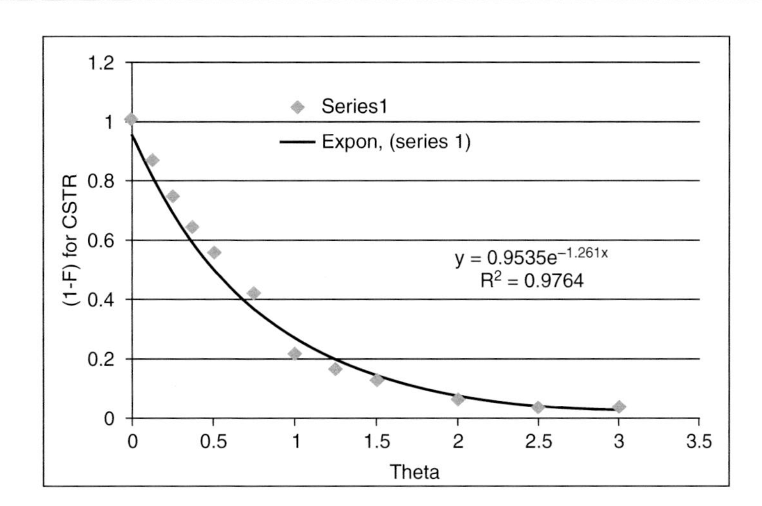

PROBLEM 4.3.6

For the same reactor as in the previous problem, draw tracer curves assuming the reactor to be (i) an ideal PFR and (ii) an ideal CSTR. Note all differences between the two reactors.

Solution

The following data were calculated for a step input of 100 mg/L of tracer concentration to an ideal PFR and an ideal CSTR.

All of the tracer concentration is expected to come out with the water at t = τ in an ideal PFR where τ = design hydraulic residence time = 2 hours. Therefore, the concentration of tracer in the effluent at t ≥ 2 h is 100 mg/L.

For an ideal CSTR and a step input, the equation derived for a step input is $C = 1 - \exp(-\theta)$ and $\theta = t/\tau$.

Time	PFR	CSTR
0	0	0.000
15	0	11.750
30	0	22.120
45	0	31.271
60	0	39.347
90	0	52.763
119	0	62.904
120	100	63.212
150	100	71.350
180	100	77.687
240	100	86.466
300	100	91.792
360	100	95.021

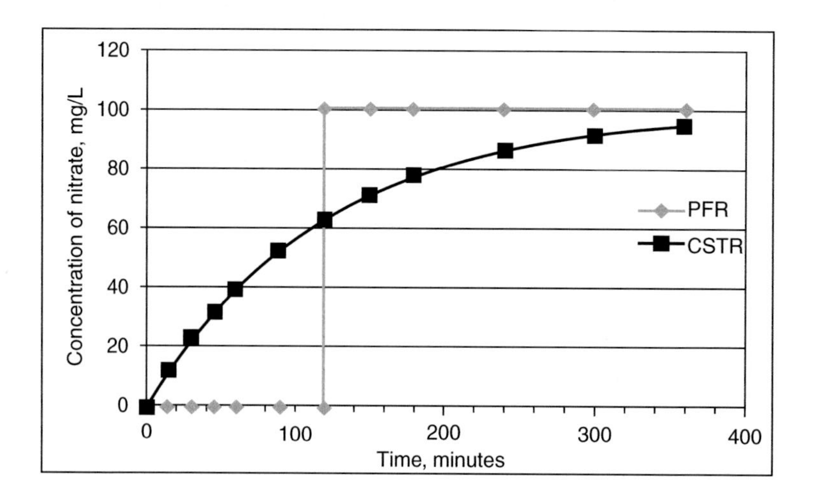

The student is referred to the following text for further details and analyzes of tracer curves: Metcalf and Eddy (2003).

4.4 BACTERIAL GROWTH AND KINETICS

Growth requirements for all organisms include carbon sources, energy sources, and essential nutrients.

Carbon sources: All organisms (including microorganisms) can be classified as autotrophs or heterotrophs. Autotrophs use inorganic carbon (mainly carbon dioxide) as their carbon source while heterotrophs use organic carbon as their carbon source.

Energy sources: All organisms rely on either chemical (oxidation-reduction) reactions or light for their energy. Chemotrophs or organisms that rely on chemical reactions for their energy can be chemoautotrophs (bacteria like methanogens or nitrifiers) or chemoheterotrophs (protozoa, fungi and most bacteria). Similarly, organisms that rely on light are termed phototrophs, and they can be photoautotrophic (plants and algae) or photoheterotrophic (some pigmented bacteria).

Bacteria reproduce by binary fission where each cell increases in size, duplicates its organelles such as DNA and eventually splits into two cells. In other words, the parent cell ceases to exist after the birth of two daughter cells. Bacterial growth can be defined in terms of generation time or doubling time.

Generation time or doubling time is the time for one cell to become two cells. Generation times for different species vary significantly depending on their environmental conditions (nutrient availability, temperature, pH) and genetic material. Most mesophilic bacteria in our environment have generation times varying from 1 to 3 hours with other species having doubling times of <10 mins to several days.

If a species has a doubling time of 1 hour, then in 24 hours, 24 generations will be born or 24 doublings of population will occur. In other words, population N can be estimated using the following equation:

the population $N = N_0 * 2^n$ 4.4.1

Where N = population at time t

N_0 = population at time t = 0

n = generations or doublings

4.4.1 Metabolic Pathways

Metabolic or biochemical pathways

Bacterial growth can occur via any one of three major biochemical pathways (Tortora, Funke and Case 2004):

Aerobic respiration: In aerobic respiration, molecular oxygen is the terminal electron acceptor (TEA) while many different organic (e.g., glucose) or inorganic compounds (e.g., ammonia, nitrite,

Fe(II), Mn(II), and sulfides) can serve as electron donors. In general, cell yields and growth rates are highest for this pathway.

Anaerobic respiration: When bacteria utilize other electron acceptors rather than oxygen for respiration in combination with various organic or inorganic compounds, the process is termed anaerobic respiration. Cell yields and growth rates are less than those for aerobic respiration but greater than that for fermentation.

Fermentation: When the same compound serves as both electron acceptor and electron donor, the process is termed fermentation. Cell yields and growth rates are lowest in this pathway. However, many important industrial processes rely on the end-products of fermentation which can serve as energy sources or other useful products. The best examples are production of alcohols, dairy products like cheese and yoghurt and biofuels.

In general, substrate (or food for microbes) is defined as the limiting nutrient for growth in a given system. Biodegradation is the transformation of substrate and has two major process components: (i) substrate utilization, and (ii) microbial growth. Microbes oxidize or consume substrate for their growth which results in the production of new cells. Figure 4.5 shows changes in substrate and biomass concentrations over time in a batch bacterial culture. A batch culture of bacteria is initiated by adding a few cells (biomass) to the wastewater or nutrient media (similar to the BOD test).

4.4.2 Bacterial Growth Phases

On addition of biomass to nutrient media in a batch culture, bacterial growth is initiated and different phases in the growth of bacteria become apparent over time as shown in Figure 4.5. Substrate concentrations change simultaneously and are shown in the same figure.

1. Lag phase: The first phase, known as the lag phase shows no net growth, i.e., there is little net increase in biomass and the growth rates and death rates of the cells are approximately equal. This phase represents the time required by the organisms to get acclimatized to the new environment before significant cell division and biomass production occurs.
2. Exponential phase: After the cells are acclimated, they multiply at their maximum rate in proportion to the concentration of the substrate, i.e., the growth-limiting nutrient.
3. Stationary phase: Eventually, the growth rate starts to slow as there are no fresh inputs of nutrients and substrate, until the growth rate and death rate are equal. The two rates may remain equal for a short or long period depending on the environmental conditions. This phase is called the stationary phase during which the concentrations of both, substrate and biomass, remain constant.
4. Death or decay phase: When the substrate is almost completely depleted and toxic end-product concentration has increased, microbial growth rates are less than their death rates. A net decrease in biomass concentration is observed, while the substrate concentration remains at its minimum value.

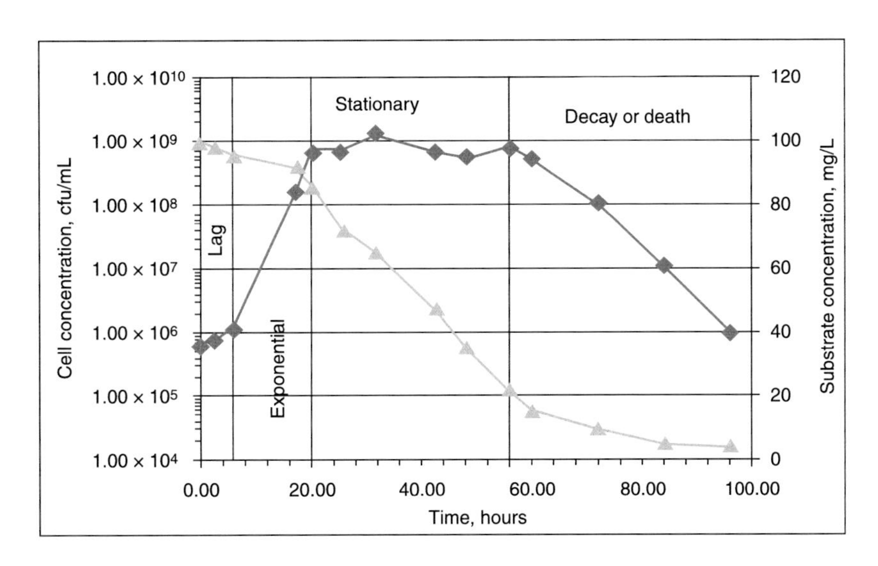

Figure 4.5 Changes in cell and substrate concentrations during microbial growth in a batch system.

4.4.3 Bacterial Growth Kinetics and Yields

Bacterial growth kinetics can be determined in batch or continuous cultures. The method for determining these constants in batch culture is provided here:

Step 1: Prepare a series of culture flasks with different substrate concentrations (S1 to S5) and inoculate each of them with the same microbial culture in the same amount. The volume of media in each culture flask should be the same for ease of comparison. Incubate the culture flasks in an incubator in the dark (to prevent algal growth) at a fixed temperature. The temperature should be maintained constant throughout the incubation period. All glassware and media should be sterilized by autoclaving prior to inoculating with the known microbial culture.

Step 2: Take samples periodically for evaluating cell concentrations (X) during the incubation period and plot them on a graph. A graph similar to Figure 4.6 can be obtained; S1 is the undiluted nutrient media and S5 is the most dilute nutrient media. As shown in the graph, each curve has a different slope in the beginning (exponential phase) and the cell population stabilizes at a later time and at a higher concentration (X) (which is also the saturation cell concentration) as the strength of the media increases.

Step 3: Determine the slope of each curve during the exponential phase; the slope is the specific growth rate μ, 1/time. The exponential phase lasts longest for S1 and least for S5. The rate of cell growth, r_g is defined as:

$$r_g = \frac{dX}{dt} = \mu X \tag{4.4.2}$$

Step 4: Plot μ versus S as shown in Figure 4.7. The maximum specific growth rate μ_m is 0.3 1/hour in this graph and K_s which is the half–velocity constant is the substrate concentration at which $\mu = \mu_m/2$, i.e., 150 mg/L in this graph. The equation of the curve in Figure 4.7 is given by the following equation formulated by Jacques Monod in 1949:

$$\mu = \mu_m \frac{s}{k_s + s} \tag{4.4.3}$$

The maximum substrate utilization rate per unit mass of microbes, *k* (mg substrate/mg cells-time) can be defined on the basis of the yield and μ_m.

$$k = \frac{\mu m}{Y} \tag{4.4.4}$$

Where Y = maximum yield coefficient which is the ratio of mass of cells formed per unit mass of substrate consumed, mg cells/mg substrate. Therefore,

$$r_g = -Yr_s \tag{4.4.5}$$

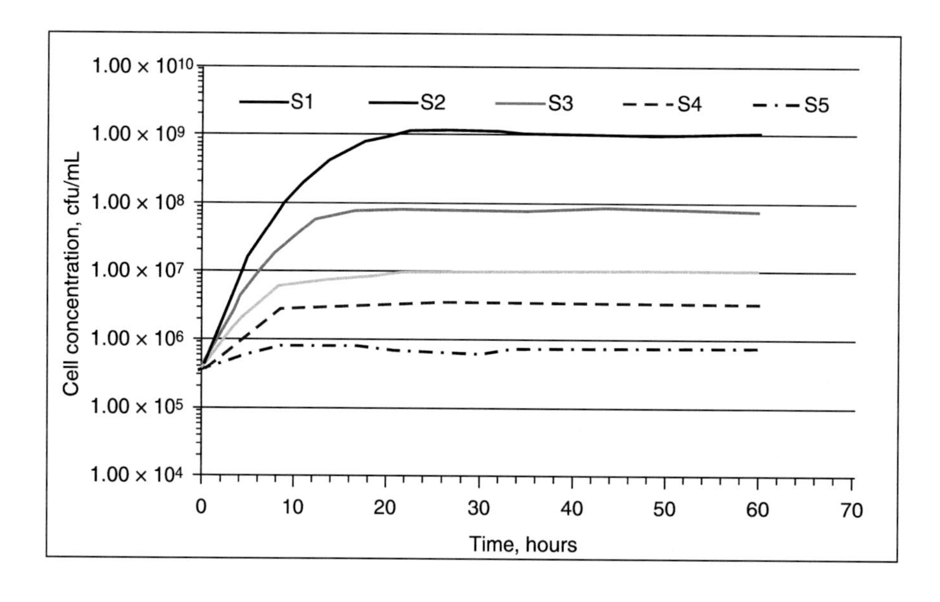

Figure 4.6 Bacterial cell concentrations in five different dilutions of the same nutrient media.

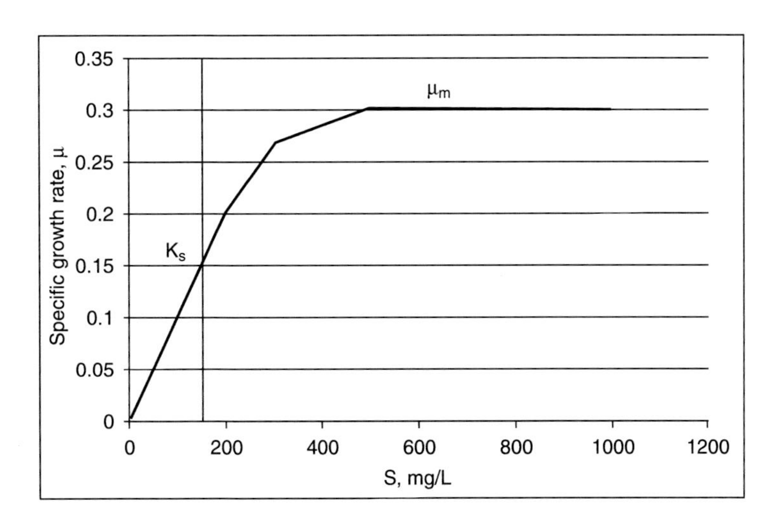

Figure 4.7 Monod kinetics or saturation kinetics for bacterial growth.

Thus, an equation for substrate utilization by actively growing microbes can be derived and the rate of utilization of the soluble substrate is given by (Metcalf and Eddy 1991):

$$r_s = -\frac{kXS}{K_s + S} \tag{4.4.6}$$

Where k = maximum specific substrate utilization rate, $\frac{g\text{-}substrate}{g\ cells - d}$

X = Biomass or cell concentration, g/m^3 or mg/L;

S = Growth limiting substrate concentration in solution, g/m^3;

K_s = half-velocity constant, substrate concentration at half the maximum substrate utilization rate, g/m^3.

Monod's equation for substrate utilization can then be rewritten as:

$$r_s = -\frac{\mu_m XS}{Y(K_s + S)} \tag{4.4.7}$$

Rate of biomass growth with soluble substrate is given by the following equation where the first term on the right hand side is for growth and the second term is for endogenous decay of biomass:

$$r_g = -Yr_{su} - k_d X \tag{4.4.8}$$

Where r_g = net biomass production rate, $\frac{g.VSS}{m^3.d}$;

k_d = endogenous decay coefficient, $\frac{g.VSS}{g.VSS.d}$

PROBLEM 4.4.1

Three flasks with the same nutrient media are inoculated with the following numbers of bacterial cells and incubated: Flask 1 – 10 cells/mL; Flask 2 – 100 cells/mL; Flask 3 – 1000 cells/mL. Given that bacteria divide by binary fission, assume a doubling time of 0.5 hr and estimate the cell numbers in each flask, 12 hrs later. Also use the exponential equation and determine k (rate constant) for these cultures.

Solution

If doubling time T = 0.5 h, cell concentrations can be calculated as N_0*2^n where n = number of generations. In this case, an incubation time of 12 hours implies 24 generations or doubling times.

Therefore, Flasks 1, 2 and 3 will be 1.68×10^8 cells/mL, 1.68×10^9 cells/mL, 1.68×10^{10} cells/mL, respectively.

The exponential growth equation is $N = N_0 \exp(k*t)$ or $\ln(N/N_0) = kt$

where N = cell concentration at time t, N_0 = initial cell concentration, t = time, hours and k = growth rate constant, 1/hour

For flask 1, $\ln(1.68*10^8/10) = 16.64 = k*12$

Solving for k = 1.387 1/hour. The growth rate remains the same in all flasks for the same conditions and the same cultures.

PROBLEM 4.4.2

A water sample was inoculated with two bacterial species with the following growth constants: Species A: $\mu_m = 1.2\ h^{-1}$; $K_s = 11.0$ mg/L; Species B: $\mu_m = 0.55\ h^{-1}$; $K_s = 1.0$ mg/L. Which species will eventually predominate (i.e., have the higher cell concentration in stationary phase) at substrate concentrations of 5, 10 and 15 mg/L.

Hint: Plot μ (growth rate) versus S, ignore decay and answer based on the graph.

Solution

Step 1: Using Monod's equation for specific growth rate μ (1/h), determine values of μ by assuming different values of S for the two different species as shown in the following figure.

As shown in the plot, at 5 mg/L – species A is growing at a slower rate than B. Therefore, species B predominates. At 10 mg/L and 15 mg/L, species A is growing faster than B, therefore A predominates, i.e., cell concentration is higher.

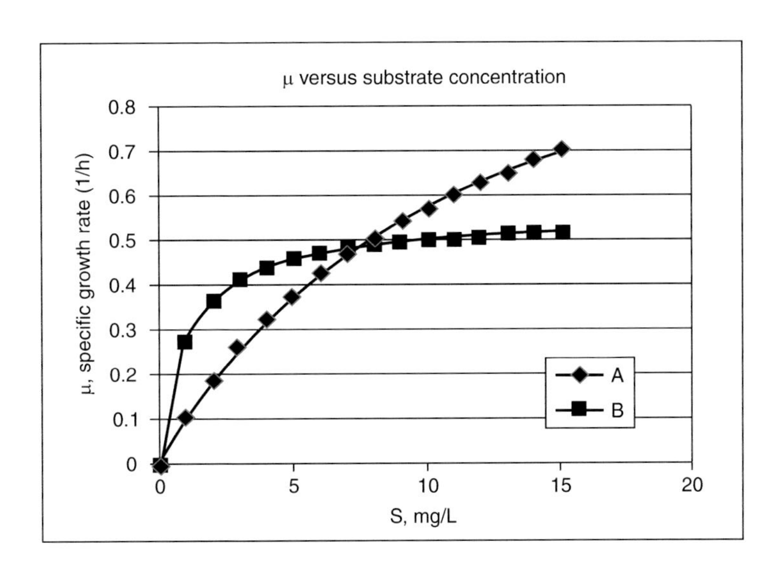

Study Outline

Mass transfer – using mass balances

Based on the law of conservation of mass, mass balances can be done for any natural or engineered system to understand its behavior. The equation of continuity can be used to quantify mass transport and transformation processes. Mass (or contaminant in our systems) can be transported by three basic processes: advection, diffusion, and/or dispersion.

Transport by advection

Transport of mass due to fluid flow is termed advection and can be analyzed using the equation of continuity. Temperature or density is generally assumed to be constant in the analyzes of aqueous systems.

Transport by diffusion

The transport of *particles* is described as *Brownian diffusion* while the transport of *dissolved compounds* is described as *molecular diffusion*. Diffusion can occur under quiescent or flowing conditions.

Brownian diffusion is a function of the thermal energy of the fluid, i.e., temperature. The Brownian diffusion coefficient can be quantified based on Stokes–Einstein equation.

Molecular diffusion is a function of the concentration gradient and is described by Fick's laws of diffusion.

Transport by eddy or turbulent diffusion

When diffusion occurs under turbulent flow conditions, the degree of mixing is much greater and the diffusion coefficient is termed eddy or turbulent diffusion coefficient.

Transport by dispersion

Deviations from laminar flow, i.e., increasing turbulence, can result in dispersion.

Mass transformation

A chemical species is said to have reacted when its chemical identity changes, i.e., mass transformation occurs. *Conservative* compounds are defined as those that do not change their chemical identity, and can be used as tracers.

Types of reactions

Reactions can be categorized as reversible or irreversible, homogeneous or heterogeneous, catalytic or non-catalytic, elementary or non-elementary, and single or multiple. Reactions can also be classified as synthesis, decomposition, or exchange reactions. Reactions can occur in series or parallel or a combination of both.

Reactions kinetics

Process selection and design depend on an understanding of the rate law and rate of reaction. Change in concentration of any chemical species can be monitored over time and analyzed to determine the order of reaction and the governing rate law. The reaction rate is a function of temperature and several other factors such as composition of the reaction mixture and the constituent concentrations. Arrhenius law is used to quantify the effect of temperature change on reaction rate constants.

Ideal reactors

Ideal reactors are well-defined in terms of mass and energy inputs, outputs, and degree of mixing. There are three types of ideal reactors: batch, completely-mixed, and plug flow. All other reactors are a combination or modification of these three basic types.

Batch reactors

In a batch reactor, there is no flow into or out of the reactor. Mass is added to the reactor, it is treated, and then discharged. In a sequential batch reactor, the cycle is repeated. Since there are no inputs or outputs of mass or energy during operation, batch reactors are most popular for determining rate laws.

Continuously stirred tank reactors

In a completely-mixed or continuously-stirred tank reactor (CSTR), it is assumed that complete mixing occurs instantaneously and uniformly throughout the reactor. The concentration of compound A leaving the reactor is equal to the concentration of the compound within the reactor, i.e., there is no spatial variation in the concentration of compound A in the reactor. The equation of continuity is used to derive expressions for steady-state and non-steady-state behavior of these systems.

Plug flow reactors

In a plug flow reactor (PFR), the reactor is much longer than it is wider; the length to width ratio is ≥ 10. Flow is assumed to be laminar, i.e., the streamlines or flow lines do not intersect, and each differential element or volume is assumed to be 'isolated' from its neighbors. Since there is no mixing of fluid elements, each fluid element behaves as a batch reactor moving at the same uniform speed as its neighbors. The concentration of compound A at any point in space within this reactor is a function of the reaction time.

Tracer curves

Tracer tests are conducted to determine the extent to which a reactor deviates from 'ideal' flow conditions and is used to assess the hydraulic performance of any reactor.

Applications of tracer studies are determining real hydraulic detention time in any reactor, assessing short circuiting, or hydraulic flow patterns in natural and engineered systems.

Bacterial growth and kinetics

Bacteria need a carbon and an energy source for their growth. Based on the carbon source, an organism can be an autotroph or a heterotroph. Autotrophs can use inorganic carbon (mainly carbon dioxide) as their carbon source while heterotrophs rely on organic carbon only. Based on energy sources, organisms can be lithotrophs that rely on chemical energy or phototrophs that rely on light energy.

Metabolic pathways

There are three major biochemical pathways: aerobic respiration, anaerobic respiration, and fermentation. Cell yields are highest for aerobic respiration and lowest for fermentation.

Bacterial growth phases

Four phases of bacterial growth can be seen in batch cultures: lag, exponential, stationary, and decay phases. All cultures may or may not show all four phases.

Bacterial growth kinetics

Bacterial growth is accompanied by substrate consumption. The loss of substrate and increase in biomass can be modeled using Monod's equation.

Study Questions

1. Note the equation of continuity and explain each term.
2. What are the differences between advection, convection, diffusion, and dispersion? What are the driving forces in each case, explain with examples.
3. What is the difference between mass transport and transformation?
4. Name the different types of reactions and explain with examples.

5. What is the difference between reaction rate and rate law?
6. Define the order of a reaction and explain with an example. For a system where the rate law is not known *a priori*, how can the order of the reaction be determined?
7. Does temperature have an effect on reaction rate? Explain with appropriate equation.
8. A contact basin is to be designed for treating drinking water. The temperature of water in winter is about 10 °C and about 30 °C in summer. What temperature should the contact basin be designed for?
9. Name the different types of ideal reactors and their defining characteristics.
10. What is a conservative compound and why is it used as a tracer?
11. What is the difference between an F-curve and a C-curve?
12. Why is the behavior of real reactors often different from that of ideal reactors?
13. What is the difference between an autotroph and a heterotroph? Explain with examples.
14. What is the difference between a chemotroph and a phototroph? Explain with examples.
15. How do bacteria reproduce? Write an equation for bacterial growth.
16. What is generation time or doubling time?
17. What are the three major metabolic pathways for bacteria? How do they differ from each other?
18. What are the four phases of bacterial growth?
19. Do all bacterial growth curves have all four growth phases? Explain your answer in detail.
20. Derive the expression for Monod kinetics for bacterial growth.

PART - II

Water and Wastewater Treatment Processes

Part II of the book deals with the treatment processes that are used in water or wastewater treatment plants. Water for domestic or municipal use is drawn from surface or groundwater sources and depending on the quality desired for drinking purposes, it is treated in a water treatment plant. After treatment, the treated water is supplied to consumers through a distribution network. Wastewater is generated by consumers after use and is collected in sewer systems from where it is transported to a wastewater treatment plant. The treated wastewater is then discharged on land, into a surface water body or into the sea. However, due to water scarcity and costs of wastewater treatment, water is now increasingly reused for irrigation or industrial purposes or recycled for other purposes. In some cases, wastewater that has been treated to tertiary or even higher levels can be used for drinking purposes. Two examples of such complete recycling of water are the public water supply systems in Singapore since 1998 and Namibia since 1968 (Pruden 2014).

Water treatment processes were mainly physico-chemical processes until a few decades ago. This was done based on an intuitive understanding that one of the major objectives of water treatment was elimination of microbes, especially pathogens, and therefore, all biological processes were kept out. However, increase in water quality requirements (removal of TOC was not necessary until regulations for disinfection by-products became effective), and the recognition that processes like filtration (porous media and membrane) and disinfection are inherently a mix of physical, chemical and biological processes has led to an understanding that any categorization is necessarily arbitrary. A specific example is the use of biofiltration for enhanced removal of TOC along with turbidity in water treatment and the recognition that biological growth occurs naturally in porous media filters and can be controlled and used to advantage. It should also be noted that physico-chemical processes are generally considered to be either primary or tertiary levels of treatment for wastewater. A list of the treatment processes, included in Chapters 5 and 6, is provided in the following table along with their points of application in water and wastewater treatment.

Treatment Unit	Water Treatment Application	Wastewater Treatment Application
	Primary Treatment	
Flow equalization	Balancing storage is the term used in water treatment and is provided within the distribution system	Flow equalization is the term used in wastewater treatment and the equalization tanks precede treatment of wastewater
Aeration	Generally for groundwater, sometimes for surface waters	Mainly in activated sludge processes
Screening	At surface water intake points	At the entry to wastewater treatment plants
Sedimentation – discrete settling (Type I)	In infiltration galleries or primary settling tanks	After screens in treatment plants, in grit chambers or primary settling tanks
	Secondary or Higher Treatment Level	
Coagulation–flocculation	A major unit process in conventional treatment; essential for most surface waters and some groundwaters	A tertiary treatment process for wastewater, mainly for nutrient removal, especially used when wastewater is to be reused or recycled
Sedimentation – flocculant settling (Type 2)	Clarifiers for settling of floc formed after coagulation	Clarifiers for biological solids from secondary treatment processes
Softening – chemical (addition of lime and soda ash)	For hard waters – generally groundwaters; chemical softening is cheaper and can be done without pretreating water	Addition of lime for phosphate removal in tertiary wastewater treatment
Softening – ion exchange	For hard waters – generally groundwaters; physical methods like adsorption and ion exchange require high-quality feed water, give very high quality treated water, and are more expensive than chemical methods	No known application except at the quarternary level of treatment; e.g., in Singapore's Nuwater treatment plant
Filtration – granular media	Generally, follows coagulation and flocculation, with or without settling.	A tertiary treatment process for wastewater; especially used when wastewater is to be reused or recycled
Filtration – membrane filters	Generally an advanced treatment process for surface waters; or a primary treatment process for groundwaters that need to be softened or desalinated	A tertiary wastewater treatment process; used especially when wastewater is to be reused or recycled

CHAPTER 5

Physico-Chemical Processes for Water and Wastewater Treatment

Learning Objectives

- *Use water supply and demand data to design balancing storage reservoirs or equalization tanks*
- *Describe the 2-film theory of gas transfer*
- *Evaluate different aeration strategies and design an appropriate aeration process for water or wastewater treatment*
- *Describe different types of screens and their applications in water and wastewater treatment*
- *Design screens and determine head loss through different types of screens*
- *Define different types of settling and the conditions under which they occur*
- *Design different types of settling tanks for primary and secondary clarification*
- *Describe the double layer model for particles and its application in coagulation processes*
- *Describe the mechanisms of coagulation and flocculation*
- *Determine hardness in water and design processes for softening water*
- *Identify types of filtration processes, evaluate their use for different water samples and design appropriate units*
- *Describe different types of membrane filters and conditions for their use*
- *Design membrane filtration processes for drinking water treatment*

5.1 BALANCING STORAGE OR FLOW EQUALIZATION

A major design unit in water and wastewater treatment is a balancing storage reservoir (water treatment) or a flow equalization tank (wastewater treatment). Since flow is never constant and is subject to seasonal and diurnal variations, fluctuations in water demand and generation of wastewater have to be accounted for in the design of these reservoirs or tanks.

Distribution reservoirs are used in water treatment plants to store treated water for balancing hourly variations in water demand and for emergencies like fires, break-downs, and repairs. These reservoirs are also called service reservoirs. The main functions of a distribution reservoir are:

1. To balance hourly variations in water demand thus allowing water treatment units and pumps to be operated at a constant rate. This in turn reduces costs and improves efficiency,
2. To maintain constant heads in the distribution mains. In the absence of storage reservoirs, pressure would fall as demand increases. Use of distribution reservoirs dampens pressure fluctuations,
3. To allow pumping of water in shifts to meet the whole day's demand without affecting water supply,
4. To supply water during emergencies like fires, breakdowns or repairs, and
5. To reduce the size of pumps, pipelines and treatment units, thereby leading to overall economy in design of water supply systems.

In water treatment, the capacity of the storage reservoir is a summation of the following (Garg 2001):

1. Balancing storage: Balancing storage or equalizing storage is the quantity of water to be stored in the reservoir for balancing or equalizing the variations caused by fluctuating water demand. Balancing storage is also referred to as the storage capacity of a balancing reservoir.
2. Breakdown storage: Breakdown storage or emergency storage is the storage required to cover emergencies like failure of pumps or any other equipment. The value of breakdown storage is difficult to assess as it depends on factors like frequency of failure and the time required for repairs. Generally, an approximate lump sum value is assumed as breakdown storage. A value of about 25 percent of storage capacity of reservoir or about to 1.5 times the average hourly supply is considered as breakdown storage for a distribution reservoir.
3. Fire demand: This is provided to supply water to extinguish fires. Sufficient amount of water must remain available in the reservoir for extinguishing fires.

In wastewater treatment, the equalization tank is designed for dampening fluctuations in flow and contaminant loading (Mihelcic and Zimmerman 2010). Fluctuations in wastewater flow are 'equalized' to provide uniform flow and/or uniform organic loading to treatment processes thereby increasing process efficiency. These tanks result in smaller size and therefore, lower costs in water and wastewater treatment. Two methods of flow equalization are used in wastewater treatment: in-line and off-line (Metcalf and Eddy 2003). In-line equalization results in very large volume tanks since the entire wastewater flow is routed through these tanks. There is much greater uniformity in flow and organic loading with this scheme as compared to off-line equalization. However, cost of construction and pumping is much higher compared to off-line equalization. Off-line equalization tanks can be constructed for diverting a part of the excess flow and then allowing it to mix with the mainstream so as to dampen fluctuations in flow or loading. The tank size with off-line equalization is much smaller and therefore, cheaper to construct and operate as compared to in-line tanks.

Two methods are commonly employed in water and wastewater treatment for designing equalization tanks: mass curve method (graphical) or analytical method.

Mass Curve Method: Hourly water demand (or use) data are collected over an extended period of time, preferably at least two years and average usage determined. Over a 24-h period, the data are plotted with time on the x-axis and cumulative demand on the y-axis (as shown in Figure 5.1). This curve is the demand curve. Another curve is obtained by joining the start and end points of the demand curve with a straight line, implying a constant rate of water supply over the 24-h period, also called the average flow rate or supply curve. The resulting figure is a mass diagram as shown in Figure 5.1 (Garg 2001).

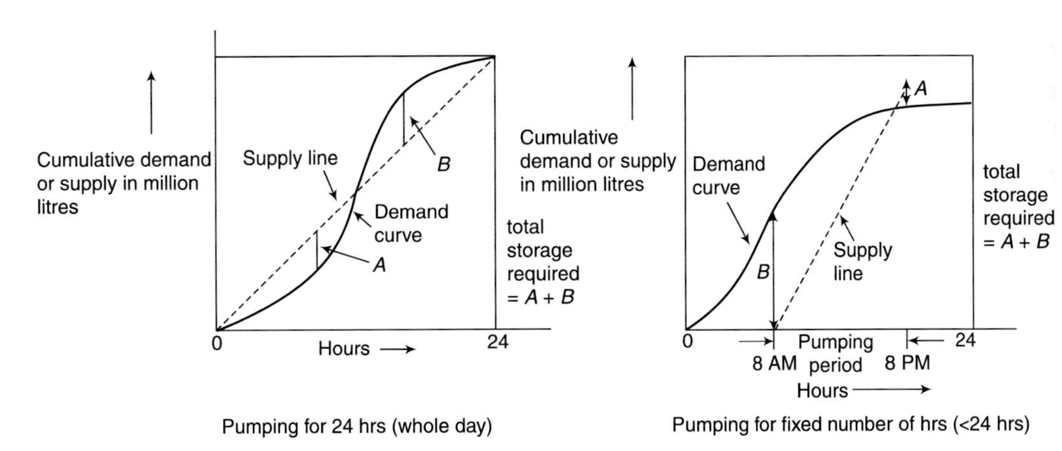

Figure 5.1 Mass diagram or demand curve for 24 hour pumping (continuous supply) and 12 hour pumping (intermittent supply), based on Garg (2001).

For determining the volume of the storage reservoir

1. Collect pumping or demand data (flowrates) for specific time intervals.
2. Determine cumulative flow.
3. Plot a hydraulic mass diagram or curve or compute analytically using a spreadsheet.

4. For the graphical method:
   a. CASE I: If water is supplied on a 24 hour basis, then draw a line connecting origin to last point on the hydraulic mass diagram.
   b. CASE II: For intermittent pumping, draw a line from point A (x = time at which pumping starts, y = 0) to point B (x = end of pumping, y = total volume or cumulative volume required for one day).
5. Draw tangents parallel to the points on the demand curve that are furthest (above and below) from the line for average flow rate or supply line. The sum of the maximum ordinates between demand and supply line as shown in Figure 5.1 is (A + B) and is equal to the volume of the storage reservoir for both cases. Depending on the nature of the data, it may happen that the demand curve lies entirely below the supply curve. In this case, there will be only one ordinate for determining the volume of the storage reservoir.
6. For the analytical method:
   a. Calculate the cumulative hourly demand and supply quantities.
   b. Calculate the difference in cumulative demand and supply. Note maximum and minimum values in demand with respect to the supply line. The sum of the absolute values of the maximum and minimum values is equal to the volume of the storage reservoir.

PROBLEM 5.1.1

A community with 10,000 people has continuous water supply at the rate of 400 Lpcd. The water is supplied by direct pumping over a 24 h period. Water demand can be divided into 4 phases as noted below. Determine the capacity of the storage reservoir.

Duration	Lpcd
5 AM to 11 AM	180
11 AM to 5 PM	80
5 PM to 11 PM	100
11 PM to 5 AM	40

Solution

The volume of the storage reservoir can be determined either analytically or graphically.

Step 1: Note L/person required during different times of the day as shown in the following table. Time starts at the beginning of the day (0 hours or midnight; in this case, you can start at 5:00 AM and create a similar table resulting in the same answer). Water required per day = 400 L/person * 10,000 persons = 40,00,000 L/day which amounts to 1,66,667 L/h. Therefore for 24 hour pumping, water has to be pumped at a constant rate of 1,66,667 L/h. The cumulative supply (A) is based on this rate and the time period of pumping.

Step 2: From midnight to 5:00 AM, water demand is 40 L/person. Therefore, water demand for the community is 40 * 10,000 = 4,00,000 L for these 5 hours. This is followed by a demand for 180 L/person in the morning from 5:00 to 11:00 AM. Cumulative demand until 11:00 AM is 180 * 10,000 + demand from previous period (4,00,000) = 22,00,000 L. Cumulative demand (B) for the remaining periods is shown in the following table.

L/person	Time, h	Cumulative supply (A), L	Cumulative demand (B), L	(A-B)
	0	0	0	
40	5	8,33,333	4,00,000	4,33,333
180	11	18,33,333	22,00,000	3,66,667
80	17	28,33,333	30,00,000	
100	24	40,00,000	40,00,000	
				8,00,000

Step 3: The volume of the storage reservoir is the sum of the 'max delta y', i.e., 4,33,333, from the demand curve to the supply curve and the 'min delta y', i.e., 3,66,667, between the two curves. Therefore, the reservoir capacity in this problem is 8,00,000 L.

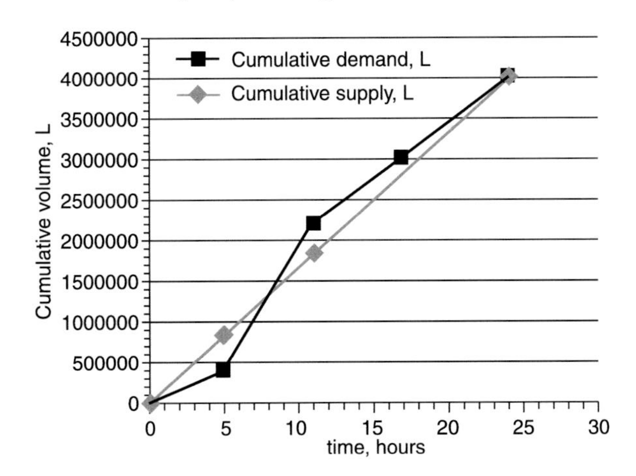

5.2 AERATION

The inclusion or addition of air into liquid or solid media is termed aeration. It has applications in both water and wastewater treatment. The most common application in drinking water treatment is for aerating sub-oxic or anoxic groundwaters. In wastewater, activated sludge processes (ASP) are common and require aeration to ensure adequate supply of oxygen for the growth of aerobic bacteria and degradation of waste (removal of BOD). Aeration is also used in other wastewater treatment processes like oxidation ponds, and extended aeration in ASP.

Aeration can be used to:

(1) increase the oxygen content of water,

(2) reduce carbon dioxide or ammonia content of water,

(3) remove methane, hydrogen sulfide and other volatile organic compounds (VOCs) like chloroform, benzene, and toluene that are responsible for taste, odor or toxicity of water, and

(4) remove iron (Fe^{2+}) and manganese (Mn^{2+}) which impart color to water by promoting their oxidation and subsequent precipitation. Compounds like arsenic (As) which may be present in reduced form and are generally associated with Fe can also be oxidized and co-precipitated leading to partial removal of As by aeration.

Aeration can be carried out by employing one of the following two methods (Peavy, Rowe, and Tchobanoglous 1985):

1. *Dispersing water into air, or*
2. *Dispersing air into water.*

In both cases, interfacial area between liquid and gas is to be maximized. This is generally done by generating bubbles or droplets of either liquid or gas (as the case may be). As shown in Problem 1.1.5, Chapter 1, decrease in size of bubbles or droplets results in higher specific surface area which helps to increase interfacial mass transfer.

Dispersing water into air

The aim here is to minimize water droplet size and thus, maximize interfacial transfer of gases. Examples of water-in-air systems include (Qasim, Motley, and Zhu 2000):

- ***Spray Aerators (Fountains)*:** These consist of a piping grid suspended over a catch basin with nozzles directed upwards. Water is sprayed into the air as droplets and is recovered when it falls back into the catch basin as shown in Figure 5.2. Often, the water is recycled.

Figure 5.2 Spray or fountain aerators (Gangtok Water Treatment Plant, Source: R. N. Sharma) (see color plate).

- ***Cascade Towers*:** They consist of a series of waterfalls (steps) that drop into a small pool. Instead of being dispersed as droplets, water is exposed to the atmosphere in thin sheets as it cascades down each step.

Figure 5.3 Cascade aerator (Gangtok Water Treatment Plant, Source: R. N. Sharma) (see color plate).

- ***Tray Towers*:** Perforated trays towers can be used for aeration. Water is delivered to the top of this stack of trays where it spreads out to increase the surface contact area. The water drips through small openings in the bottom of each tray. As water falls from one tray to another, air is entrained in it or VOCs are removed from it. Tray aerators are most often used for oxidation of iron and manganese or the removal of volatile organic compounds (VOCs).
- ***Packed Bed Aerator*:** The distributor introduces water evenly across the top of a tower packed with plastic, ceramic or metal objects engineered to maximize air-water contact. Air is pushed or drawn upward through the tower against the direction of water flow. A pump at the base collects and removes the treated water.

Since most of these processes use gravity as their driving force, they are also known as gravity aerators.

Dispersing air into water

In dispersing air into water, the aim is to minimize the size of the air bubble so as to maximize interfacial area. Air-in-water systems most often consist of tanks through which water flows. Air is injected through a porous bottom or through spargers near the bottom. Since it takes less energy to disperse air rather than water, smaller, less complicated equipment is required. These types of aeration devices have found greater use in wastewater treatment than in potable water treatment. Air-in-water aeration can be carried out using:

- ***Diffusers or Spargers*:** The air is bubbled into water through diffusers, tubes or spargers. Diffusers are either fine or coarse diffusers depending on the diameter of the nozzles.
- ***Mechanical Aerators*:** These aerators work by vigorously agitating source water with mechanical mixers. As the water is churned, it is infused with air.

The two-film theory of mass transfer is used to describe and derive the equation for mass transfer in two-phase systems, i.e., liquid-gas systems (Cornwell 1990).

Major assumptions in this theory are:

i. Mass transfer occurs through interfacial layers.
ii. The rate-limiting step is the one that offers the greatest resistance to mass transfer.
iii. In stagnant systems, the process is diffusion-controlled while in mixed or flowing systems, the process is controlled by the degree of turbulence and eddy diffusion. As turbulence increases, the thickness of the interfacial layers decreases.

Milestones in aeration theory and concepts (Hendricks 2011)

1855 – A. E. Fick published his first diffusion paper

1924 – 2-film theory presented by W. K. Lewis and W. G. Whitman

1924 – Streeter–Phelps derivation for stream aeration or 'self-purification capacity of streams' was presented

1935 – Higbie proposed the idea that film thickness decreased as turbulence increased

1951 – Danckwertz elaborated on the above

1958 – O'Connor and Dobbins applied the above ideas to stream aeration

1959 to 1962 – Bewtra derived oxygen transfer rates for coarse and fine diffusers

1963 – Kalinske derived oxygen transfer rates for turbine aerators

1968 – Onda correlations published to estimate column height for stripping VOCs using packed beds

Rate of mass transfer in the following cases is either:

i. Gas film-controlled: for highly soluble gases like ammonia and sulphur dioxide, there is greater resistance in movement through the gas phase (Table 5.1),
ii. Liquid film-controlled: for sparingly soluble gases like oxygen, hydrogen, methane and nitrogen, there is more resistance in the liquid film layer, or
iii. Mixed film-controlled: for gases with intermediate solubility like hydrogen sulphide.

Table 5.1 Solubility of pure gases in water at 20 °C and 1 atm pressure (data from Table H.5 (Hendricks 2011))

Gas	C_s (mg/L)	Henry's constant, atm[44]
Hydrogen	1.60	
Nitrogen	19.01	
Methane	23.18	3.8×10^4
Carbon monoxide	28.40	
Oxygen	43.39	4.3×10^4

(Contd.)

44 H (atm) = p/(c*P) where p = mol gas/ mol air; c = mol gas/ mol water; P = total pressure in atm, usually 1; c for water = 1000 g/L/(18 g/mol) = 55.6 mol/L (AWWA 1990).

Gas	C_s (mg/L)	Henry's constant, atm[44]
Ozone	482.00	5.0×10^3
Carbon dioxide	1,688.00	1.51×10^2
Radon	2,347.00	
Hydrogen sulphide	3,846.00	5.15×10^2
Chlorine	7,283.00	
Sulphur dioxide	1,12,800.00	
Ammonia	5,29,000.00	

Design of aeration systems

For aeration, gas transfer rate is proportional to the exposure area per unit volume, time of exposure, concentration gradient, temperature, gas and liquid flow rates and chemical characteristics. The objective of aeration is to maximize transfer of gases either into the liquid or out of it.

Factors that influence gas transfer rate are:

1. **Specific surface area of gas bubbles or liquid droplets:** Maximizing gas transfer requires maximizing interfacial area between liquid and gaseous phases. Specific surface area or the surface area to volume ratio for spheres can be calculated and is inversely proportionate to the radius of the sphere, i.e., 3/R or 6/D where R = radius of droplet or bubble and D = diameter of droplet or bubble. This implies that interfacial area or specific surface area increases as the radius of the sphere decreases. For achieving design objectives, it is therefore necessary to reduce the size of the liquid droplets or gas bubbles to maximize interfacial area and gas transfer.

2. **Contact time or time of exposure:** As contact time increases, the mass of a compound that is transferred from one phase to another phase increases. Therefore, the actual time taken by the gas to be transferred from gas phase to liquid phase or vice versa needs to be maximized. If contact time decreases, overall gas transfer decreases. However, increasing contact time results in higher tank volume making it an expensive proposition. Therefore, a reasonable contact time is used in design depending on the nature of the contaminant to be removed and final concentration to be achieved. Generally, the final concentration is defined by standards or regulations.

3. **Concentration gradient (C_s-C_0):** The driving force for gas transfer is a concentration gradient. As long as the concentration of the contaminant in water is greater than or less than its saturation concentration in water, i.e., it is not in equilibrium with the gas phase, gas transfer will take place.

4. **Gas and liquid flow rates:** The transfer of gas from liquid or into it is determined by the flow rates of gas (generally air) and liquid (generally water) through the aeration tanks or towers.

5. **Temperature:** The overall gas transfer coefficient decreases with decrease in temperature, but saturation concentration will increase with decrease in temperature. Therefore, the net impact of temperature on overall mass transfer cannot be predicted without actual data.

6. **Chemical characteristics of the solute:**

   a. **Saturation concentration of the solute:** The nature of the solute will impact its transfer into or out of the liquid phase.

   b. **Henry's constant for the solute:**

$$\text{Henry's constant} = \frac{\text{moles fraction of component in gas phase}}{\text{moles fraction of component in aqueous phase}} = P_g/C_s$$

   c. **pH for some solutes:** The speciation and solubility of some gases like carbon dioxide, hydrogen sulphide and ammonia are strongly affected by pH. However, other gases like oxygen and methane which are sparingly soluble in water are not affected by pH, and their gas transfer rates will not be affected by pH either.

Based on the above factors, the rate of change of concentration can be defined as:

$$\frac{dC}{dt} = K_L a\,[C_S - C] \qquad 5.2.1$$

Integration of the above equation from t = 0 to t, and $C = C_0$ at t = 0 to some concentration C at time t, leads to the following expression:

$$[C_S - C] = [C_S - C_0]*\exp[-K_L a*t] \qquad 5.2.2$$

where

dC/dt = rate of change in concentration, mg/L-s

K_L = overall mass transfer coefficient, m/s

A/V = a = specific surface area, 1/m

C_s = saturation concentration, mg/L

C_0 = concentration at t = 0, mg/L

t = time, minutes or any other unit of time

Values for the overall mass transfer coefficient, K_La are generally found in various reference books for a temperature of 20 °C and 1 atmospheric pressure. A temperature correction equation is used for correcting K_La values for other temperatures using a temperature correction coefficient, θ = 1.02 and the following equation:

$$K_L'a = K_L a * \theta^{(T-20)} \qquad 5.2.3$$

PROBLEM 5.2.1

Calculate the saturation concentration in water for oxygen and carbon dioxide at 20 °C at mean sea level. Assume atmospheric pressure at mean sea level (MSL) is 1 atm, partial pressure of oxygen is 0.21 atm and carbon dioxide concentration in the atmosphere is 400 ppm (which was the case in 2015)

Solution

Oxygen

Henry's constant at 20 °C = 4.3×10^4 atm

Based on the data provided in Table 5.1, the concentration of pure oxygen in water at 20 °C and 1 atm of oxygen is 43.39 mg/L. However, the partial pressure of oxygen in the atmosphere is 0.21 atm. In other words, 21 percent of the atmosphere is oxygen and the saturation concentration has to account for this, i.e., 43.39*0.21 = 9.11 mg/L. Oxygen is sparingly soluble in water and does not react with water or other species in water accounting for its low solubility.

Carbon dioxide

Similarly, the partial pressure of carbon dioxide in the atmosphere in 2015 was 400 ppm. Therefore, based on values in Table 5.1, the saturation concentration of carbon dioxide in water is 1,688 mg/L*4*10^{-4} = 0.6752 mg/L. This number does not account for the effect of pH, the formation of carbonic acid in water and its dissociation into carbonate and bicarbonate forms which can lead to much higher concentrations of carbon dioxide in water.

PROBLEM 5.2.2

Calculate the saturation concentration in water for oxygen and carbon dioxide at 20 °C at an elevation of 2000 m. Use the regression equation for relating atmospheric pressure to elevation (meters) (Hendricks 2011): pressure (kPa) = $101.325 - 0.011944x + 5.3142*10^{-7}*x^2 - 1.3476*10^{-11}*x^3 + 8.2464*10^{-15}*x^4 - 2.3906*10^{-18}*x^5 + 2.0382*10^{-22}*x^6$

where x = elevation in meters. 1 atm = 101.325 kPa

Solution

If atmospheric pressure at MSL is 1 atm, then at an elevation of 2000 m, the atmospheric pressure is calculated as 0.785 atm using the above regression equation.

For determining saturation concentration of oxygen:

Partial pressure of oxygen at 2000 m is 0.21 *0.7952 atm = 0.167 atm.

From Table 5.1, the saturation concentration of pure oxygen in water is 43.39 mg/L. Therefore, the saturation concentration of atmospheric oxygen in water is 43.39*0.1626 mg/L = 7.05 mg/L. These calculations indicate lower C_s values for DO based on elevation only. However, higher elevations are generally associated with lower temperatures and therefore, higher C_s values for DO.

For determining saturation concentration of carbon dioxide:

Partial pressure of carbon dioxide at 2000 m is $4*10^{-4}*0.774$ atm = $3.1*10^{-4}$ atm.

From Table 5.1, the saturation concentration of pure carbon dioxide in water is 1,688 mg/L. Therefore, the saturation concentration of atmospheric carbon dioxide in water at an elevation of 2000 m is $3.18*10^{-4}*1{,}688$ mg/L = 0.537 mg/L.

PROBLEM 5.2.3

A groundwater source is contaminated by volatile organic compounds. Compound X is the main contaminant with a concentration of 500 micro-g/L in the water. It has a saturation concentration at 20 °C of 5 micro-g/L. K_La (overall mass transfer coefficient) at 20 °C is 0.05 1/min. Determine the minimum time required to achieve a minimum concentration of 10 micro-g/L. If the minimum water temperature in winter is 5 °C, determine the minimum time required to achieve the same concentration in winter. Temperature correction coefficient is 1.02. What volume of aeration tank is required, if the flow to be treated is 100 million liters per day (MLD)?

Solution

Concentration of contaminant, C_0 = 500 μg/L

Saturation Concentration, C_s = 5 μg/L

Concentration at time t, C_t = 10 μg/L

Mass transfer coefficient, K_La (at 20 °C) = 0.05 l/min

Temperature correction coefficient θ = 1.02

Time required when concentration C_t = 10 μg/L is determined using equation 5.2.2:

t = 91.90 min, i.e., 92 min

Temperature, T = 5 °C

Temperature correction coefficient, θ = 1.02

Mass transfer coefficient K'_La at 5 °C is determined by using equation 5.2.3:

$$K'_L a = K_L a * \theta^{(T-20)}$$

$$K'_L a = 0.037151 \text{ 1/min}$$

Calculating time required for T = 5 °C

t = 123.69 min

t = 124 min

Design the aeration tank for the minimum temperature of 5 °C so as to meet standards in winter as well as in summer.

Volume of tank = 69.44 m^3/s*124 min = 8611.11 m^3

PROBLEM 5.2.4

A water treatment plant has to treat a flow rate of 50 MLD. Design an aeration tank with 5 compartments of depth 2.5 m, and length-to-width ratio of 5:1. Detention time in tank should be 20 min. What are the total air requirements if 0.5 m^3/min of air-m^3 of tank are to be supplied?

Solution

A schematic of the aeration tanks is shown here:

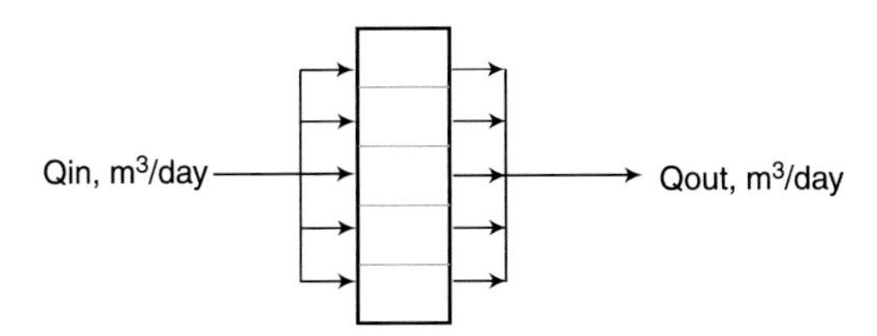

Flow rate (Q) = 50 MLD

No. of compartments (n) = 5

Depth (D) = 2.5 m

Length (L) to width (W) ratio = 5:1

Detention time (t) = 20 min

Supplied air (A) = 0.5 m^3/min-m^3 of tank

Flow rate in each compartment (q) = Q/n = 10 MLD = 10^7 L/d * (1 d/ 24 h) * (1 h/60 min) = 6.944 m^3/min

Volume of each compartment (V) = q*t = 138.88 $m^3 \approx 140\ m^3$

Now, V = L*W*D

Since L:W = 5: 1, therefore, L = 5W

$$V = 5W^2 * D \text{ where } D = 2.5\ m^3$$

Solving for W results in W = 3.35 m and L = 16.75 m

Air requirement (m^3 of air/min) = Volume of tank*daily supplied air = V*air supplied= 70 m^3 of air/min

Daily air requirement = 10^7 L/d * (1 d/ 24 h) * (1 h/60 min) = 1,00,800 m^3/day.

5.3 SCREENING

It is the first unit operation in surface water treatment plants and wastewater treatment plants. A screen is a device with openings used for removing large objectionable solids from influent water or wastewater.

Types of screens (Metcalf and Eddy 2003):

1. Coarse screens with an opening size of 6–150 mm
2. Fine screens with an opening size of <6mm
3. Microscreens with an opening size of <50μm

Coarse Screens

Coarse screens consist of parallel bars or rods and are shown in Figures 5.4 to 5.6. They are also called bar racks or bar screens. They remove coarse materials which can damage equipment like pumps or reduce the efficiency of treatment processes.

Figure 5.4 Vertical bar screen in Bhadreswar Wastewater Treatment Plant, 2010.

Figure 5.5 Inclined bar screen in Bhadreswar Wastewater Treatment Plant, 2010.

Screens can be cleaned either manually or mechanically. The screens shown above are static or fixed screens and need to be cleaned manually. Photos in Figure 5.6 show a moving mechanical screen installed in a wastewater treatment plant in Pune where the screenings (screened materials) are collected at the top as the belt or screen moves in the upward direction.

(a)

(b)

Figure 5.6 Moving screen in a wastewater treatment plant in Pune: (a): Front view, screen is moving from bottom to top; (b). Back view of top of the screen where the screened materials are collected (see color plate).

Types of mechanically cleaned screens:

1. Chain driven screen
2. Reciprocating rake screen
3. Catenary screen
4. Continuous belt screen

Design of Coarse Screens

Key design parameters for coarse screens are flow velocity, mesh size or spacing between bars, and head loss through screens and appurtenances. To prevent deposition of solids in the channel, a minimum approach velocity of 0.4 m/s is prescribed. To minimize passage of debris through the screens, velocity through the screen should not exceed 0.9 m/s (Metcalf and Eddy 2003). Generally, a minimum bar width of 10 mm and minimum spacing between the bars of 10 mm is assumed. Therefore, the area of the open spaces between the bars is approximately 50 percent of the channel cross-sectional area. In such cases, the maximum velocity in the channel cannot exceed 0.45 m/s (to ensure that maximum velocity through the screen is <0.9 m/s).

Head loss through vertical screens

Head loss through vertical screens can be calculated using Bernoulli's equation:

$$h_L = \frac{1}{2g}\left[\frac{v_{sc}^2 - v_{us}^2}{C_d}\right] \qquad 5.3.1$$

The above equation is often simplified to the following form by assuming that the approach velocity is zero; the assumption is not realistic but is in keeping with a conservative design approach since the value for head loss with equation 5.3.2 will be higher than that with equation 5.3.1:

$$h_L = \frac{1}{2g}\frac{v_{sc}^2}{C_d} \qquad 5.3.2$$

h_L = head loss through screen, m

g = gravitational acceleration, 9.81 m/s^2

v_{sc} = flow velocity through screen, m/s

v_{us} = flow velocity through channel upstream of screen or approach velocity, m/s

C_d = coefficient of drag; values ranges from 0.6 to 0.9, dimensionless

A = submerged area of openings, m^2

Head loss through inclined bar screen

An inclined bar screen in an open channel is shown in Figure 5.7. The screen is perpendicular to the flow direction and inclined to the bottom of the channel at an angle θ.

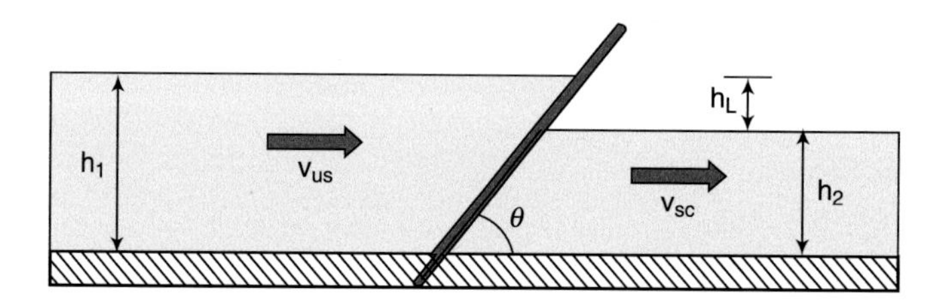

Figure 5.7 Schematic of an inclined screen.

Head loss through a bar screen inclined at an angle θ to the horizontal can be calculated by using the continuity equation, and Bernoulli's or Kirschmer's equations.[45]

The continuity equation implies that flow through the open channel upstream of the screen is equal to the flow through the screen openings, i.e., $Q = A^* v_{us} = A_{sp}{}^* v_{sc}$

Where

Q = flow rate, m^3/s

A = cross-sectional area of open channel upstream of screen, m^2

v_{us} = approach velocity or velocity in channel upstream of screen, m/s

A_{sp} = area of open spaces in screen, m^2

v_{sc} = velocity through screen, m/s

45 http://site.iugaza.edu.ps/frabah/files/2011/09/2.-Preliminiary-treatment-.pdf.

Head loss (h_L) through an inclined screen is determined using Kirschmer's equation where β is a shape factor which ranges from 2.42 for sharp-edged rectangular bars, 1.83 for rectangular bars with semi-circular face on upstream side, and 1.79 for round bars (Karia and Christian 2006).

$$h_L = \beta\left(\frac{w}{b}\right)^{\frac{4}{3}} * h_{us} * \text{Sin}\,\theta \qquad 5.3.3$$

$$h_{us} = \frac{1}{C_d}\frac{(v_{sc}^{\ 2} - v_{us}^{\ 2})}{2g} \qquad 5.3.4$$

W = maximum width of bars facing flow, m

b = minimum width of clear opening, m

h_{us} = head loss through channel upstream of screen, m

where C_d = coefficient of discharge = 0.6 to 0.9, often 0.7 is assumed.

g = gravitational acceleration, 9.81 m/s^2

A depth-to-width ratio of 1.5 for the screen channel is considered to be efficient.[46]

Fine Screens

Fine screens are used in preliminary treatment (i.e., after coarse screens), primary treatment (*in lieu* of primary clarifiers) and treatment of combined sewage. Fine screens used for preliminary and primary treatment can be of static, rotary drum or step type.

Microscreens

They are mainly used to remove fine solids left in the secondary effluent or algal cells from water treatment plant influent. Microscreens are made of fabric with micron-sized openings (Hendricks 2011). The removal mechanism is straining rather than sieving which is the mechanism for coarse and fine screens. This results in the formation of a mat on the surface and needs frequent cleaning. Cleaning is done by backwashing with water jets from behind the screens which loosens the mat deposited on the screen.

Screenings

The solids removed by the screens are called screenings. Coarse screenings include rocks, leaves, plastic, and rags. Fine screenings include grit, and food waste among various other things. Due to the predominantly organic nature of fine screenings, their moisture content is higher and their specific gravity is lower than coarse screenings. The screenings are either discharged directly or transported via belt conveyors or pneumatic ejectors into a screenings grinder or screenings compactor. The screenings compactor dewaters and reduces the volume of screenings. Disposal of screenings is done by burial on the plant site, incineration or in landfills. Sometimes, they are ground by grinders and are returned to the wastewater.

46 http://site.iugaza.edu.ps/frabah/files/2011/09/2.-Preliminiary-treatment-.pdf.

The amount of solids collected as screenings was exponentially related to the spacings between the bars. Using data from 133 wastewater treatment plants, a general regression equation was determined for the amount of screenings with respect of bar spacings (Hendricks 2011):

Average y = 0.11827*exp(–0.064587x)

Maximum y = 0.21533*exp(–0.060845x)

where y = average screenings collected, m^3 solids/ million L of wastewater,

x = bar spacing, mm

In practice, bar spacing of 20 to 40 mm is commonly used in wastewater treatment plants.

PROBLEM 5.3.1

Approach velocity in a screen channel should be >0.4 m/s to prevent deposition of solids in the channel and velocity through the screen should be <0.9 m/s to prevent passage of debris through screen. For flow rates of 10 or 50 MLD, determine the channel dimensions assuming depth of flow and width of channel are equal and approach velocity is 0.4 m/s. Depth of channel should be at least 0.5 m higher than depth of flow in channel.

Solution

For a flow rate, Q = 50.00 MLD = 0.58 m^3/s

For velocity of approach = 0.40 m/s

Based on Q = A*v where A = area of channel and v = approach velocity

Area of channel = 1.45 m^2

Depth of flow = width of channel = 1.20 m

Depth of channel is 0.5 m greater than the depth of flow = 1.7 m

Similarly, for a flow rate, Q = 10.00 MLD = 0.116 m^3/s

For velocity of approach = 0.40 m/s

Area of channel = 0.29 m^2

Depth of flow = width of channel = 0.54 m

Depth of channel is 0.5 m greater than the depth of flow = 1.04 m

PROBLEM 5.3.2

Determine head loss through a screen with square bars of 10 mm width spaced at 10 mm intervals. Flow rate through the screen is assumed to be 50 or 10 MLD. Assume coefficient of discharge is 0.7 and solve by (A) assuming an approach velocity of 0.45 m/s in screen channel and (B) assuming approach velocity is zero. Assume depth of flow is equal to width of channel. Note depth of channel should be at least 0.5 m above the depth of flow in channel.

Solution

There are 4 different cases for which head loss is to be calculated and it is simpler to set up a spreadsheet for doing so as shown below.

Step 1: For the two flow rates of 50 and 10 MLD, determine flow rate in m^3/s. Using the approach velocity given in Case A, determine the dimensions of the channel assuming depth of flow = width of channel.

Step 2: Calculate the number of bars to be used across the channel by dividing the width by 20 mm and subtracting 1 from the answer. Then determine area of bars facing flow and subtract this area from the channel cross-sectional area to determine area of the openings between the bars.

Step 3: Determine velocity through the screen. Check if <0.9 m/s.

Step 4: Use equations 5.2.1 and 5.2.2 to determine head loss for the two different approach velocities and flow rates as shown in the following table.

Flow rate, Q	50.00	10.00	MLD
	0.58	0.12	m^3/s
A. For velocity of approach (assumed)	0.45	0.45	m/s
Cross-sectional area of channel, A = DW	1.29	0.26	m^2
If depth of flow (D) = width of channel (W)	1.13	0.51	m
Cross-sectional area of channel, A = DW	1.29	0.26	m^2
Width of one bar	10.00	10.00	mm
Spacing between bars	10.00	10.00	mm
Number of bars	55.70	24.36	bars
Area of bars	0.63	0.12	m^2
Opening area	0.65	0.13	m^2
Velocity through screen, v_{sc}	0.88	0.87	m/s
Head loss through the screen	0.042	0.040	m
B. For velocity of approach (assumed)	0.00	0.00	m/s
Head loss through the screen	0.057	0.055	m

PROBLEM 5.3.3

A 0.5 m wide screen has bars of 10 mm width spaced at 10 mm intervals. Flow rate through the screen is 10 MLD and depth of flow is 0.52 m. Assume coefficient of discharge is 0.7. Use Kirschmer's equation for accounting for head loss through inclined bar screens, assuming the bar screen is inclined at a 45 degree angle with the horizontal. Compute for bar shape factor of 2.42 for sharp-edged bars and 1.83 for round edged square bars.

Solution

Flow rate, Q = 10 MLD = 10000000 L/d = 10000 m^3/d = 0.116 m^3/s

Cross-sectional area of channel, A = 0.26 m^2

Velocity in channel upstream of screen, v_{us} = **0.446 m/s**

Number of bars = 24 bars

Area of bars = 0.125 m^2

Clear area (area of spacings) = 0.135 m^2

Velocity through screen, v_{sc} = **0.856 m/s**

Kirschmer's equation: head loss through inclined screen, $h_L = \beta^* h_{us}^* \mathrm{Sin}\ \theta^*(W/b)^{(4/3)}$

β is the shape factor and is 2.42 for sharp-edged bars; 1.83 for round-edged bars

W = width of bars facing opening, m; in this example W =10 mm

b = minimum width of opening, m = 10 mm

Head loss in channel upstream of screen, use Bernoulli's equation (5.3.1):

$$h_{us} = 0.035 \text{ m}$$

Head loss through screen for β = 2.42 is **0.06 m**

Head loss through screen for β = 1.83 is **0.046 m**

5.4 SETTLING

Particles in the environment vary in shape, size and specific gravity resulting in great differences in their behavior and interaction with other particles. Table 5.2 summarizes the sizes and specific gravities of some particles commonly found in aquatic systems.

Table 5.2 Sizes and specific gravities of particles found in aquatic systems

Particles	Size, microns	Specific gravity	Nature of particles
Virus	0.03 to 0.8[48]	1.25[47]	Colloidal
Bacteria	0.2 to 10[48]	1.25[47]	Colloidal
Clay	<2[50]	2.6 to 2.75[50]	Colloidal
Silt	2 to 60[50]	2.6 to 2.75[50]	Colloidal or Discrete
Soil, coarse	60 to 60,000[50]	2.6 to 2.75[50]	Discrete
Sand, filter	100 to 1200[49]	2.5 to 2.65[49]	Discrete
Flocculant particles	>1000	1 to 2.6 depending on size and compactness[49]	Discrete

47 McIntosh and Selbie 1937.

48 Madigan et al. 2015.

49 JMM Inc. 1985.

50 http://environment.uwe.ac.uk/geocal/SoilMech/classification/soilclas.htm.

Removal of particles by gravity is termed sedimentation or settling or clarification. The settling of particles depends on their shape, size and specific gravity and their concentration in suspension.

Particles can be either discrete or flocculant particles.

- Discrete particles are defined as particles that do not change size, shape and specific gravity over time.
- Flocculating particles are defined as particles that change size, shape and specific gravity over time as they aggregate or coalesce.

Particles dispersed in any fluid constitute a suspension, in contrast to dissolved compounds which constitute a solution. **Suspensions** can be either dilute or concentrated.

- Dilute suspensions: If the concentration of particles in a suspension is insufficient to displace water as the particles settle then the suspension is termed a dilute suspension. In general, suspensions of particles with less than 1000 mg/L are considered dilute.
- Concentrated suspensions: If the concentration of particles in a suspension is sufficient to displace water as the particles settle, it is defined as a concentrated suspension; this frequently means suspensions with >1000 mg/L of particles.

Settling characteristics of particles are categorized as Type 1 to 4 depending on the nature of particles and their suspensions.

***Type 1*:** This applies to discrete particles in dilute suspensions. The behavior of these particles can be described by Newton's equation. Stokes' equation describes the specific case of settling of spherical particles under laminar flow conditions.

***Type 2*:** This applies to flocculant particles in dilute suspensions. Since these particles change character over time, their behavior cannot be predicted *a priori*. Experimental settling curves have to be generated to understand the behavior of these particles.

***Type 3*:** This applies to particles settling in concentrated suspensions and is also called hindered or zone settling. Particles are in a fixed position relative to each other; the entire mass of particles settles as a whole, i.e., as a zone. The rate of settling is a function of the concentration of particles.

***Type 4*:** Also called compression settling, this applies to particles settling in concentrated suspensions. However, due to higher concentrations, further settling takes place only by compression (compaction) of the previously settled mass.

5.4.1 Discrete Settling (Type 1)

Type 1 or discrete settling: This type of settling applies to discrete particles and can be modeled based on Newton's Law (1687) (Whiten and Ozer 2015). A particle settling due to gravity has three forces acting on it: [i] gravity - in the downward direction and [ii] buoyancy in the upward direction. As the particle starts falling in the medium, [iii] frictional or viscous drag forces act in the direction opposite to its direction of movement (Peavy, Rowe, and Tchobanoglous 1985).

Gravitational force: $F_g = \rho_p g V_p$

Buoyancy force: $F_b = \rho_w g V_p$

Net force on stable particle: $F_{net} = F_g - F_b = (\rho_p - \rho_w) g V_p$ 5.4.1

Where ρ_p = density of particle, ρ_w = density of water, V_p = Volume of particle, g = acceleration due to gravity.

The viscous drag force acting on the moving particle is given by

$$F_d = \frac{C_d A_p \rho_w v_s^2}{2} \quad 5.4.2$$

C_d = coefficient of drag, dimensionless

A_p = projected area or cross-sectional area of particle in flow direction (m^2),

v_s = settling velocity of particle (m/s).

Solving for equations 5.4.1 and 5.4.2 for spherical particles, $V_p/A_p = (4/3)\pi^*(d_p/2)^3/(\pi^*(d_p/2)^2) = 2d_p/3$, we get

$$v_s = \sqrt{\frac{4}{3}\left(\frac{\rho_p}{\rho_w} - 1\right)\frac{gd_p}{C_d}} \quad 5.4.3$$

where $C_d = \frac{24}{Re}$ for laminar flows (Re<0.2) 5.4.4

$$C_d = \frac{24}{Re} + \frac{3}{\sqrt{Re}} + 0.34 \text{ for transitional flows } (0.2 \leq Re \leq 1000) \quad 5.4.5$$

$C_d = 0.445$ for turbulent flows ($1000 < Re < 2 \times 10^5$) 5.4.6

and Reynolds number (Re) is given by $Re = \frac{\text{Ø} \rho_w V_s d_p}{\mu}$ 5.4.7

where d_p = diameter of particle (m), μ = dynamic viscosity of water (N-s/m^2)

Ø = sphericity of the particle (how much does it deviate from a perfect sphere) ranges between 0 and 1.0. A perfect sphere has a sphericity of 1.0 and a shape factor of 6 as explained below.

Sphericity of a particle = surface area of a sphere with the same volume as the particle of interest divided by the surface area of the particle (Geldart 1990)

The above equation can be written as $\text{Ø} = \frac{d_{sv}}{d_v}$ 5.4.8

Where d_v = volume diameter of a particle and d_{sv} = surface-volume diameter for an individual particle is defined as the diameter of a sphere which has the same ratio of external surface area to volume as the particle = $\frac{1}{\sum \frac{x_i}{d_i}}$

And x_i = mass fraction of particles in the *i*th size range with average diameter, d_i

Shape factor is 6/Ø and is taken as 2 for sand, 2.25 for coal, 4.0 for gypsum and 22 for graphite flakes (Gregory and Zabel 1990).

Table 5.3 Particles and deviations from sphericity (Geldart 1990)

Particles	Volume diameter, d_v (micron)	Surface-volume diameter, d_{sv} (micron)	Wadell's sphericity, θ
Glass spheres	514	514	1.0
	787	772	0.98
	1,204	1,178	0.98
Sand	384	296	0.77
	514	455	0.885
	832	599	0.72
	989	698	0.705
Plastic cuboids	3,664	2,387	0.65
	3,700	2,800	0.757
Limestone	3,254	2,280	0.7
	3,492	2,280	0.65

For laminar conditions and particles that are perfect spheres, i.e., Ø = 1.0 and substituting $C_d = 24/Re$ and Re, we get Stokes' Law (1851) for settling velocities of discrete particles:

$$v_s = \frac{(\rho_p - \rho_w) g d_p^2}{18\mu} \quad 5.4.9$$

Stokes' law is applicable only for laminar flow and spherical particles. The standard drag curve shown in Figure 5.8 is based on the above equations and following values for the constants:

Different diameters of spherical particles were assumed ranging from 0.1 micron to 48.2 mm which correspond to Reynolds number values of 1.8×10^{-12} to 2×10^5. The particles had a specific gravity of 2.65 with a shape factor of 1. Water temperature was assumed to be 20 °C with a density of water of 998 kg/m^3 and dynamic viscosity of 1.0×10^{-3} $N\text{-}s/m^2$. The corresponding C_d values were calculated and plotted with respect to Re in the above curve and the flow regimes defined as laminar (Re < 0.2), transitional and turbulent (Re > 1000) based on cut-offs mentioned in the literature (Whiten and Ozer 2015). There is a sudden drop in the value of the drag coefficient for Re > 2×10^5 and Newton's equation is no longer applicable after that.

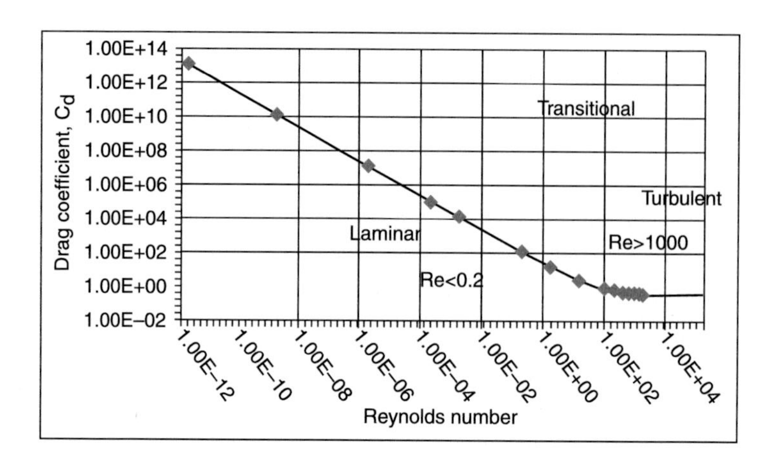

Figure 5.8 A standard drag curve for discrete particles in a range of Re from 10^{-12} to 2×10^5.

Settling column analysis for Type 1 settling

Analysis of the settling behavior of discrete particles can be done using settling column analysis as described below:

- A dilute homogeneous suspension of particles is poured into a very tall column with a single port at the bottom. The height of the water column is noted.
- Samples are removed from the port at periodic time intervals.
- The concentrations of the suspended-solids in the samples are determined.
- Based on the above data, the mass fraction of particles remaining in suspension over time (x_i) and their corresponding settling velocities (v_i) are determined.
- Plotting a graph of x_i versus v_i and using the equation derived below, the removal efficiency of a sedimentation or settling tank can be determined.

Any settling basin is designed for the complete removal of all particles having terminal settling velocities $\geq v_s$ which is the design settling velocity. Thus, particles with settling velocity $\geq v_s$ are completely removed (100 percent removal). Assuming a uniform distribution of particles over depth, we can determine the fraction of particles with settling velocities $v_p < v_s$ that are removed in proportion to the ratios of their velocities.

Thus, the net removal efficiency of the settling tank is given by $(1 - X_0) + \int_0^{X_0} \frac{v_i}{v_s}$ 5.4.10

Where $1 - X_0$ = fraction of particles with settling velocity $\geq v_s$, where X_0 corresponds to v_s.

PROBLEM 5.4.1

Determine the settling velocities of spherical particles with diameters 1 micron, 10 micron, 100 micron, 1 mm and 15 mm based on Stokes' law. Assume water temperature to be 20 °C with a density of water of 998 kg/m^3 and dynamic viscosity of 1.0×10^{-3} N-s/m^2. Assume particles have a specific gravity of 2.65 at the same temperature. Note where Stokes' law is no longer applicable and why. Use Newton's law to determine settling velocity in transitional flow regimes.

Solution

For extremely small-sized particles with density that is much greater than water, it is safe to assume that Stokes' equation applies. Determine settling velocities using Stokes' equation for all cases and check for Reynolds number (Re). If Re < 0.2, laminar condition apply and Stokes' law is valid.

For Re > 0.2, particle setting velocities are determined iteratively as shown in the following flowchart and the method followed is explained here:

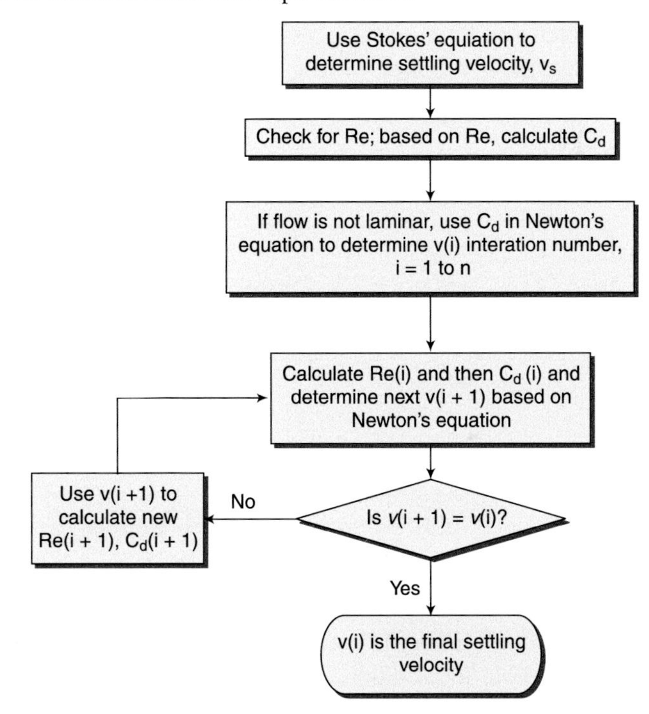

Flow chart for determining settling velocities using Newton's equation.

Step 1: For each diameter, determine settling velocity v_s based on Stokes' law

Step 2: Calculate Re based on each v_s and determine flow regime based on Re values:

Re < 0.2 - laminar, Stokes' equation directly;

$0.2 <$ Re < 1000 - transitional flow; based on Stoke's law, Re was 0.895. Therefore, use equation [5.4.5] to determine C_d. For a diameter of 0.1 mm, in the first iteration v_s was found to be 8.98×10^{-4} m/s, Re = 0.895, i.e., flow is transitional and C_d was 30.3.

Re > 1000 – turbulent flow; use C_d =

For Re $> 2 \times 10^5$, Newton's law is no longer applicable.

Step 3: If Re > 0.2, the flow regime is no longer laminar and Stokes' law is no longer applicable. Use C_d from Step 2 in Newton's equation to determine a new settling velocity, v(i).

Step 4: Calculate Re (i) and C_d (i) based on v(i) and use these values to calculate a new v(i + 1) using Newton's equation.

Step 5: Is v(i + 1) equal to vi? If yes, vi is the final settling velocity. If no, calculate new Re(i + 1) and Cd (i + 1) to determine new vi. Continue until v(i + 1) = vi. For example, using v1, calculate Re2 and Cd2 to determine new v2. Is v2 = v1? If no, use v2 to calculate Re3 and Cd3 and determine v3. Is v3 = v2? If yes, stop. If no, continue calculating new vi. For a particle diameter of 0.1 mm, Re for the first iteration using Stokes' equation are shown in the table below and the final Re = 0.795, final Cd = 33.9 and final settling velocity = 7.99×10^{-3}

Diameter of particle, micron	Diameter of particle, m	Reynolds number	Flow regime	Settling velocity, m/s	Is Stokes' law applicable?
1	0.000001	8.95×10^{-7}	Laminar	8.98×10^{-7}	Yes
10	0.00001	8.95×10^{-4}	Laminar	8.98×10^{-5}	Yes
60	0.00006	1.93×10^{-1}	Laminar	3.23×10^{-3}	Yes
100	0.0001	8.95×10^{-1}	Transitional	7.99×10^{-3}	No
1000	0.001	8.95×10^{2}	Transitional	1.75×10^{-1}	No
1100	0.0011	1.19×10^{3}	Turbulent	1.89×10^{-1}	No
10,000	0.01	8.95×10^{5}	Turbulent	Newton's law does not apply	
15,000	0.015	3.02×10^{6}	Turbulent	Newton's law does not apply	

PROBLEM 5.4.2

Settling column analysis was run on used backwash water from the IIT Kharagpur water treatment plant. The column used was 45 cm in length, and 2000 mL volume. Water samples were taken at a depth of 9 cm from the bottom of the column at varying time intervals and analyzed for total solids (TS) concentrations. Also, note that the column depth falls by 9 mm at each sampling time interval. The data are summarized in the table below. For a loading rate of 10 m/d (10 m^3/m^2-d), determine the theoretical removal efficiency.

Time, min	0	30	60	90	120	150	180
Conc., mg/L	953	570	420	400	380	370	350

Solution

Step 1: Set up a spreadsheet or table with the following columns: sampling time (column 1), concentration (mg/L) – column 2, depth of water column, mm (column 3), mass fraction of solids remaining (TS) in each sample (column 4), settling velocity (column 5) and area above curve (column 6).

Depth of water column, mm = 450 –sample number*9 mm

Mass fraction remaining, $X_i = 1 - (C_0 - C_i)/C_0$

Settling velocity (v), mm/min = Depth of water column/time

Area above curve = $\Delta X * \Delta v$ is determined by taking small intervals on the y-axis and calculatingm graphically. Another method is to take the mid-point between two v_i values and multiply that by the corresponding ΔX for each incremental area above the curve.

Step 2: Determine the design settling velocity based on the loading rate given. Determine area above the curve and to the left of the design settling velocity.

Step 4: Plot graph of settling velocity versus mass fraction solids remaining.

Step 5: Use equation 5.4.10 to determine removal efficiency for the given loading rate.

1	2	3	4	5	6
Time, min	**Conc., mg/L**	**Depth in column at different times, mm**	**Mass fraction remaining**	**Setting velocity (v_i), mm/min**	**Area above curve = $\Delta X*v_i$**
0	953	450			
30	570	441	0.598	12	
			0.465	6.94	
60	420	432	0.441	6	0.160
90	400	423	0.420	4	0.105
120	380	414	0.399	3	0.073
150	370	405	0.388	2.4	0.028
180	350	396	0.367	2	0.046
			0.000	0	0.367
					Total area = 0.780

v_0 = 10 m/d

v_0 = 6.94 mm/min

Corresponding mass fraction that is 100% removed = X_0

Since v_0 = 6.94 mm/min, interpolate to determine X_0 either analytically or graphically.

X_0 = 0.465 which is 100% removed.

To determine fraction that is partially removed in proportion to its velocity, v_i

Slope between v_i = 6 and 12 = 0.026 and partial fraction = area above curve/v_0 = 780/6.94

Removal efficiency = $1 - X_0$ + area above curve/v_0 = 0.6469 = 64.69%

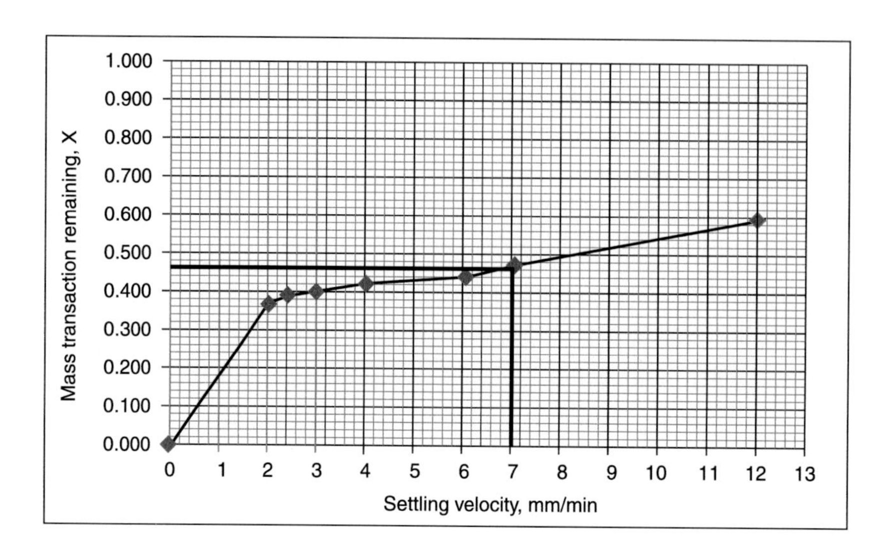

5.4.2 Design of Settling Tanks

Settling basins (tanks) employed for solids removal in water and wastewater treatment plants are classified as rectangular, circular or solids-contact clarifiers.

Rectangular Settling basin or tank

Rectangular settling tanks can be divided into four different functional zones as shown in Figure 5.9.

- ***Inlet Zone*:** Flow at the inlet pipe of the tank results in turbulence which is not conducive to settling. Baffles with holes are generally provided for flow distribution and for separating the turbulent inflow from the settling zone which should have laminar flow. The holes in the baffles allow the flow to be evenly distributed across the width of the tank, i.e., the cross-section of the tank.
- ***Settling Zone*:** Settling of discrete particles or solids occurs in this zone. Flow in this zone has to be laminar for achieving maximum removal efficiency.
- ***Sludge Zone*:** Solids removed from the water column or settling zone settle into the sludge zone. Scrapers for removing settled solids are provided in this zone and in Figure 5.9a are shown on a chain or a belt. The floor of the tank is provided with a slope for allowing solids to be collected in a sludge hopper at one end, usually close to the inlet end.
- ***Outlet Zone*:** Clarified liquid is discharged through the outlet zone over weirs. These weirs are usually V-notch type as shown in Figure 5.10. The width of the tank is the minimum weir length. If greater weir length is required, inboard weir arrangement can be provided by constructing collection troughs parallel to the width of the tank.

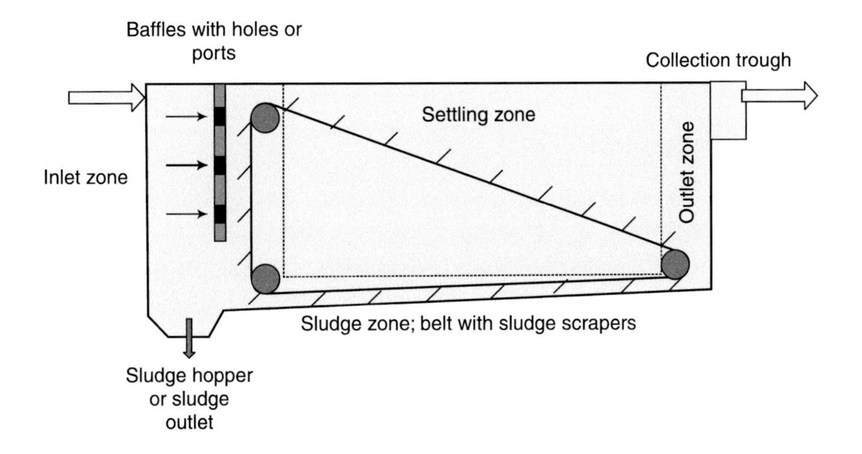

Figure 5.9a Rectangular settling tank with sludge scrapers mounted on a chain ease.

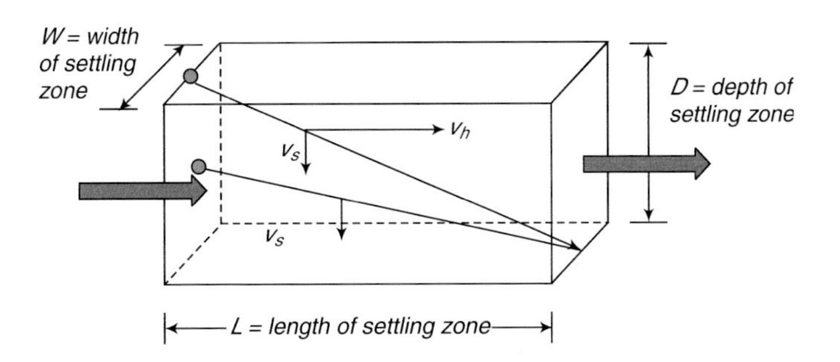

Figure 5.9b Schematic of settling zone in a rectangular settling tank.

Figure 5.10 Series of V-notches in a settling tank to prevent flow of solids into the collection trough or channel (see color plate).

The trajectory of a particle settling with velocity v_s is shown in Figure 5.9b. The path is a straight line because both, the horizontal (v_h) and vertical velocities (v_s) are assumed to be constant. The vertical velocity, v_s (which is also Stokes' setting velocity) = D/T where D = depth of the tank and T = time taken to settle to bottom of tank.

The horizontal velocity or the forward velocity is given as $v_h = Q/A_{cs}$ where A_{cs} is the cross-sectional area of the tank and is equal to W * D where W is the width of the tank and D is the depth of the tank. Detention time (or retention time) is defined as the time for which water remains in the settling tank. Mathematically, detention time T(h) = V (m^3)/Q(m^3/h) = LWD/Q where V is volume of the tank.

Any particle entering at the top of the tank will be removed if its settling velocity, $v_i \geq v_s$.

Equating the two equations for $T = D/v_s = LWD/Q$ we get $v_s = Q/LW$.

This can be written with another symbol, $q_0 = Q/A_s$

Where $q_0 = v_s$ = surface overflow rate (SOR) or design settling velocity, m/d

Q = the volumetric flow rate, m^3/d (flow rate has units of m^3 per unit time),

A_s = Surface area = L * W, m^2

SOR is the most important design parameter since it represents the critical (or design) particle settling velocity for complete (or 100 percent) removal. Particles with settling velocities $v_i < v_s$ will be removed in proportion to their velocities, i.e., removal is proportionate to v_i/v_s (Peavy, Rowe, and Tchobanoglous 1985)

Weir overflow rate (WOR) is defined as $\mathbf{Q/L_w}$ where L_w is the length of weir. It is an important factor in the design of rectangular tanks. For simple weirs, weir length may be taken equal to the width **W** of the tank. If the width of the tank is less than the required weir length, additional weir length can be provided using an inboard weir arrangement.[51]

Circular Settling Basin

Circular settling tanks have the same functional zones as rectangular basins, but the flow regime is different and is shown in Figure 5.11. There is a central pipe through which water enters the tank at the top and flows radially outwards. The horizontal velocity of water continuously decreases as the distance from the center increases. A discrete particle with settling velocity v_s experiences a continuous change in its absolute velocity because of its decreasing horizontal velocity.

51 http://nptel.iitm.ac.in/courses/Webcourse-contents/IIT-KANPUR/wasteWater/Lecture%206.htm.

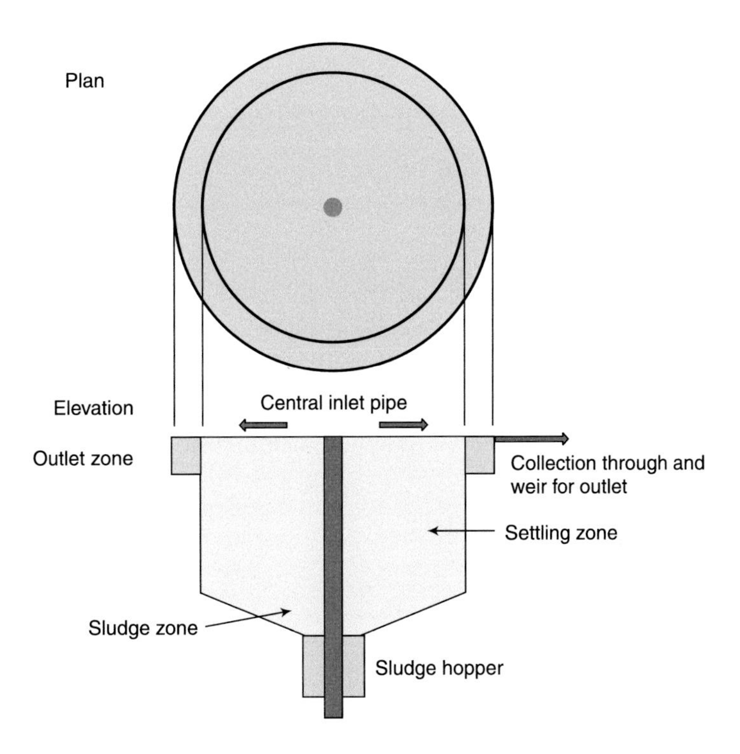

Figure 5.11 Schematic of circular settling tank.

The design of a circular settling basin is also based on overflow rates and detention times. Weir overflow rates and horizontal velocity are not considered in the design of circular tanks but they have to be checked to ensure that they fall within an acceptable range.

Design assumptions used for rectangular and circular settling basins are presented below:

- Dimension of rectangular basins
  - † length: 15–90 m
  - † width: 3–24 m
  - † depth: 3–5 m
  - † Length to width ratio: >5
- Dimensions of circular basins
  - † depth: 3–5 m
  - † diameter: 4–60 m
- Detention time
  - † plain sedimentation: 3–4 h
  - † coagulated sedimentation (clarifier): 2–2.5 h

- Surface overflow rate

  † for plain sedimentation: 12,000–18,000 L/d/m^2 tank area

  † for thoroughly flocculated water: 24,000–30,000 L/d/m^2 tank area
- Velocity of flow: Not greater than 30 cm/min for horizontal flow

PROBLEM 5.4.3

Design a long rectangular basin for type-1 settling for a city that uses 24 MLD of treated water. Settling column tests have shown that an overflow rate of 20 m/d provides satisfactory removal at a depth of 3 m. If the following conditions apply to the design of a single settling unit for discrete particle settling, determine the number of units and their dimensions. For dilute suspensions of discrete particles:

a. Depth of basin ≤3 m; L:W ratio ≤ 4, and L ≤ 12 m,

b. Detention time ≤ 4 h,

c. Horizontal velocity ≤ 36 m/h and weir overflow rates ≤ 14 m^3/m-h.

Solution

Parameter	W = 4	W = 3	Units	Check
Flow, Q	24000.00	24000.00	m^3/d	
Surface Overflow rate, q_0	20.00	20.00	m/d	
Surface area, As = Q/q_0	1200.00	1200.00	m^2	
If L:W = 4, and L = 12, then W has to be 3m; if L:W = 3, then W = 4m	**W = 4**	**W = 3**		
then number of this unit = Total area/(4 or 3W^2)	25.00	33.33	units	
Design units (round off to higher integer value)	25.00	34.00	units	
Depth	3.00	3.00	m	
Detention time = V/Q	0.15	0.15	d	
	3.60	3.60	h	OK
Horizontal velocity = Q/A_{cs} of each unit	80.00	80.00	m/d	
	3.33	3.33	m/h	OK
Flow through each unit (Q)	960.00	705.88	m^3/d-unit	
Weir overflow rate	9.60	6.92	m^3/m-h	OK
Actual volume or number of units will be more than calculated here; minimum 20% additional capacity is required for cleaning and maintenance.				

5.5 COAGULATION–FLOCCULATION

Coagulation, flocculation and settling are used in water treatment with the primary objective of turbidity removal. They are also used as tertiary treatment processes in municipal wastewater treatment, generally for nutrient removal (mainly phosphorus).

Turbidity is a measure of the concentration of colloidal particles in suspension due to the tendency of these particles to scatter light. Turbidity is measured in terms of nephelometric turbidity units (NTU). Colloidal particles can range in size from 1 nm to 1 micron, and include both organic and inorganic particles. Examples of colloidal particles include silt and clay particles, viruses, bacteria, NOM, and large proteins.

Under normal or usual environmental conditions, all particles are negatively charged resulting in net repulsive energy of interaction. This energy barrier prevents particles from coming close together and forming aggregates or floc which can settle by gravity. These particles are termed 'stable particles' since they do not settle or aggregate and remain in suspension indefinitely.

Coagulation is the *chemical conditioning* of stable particles. In this process, particles are destabilized and their physico-chemical characteristics are modified so as to allow aggregation as described in the following sections.

Flocculation is the *physical conditioning* of particles. In this process, gentle mixing of the destabilized particle suspension is done so as to accelerate inter-particle contact, which in turn promotes aggregation and settling of the aggregated particles or 'floc'.

The coagulation–flocculation process facilitates:

- Removal of colloidal particles like bacteria, soil, sand, silt and clay particles.
- Simultaneous removal of associated dissolved compounds like natural organic matter (NOM), heavy metals, color-, taste- and odor-causing compounds, and synthetic organic compounds (SOCs) like pesticides.

5.5.1 The Stability of Particles

Colloids are particles that do not settle under gravity and remain suspended in fluid media (Crittenden et al. 2012). These particles are termed 'stable' and their stability is attributed to two major reasons: electrostatic stabilization and steric stabilization.

Electrostatic stabilization: Under most natural conditions (especially where pH is between 6 and 8), colloidal particles have negatively charged surfaces, resulting in net electrostatic repulsion between particles (Amirtharajah and O'Melia 1990). These forces render these particles 'stable' in suspension since they are fixed in space with respect to each other and will not settle spontaneously.

The negatively charged surfaces of particles are due to the chemical nature or composition of the particles and the solution chemistry. For example, the surface charge on solid particles like silica depends on the pH of the solution. At pH >2, silica is negatively charged due to the following reactions:

$$SiOH_2^+ \leftrightarrow SiOH + H^+$$

$$SiOH \leftrightarrow SiO^- + H^+$$

As pH increases, the concentration of protons in solution decreases and the above reactions shift increasingly to the right resulting in an increase in negatively charged species of silica. Similarly, large organic compounds like proteins which have two major functional groups – carboxyl and amino groups are negatively charged at pH > 4 due to loss of protons from these groups.

Another reason for surface charge on colloidal particles is the adsorption of ions on these particles. If ions like Ca^{2+} or phosphate (PO_4^{3-}) adsorb to the surfaces of these particles, they can impart positive or negative charges, respectively, to the colloidal particles.

Clay particles are layered crystal structures with SiO_2 in tetrahedral arrangement and no net surface charge. However, imperfections in this arrangement due to the presence of impurities like Al, Fe(II) or Mg can result in a net negative charge on the surfaces of these crystalline particles.

Steric stabilization: Colloidal particles can also be stable due to space-occupying or steric effects. Often, organic polymers like natural organic matter (NOM) adsorb to colloidal particles. Parts of these long-chain compounds have hydrophobic and hydrophilic portions which result in the formation of loops and tails where the polymers adhere to the particles. The hydrophilic portions remain in aqueous solution while the hydrophobic portions remain adhered to the particles. These loops and tails prevent colloidal particles from coming close to each other and forming aggregates. This phenomenon is important in the manufacture of paints where maintaining the stability of colloidal pigments in suspension impacts the quality of the paint in the long-term.

It is important to note that the stability of colloidal particles is strictly a surface phenomenon.

5.5.2 Coagulation Theory

There are four different mechanisms by which particles can be coagulated (Amirtharajah and O'Melia 1990).

1. ***Double layer compression*:** Due to the negative surface charge of particles, two electrical layers are formed around any particle. A schematic of the double layer model is shown in Figure 5.12 (Elimelech et al. 1995):
   - A fixed, rigid (Stern) layer of counter-ions is formed at the surface and remains attached to the particle even when it moves. In this case, the counter ions are positively charged since the particles are negatively charged.
   - A diffuse layer is formed by counter- and co-ions depending on the ionic strength (I), pH, and particle charge. When two colloidal particles approach each other, their diffuse layers overlap and interact resulting in a net repulsive force between them. This repulsive potential energy increases with decrease in separation distance and prevents aggregation of particles.

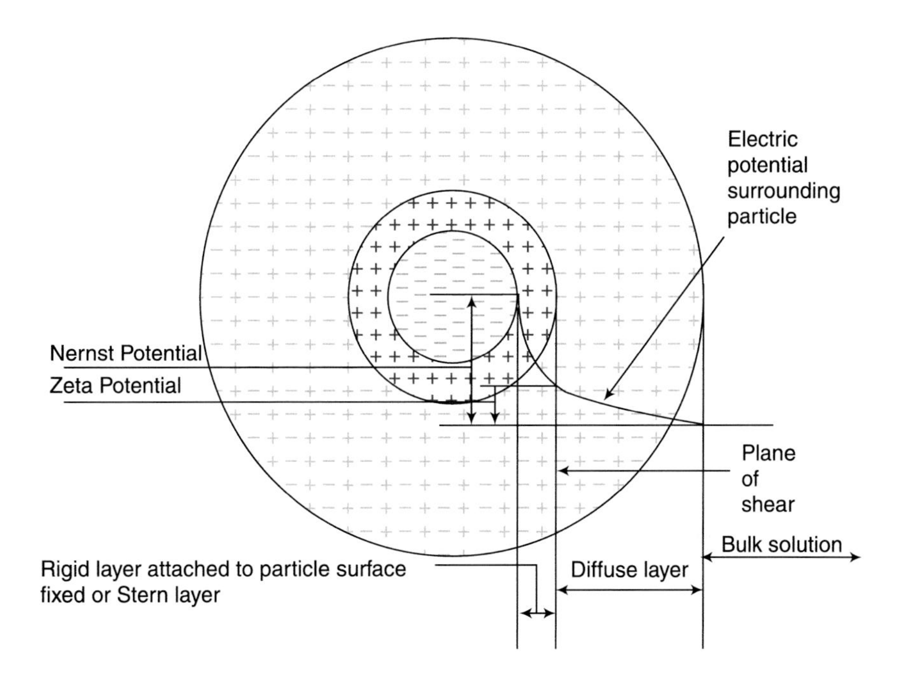

Figure 5.12 Double layer model of a particle.

The thickness of the double layer can be determined based on the following equation (Crittenden et al. 2012):

$$\kappa^{-1} = 10^{10}\left[\frac{2*1000*e^2*N_A*I}{ss_0 kT}\right]^{-1/2} \quad 5.5.1$$

κ^{-1} (1/kappa) = double layer thickness, Å

10^{10} = length conversion, Å/m

1000 = volume conversion, L/m^3

e = electron charge, 1.60219 *10^{-19} Coulombs

N_A = Avogadro's number, 6.02205*10^{23} molecules/mol

I = ionic strength, ½ $\sum z^2C$, mol/L

z = magnitude of positive or negative charge on ion

C = molar concentration, mol/L

k = Boltzmann constant, 1.38066*10^{-23} J/degree K

T = absolute temperature, degree Kelvin (273.15 + degree C)

s = permittivity relative to a vacuum (s for water is 78.54, unitless)

s_0 = permittivity in a vacuum, 8.854188*10^{-12} C^2/J.m where C^2/J ≡ N/V^2

The ionic strength of a solution increases when a coagulant salt is added due to the high concentrations of counter-ions and co-ions in solution. Further, if the ionic strength of the solution is increased, the concentration of ions increases resulting in a more compact diffuse layer with decreased thickness. When the diffuse layer is sufficiently compressed, the particles can come closer to each other. Under high ionic strength conditions, electrostatic repulsion decreases and weak but attractive van der Waal's forces that exist between all particles become dominant. This allows particles to aggregate and form floc and settle. An insight into the bond energies for different types of interactions is provided in Table 5.4.

2. ***Adsorption and charge neutralization:*** Coagulants like alum and ferric chloride produce oxide and hydroxide species in aqueous solution. Since colloidal particles are generally negatively charged under normal environmental conditions (at pH ranging from 5 to 9) the cationic species of these coagulants adsorb to the negatively charged surfaces leading to charge neutralization. Once the surface charge is neutralized, there is no electrostatic repulsion between particles to keep them stable with respect to each other. Attractive van der Waal's forces can then lead to aggregation and settling of particles. However, an overdose of coagulant can lead to restabilization of particles due to charge reversal of the particle surface from negative to positive.

   The Schulze–Hardy rule (1900) states that destabilization of colloids by an indifferent electrolyte is due to counter-ions and coagulation effectiveness increases with ion charge (Nowicki and Nowicka 1994). Based on coagulation theory, there is a critical coagulant concentration (CCC) which is the minimum concentration of a particular electrolyte required for rapid coagulation of a solution or suspension. In other words, the transition from kinetic stability to rapid aggregation occurs over this narrow range of electrolyte concentrations termed critical coagulation concentration (CCC) (Elimelech et al. 1995).

   At large zeta potentials: $CCC = constant/\ z^6$ 5.5.2

   While at low zeta potentials: $CCC = constant/\ z^2$ where z = valence of the counter ion. Therefore, concentrations of Na^+, Ca^{2+} and Al^{3+} required to destabilize a negatively charged colloid vary in the approximate ratio of 729: 11.4: 1 based on the above equation:

   CCC for Na, Ca^{2+} and Al^{3+} is $1/1^6: 1/2^6: 1/3^6 = 1: 1/64: 1/729 = 729: 11.4: 1$

Table 5.4 Bond energies for different types of interactions (Brock 1990)

Bond	Example	Bond energy, kcal/mole	Bond energy, kJ/mole
Single covalent bonds			
	H-H	104	435
	C-H	98	410
	C-O	88	368
	C-N	70	293

Bond	Example	Bond energy, kcal/mole	Bond energy, kJ/mole
	C-S	62	259
Double covalent bonds			
	C = C	147	615
	C = O	168	703
	O = O	96	402
Triple covalent bonds			
	C ≡ C	192	803
	C ≡ N		954
Other interactions			
	Benzene		518
	H bonds		5 to 30
	Hydrophobic interactions	2	8.368
	van der Waal's forces		4 to 8
	Ionic bonds measured in terms of lattice energy for NaCl		769

Dilute solution in nature – low ionic strength

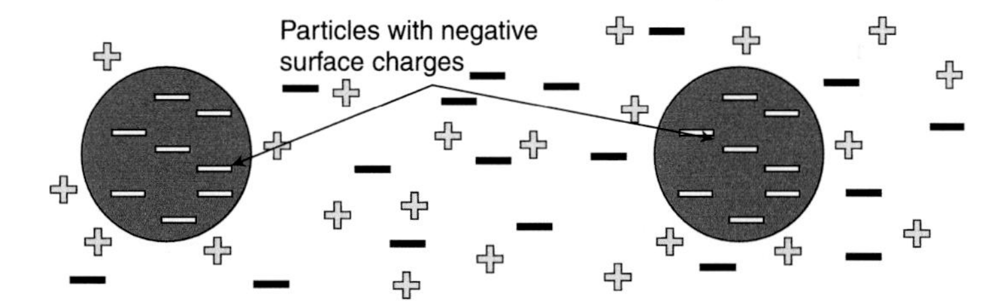

After addition of coagulants to solution – high ionic strength

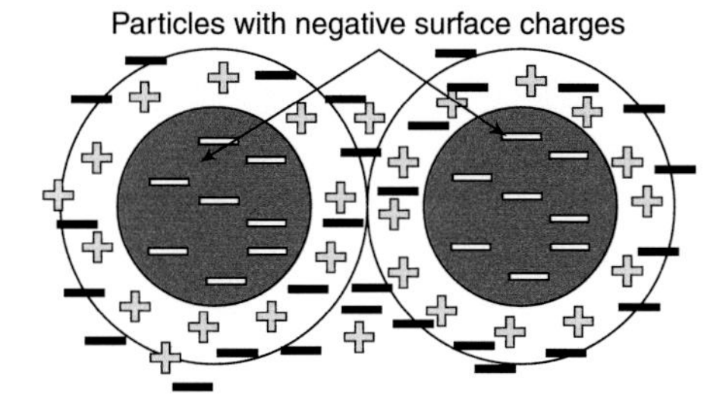

Figure 5.13 Adsorption and charge neutralization leading to coagulation.

Estuarine regions are good examples of natural, spontaneous coagulation. Freshwater streams or rivers often have high turbidity due to their burden of silt, clay and other stable particles and low ionic strength. Under normal conditions, these colloidal particles remain stable and do not settle by themselves. The area where these streams or rivers flow into the sea is tidally influenced. There is a sudden increase in ionic concentration at the mouth of the river due to ingress of seawater containing high salt concentration. The salts in sea water are capable of neutralizing the negative surface charge of the stable clay and silt particles allowing these particles to come closer together. The net interaction energy is now attractive rather than repulsive, and van der Waal's attractive forces between neutral particles dominate over electrostatic forces. This allows the particles to aggregate and settle in estuarine regions leading to the formation of deltas.

3. ***Sweep Coagulation*:** If the coagulant concentration is very high, oxides and hydroxides of the coagulant cation will precipitate. Examples include formation of Al and Fe hydroxides when alum or ferric salts are added to water as coagulants. These precipitates are capable of settling by gravity and any colloidal particles in their path get 'swept' by the precipitate, they are entrapped in the precipitate floc and are removed from suspension. Sweep floc formation is important for the removal of turbidity, color and TOC from drinking water sources.
4. ***Inter-particle bridging*:** Synthetic organic polymers have been used as coagulant aids since the 1960s. Each of these polymer molecules attaches itself to several colloidal particles resulting in particle-polymer-particle aggregates. This results in the formation of large aggregates of particles bridged together by polymers. These aggregates or floc are then large and/or heavy enough to settle by gravity.

Coagulation and flocculation are accomplished in two separate steps: rapid mixing and slow mixing, respectively.

Rapid mixing of coagulants is essential for achieving uniform dispersion of coagulant species in the particle suspension. Rapid mixing can be done by injection of coagulant on the suction side of pumps, upstream from hydraulic jumps or in flow through basins. Most modern designs use either mixing tanks with back mix impeller or inline flash mixers or direct injection of coagulant into the water main.

Some of the important coagulants used in water treatment include alum (hydrated aluminum sulphate salts), ferric salts like ferric chloride and ferric sulphate, activated silica, and organic polymers. Optimum dose of coagulant for a specific type of water has to be determined experimentally using jar tests and the method is described in the following Section (Section 5.5.3).

5.5.3 Flocculation

The objective of flocculation is to enhance inter-particle contact and promote aggregation of the destabilized colloids generated during coagulation. The greater the number of collisions (or contacts) between particles, the greater the probability that the particles will adhere to each other, aggregate and settle. It is essential that mixing is done slowly or gently to ensure that floc formed does not break due to shear.

Three physical processes that lead to inter-particle collisions are:

1. **Perikinetic flocculation or Brownian diffusion**

   Water molecules vibrate in their positions depending on their temperature or thermal energy. This leads to random motion of particles in water due to collisions with water molecules. The driving force is thermal energy of fluid = kT where k = Boltzmann's constant and T = absolute temperature, degree K.

2. **Orthokinetic flocculation or fluid shear**

   The driving force for fluid shear is the velocity gradient (G) in either laminar or turbulent fluid fields. In other words, differences in velocities of different fluid elements in a system can lead to inter-particle collisions. Flocculation tanks are designed based on a dimensionless parameter, Gt, where G is a measure of the relative velocity of two particles of fluid and the distance between them and t is time. Gt values range from 10^4 to 10^5 for t ranging from 10 to 30 min (Hendricks 2011).

3. **Differential settling**

   Since most real suspensions are heterogeneous in nature, i.e., particles are of different sizes, their settling velocities (even when they are governed by Stokes' law) are also different. This vertical transport of particles by gravity at different velocities leads to inter-particle collisions which can result in aggregation of particles.

Jar tests for determining optimum coagulant dose

Jar tests are required to determine the optimum coagulant dose for turbidity removal in a water source. Common coagulants used are alum, ferric chloride and ferric sulphate. These coagulants are cheap and readily available. Since the 1960s, synthetic organic polymers have been used frequently for turbidity removal. They are often more effective than inorganic coagulants but are more expensive and not easily available.

Instruments required: Flocculator or jar test apparatus, turbidity meter, and pH meter.

Stock Solution of coagulant: A stock solution of the coagulant is prepared by dissolving a known amount of coagulant in 1000 mL distilled water. The jar test apparatus generally has room for 6 jars. Six to twelve jars are prepared with varying concentrations of coagulant, and are placed in the jar test apparatus. The coagulation–flocculation–settling process is simulated in jar tests in a 3-step process (Narayan and Goel 2011). The solutions are rapid mixed at 100 rpm for 2 min, gently mixed at 35 rpm for 15 min and allowed to settle for 1 hour. Supernatant samples are removed from the top of the solutions without stirring or disturbing the contents of the beaker. Turbidity of the supernatant samples is measured before and after filtration through a coarse filter paper. These results provide turbidity removal efficiencies after coagulation–flocculation–settling and after coagulation–flocculation–settling–filtration, respectively.

5.5.4 Design considerations

Rapid mixing tanks are designed for mixing the coagulant into the water to be treated. They have detention times of approximately 0.5 min. Ideally, the detention time should be 2 min. Impellers or in-line blenders can also be used for the purpose of rapid mixing. Best G values for impellers are 700 to 1000 s^{-1} and 3,000 to 50,000 s^{-1} for in line blenders (Amirtharajah and O'Melia 1990).

Slow mixing is provided in flocculation tanks for floc formation. Detention time in flocculation tanks is approximately 0.5 hours. If the mixing speed is too fast, it will break the floc. Therefore, very slow and gentle mixing is provided to maximize number of particle collisions and probability of particle capture and aggregation. Optimum speed, just like optimum coagulant dose, has to be determined experimentally.

Design parameters for designing flocculation tanks or basins include mixing time (t, seconds) and velocity gradient (G, 1/s). Rapid mix tanks operate best at G-values ranging from 700 - 1000 (1/s) with detention times of approximately 2 min (Peavy et al. 1985).

$$G = \sqrt{\left(\frac{P}{V\mu}\right)} \tag{5.5.3}$$

where G = velocity gradient, 1/s, P = power input, W (N-m/s), V = volume of mixing basin, m^3, μ = dynamic viscosity (N-s/m^2).

Factors affecting the coagulation–flocculation process include:

- ***Velocity gradient, G*:** It increases the probability of collisions between particles. Increase in collision rate will enhance aggregate formation.
- ***Time*:** Increasing contact time increases the probability of particle interaction and aggregate formation.
- ***pH*:** The pH of a solution affects the coagulation process as it determines the formation of coagulant species and the subsequent destabilization of particles.

G*t is a dimensionless design parameter which controls flocculation, where G is the velocity gradient and t is contact time. G values in practice range from 10 to 100 1/s so that G*t values for flocculation range from 10^4 to 10^5 (Hendricks 2011).

Recently developed designs are a combination of rapid mixing, flocculation and settling in the same tank. The G value can be calculated as follows:

P = Dv_p; where D is drag force, v_p is velocity of paddle tip relative to water (about 75 percent of actual speed), and P is power input.

D = $C_d A_P \rho v_p^2$ where C_d is dimensionless coefficient of drag, A_p is area of paddle blades (total area of wooden slats)

Using the above equations, we get

$$G = \sqrt{\left(\frac{C_D A_p \rho v_p^3}{2V\mu}\right)} \quad 5.5.4$$

For low turbidity and color waters, coagulant aids like activated silica, or kaolinite are added along with alum or any other coagulant to improve the extent and rate of flocculation. By increasing initial turbidity in terms of particle concentration, the number of inter-particle collisions increases leading to better flocculation and settling.

PROBLEM 5.5.1

The optimum coagulant dose of alum was 40 mg/L and for ferric chloride, it was 20 mg/L to achieve the desired turbidity levels. If no alkalinity is present in the treated water, and hydroxides of the above salts are being formed, determine the minimum change in pH that will occur due to addition of these coagulants.

Solution

Alum or aluminium sulphate

$Al_2(SO_4)_3.14H_2O$ <–> $2Al(OH)_3 + 3SO_4^{2-} + 7H^+ + OH^-$

Molecular weight of alum = 594 g/mol or mg/mM

Molal conc of alum = 0.06734 mM alum

Excess $[H^+]$ generated, i.e., $[H^+] - [OH^-]$ = 6 mM/mM alum

$[H^+]$ generated = 0.40404 mM

Final pH (assuming initial pH = 7, i.e., $[H^+] = 10^{-7}$ M which can be ignored since the amount of H^+ generated is several orders of magnitude higher) = 3.39 units.

Ferric chloride

$FeCl_3 + 3H_2O$ <–> $Fe(OH)_3 + 3Cl^- + 3H^+$

Molecular weight of ferric chloride = 162.5

Molal conc of ferric chloride = 0.12 mM

$[H^+]$ = 3 mM/mM ferric chloride

$[H^+]$ generated = 0.37 mM

Final pH = 3.43 units

PROBLEM 5.5.2

Most natural waters contain varying amounts of alkalinity, measured as $CaCO_3$. $CaCO_3$ is relatively insoluble and soluble alkalinity is generally represented as $Ca(HCO_3)_2$ - Calcium bicarbonate. Write balanced reactions for each of the three coagulants, i.e., alum, ferric sulphate and ferric chloride. Determine the amounts of alkalinity consumed in each reaction (express your answers in terms of mole/mole of coagulant and mg/mg coagulant). Also, determine the amount of solids produced per mole (or mg) of coagulant added.

Solution

For alum or aluminium sulphate:

$Al_2(SO_4)_3 .14H_2O + 3Ca(HCO3)2 \leftrightarrow 2Al(OH)_3 + 3CaSO_4 + 14H_2O + 6CO_2$

Alternatively,

$Al_2(SO_4)_3 .14H_2O + 3CaCO_3 \leftrightarrow 2Al(OH)_3 + 3CaSO_4 + 11H_2O + 3CO_2$

MW of $Ca(HCO_3)_2$ = 162

MW of $Al_2(SO_4)_3.$ 14 H_2O = 594

From the stoichiometry of the reaction, moles of alkalinity consumed per mole of coagulant = 3 moles of $Ca(HCO_3)_2$ or $CaCO_3$/ moles of alum

Also, mg of alkalinity consumed per mg of coagulant = 0.82 mg $Ca(HCO_3)_2$/mg alum = 0.51 mg $CaCO_3$/mg alum

Molecular weight of $Al(OH)_3$ = 78

Moles of $Al(OH)_3$ formed = 2.00 mol/mol of alum = 0.26 mg aluminium hydroxide/ mg alum

For ferric sulphate:

$Fe_2(SO_4)_3 + 3Ca(HCO_3)_2 \leftrightarrow 2Fe(OH)_3 + 3CaSO_{4\downarrow} + 6CO_2$

Alternatively,

$Fe_2(SO_4)_3 + 3CaCO_3 + 3H_2O \leftrightarrow 2Fe(OH)_3 + 3CaSO_4 + 3CO_2$

MW of $Fe_2(SO_4)_3$ = 400

MW of $Ca(HCO_3)_2$ = 162

From the stoichiometry of the reaction, moles of alkalinity consumed per mole of coagulant = 3.00 moles of $Ca(HCO_3)_2$ /moles of ferric sulphate

Also, mg of alkalinity consumed per mg of coagulant = 1.22 mg $Ca(HCO_3)_2$/ mg ferric sulphate = 0.75 mg $CaCO_3$/ mg ferric sulphate

Molecular weight of $Fe(OH)_3$ = 107

Moles of $Fe(OH)_3$ formed = 2.00 mol/mol of ferric sulphate = 0.54 mg ferric hydroxide/ mg ferric sulphate

For ferric chloride:

$2FeCl_3 + 3Ca(HCO_3)_2 \leftrightarrow 2Fe(OH)_3 + 3CaCl_2 + 6CO_2$

Alternatively,

$2FeCl_3 + 3CaCO_3 + 3H_2O \leftrightarrow 2Fe(OH)_3 + 3CaCl_2 + 3CO_2$

Molecular Weight of $Ca(HCO_3)_2$ = 162

Molecular Weight of $FeCl_3$ = 162

From the stoichiometry of the reaction, moles of alkalinity consumed per mole of coagulant = 1.50 moles of $Ca(HCO_3)_2$/moles of ferric chloride

Also, mg of alkalinity consumed per mg of coagulant= 1.50 mg $Ca(HCO_3)_2$/ mg ferric chloride = 0.92 mg $CaCO_3$/ mg ferric chloride

Molecular Weight of $Fe(OH)_3$ = 107

Moles of $Fe(OH)_3$ formed = 1.00 mol/mol of ferric chloride = 0.66 mg ferric hydroxide/ mg ferric chloride

PROBLEM 5.5.3

A water sample contains turbidity of 30 NTU which is approximately equal to 50 mg/L of clay and the optimum dose of alum is 40 mg/L while that of ferric chloride is 20 mg/L. Determine the amount of sludge produced in achieving the desired level of turbidity (5 NTU), report answer in terms of sludge produced (mg/L per unit of NTU removed).

Solution

$Al_2(SO_4)_3.14H_2O + 3Ca(HCO_3)_2 \leftrightarrow 2Al(OH)_3 + 3CaSO_4 + 14H_2O + 6CO_2$

Alum (40 mg/L) = 0.06734 mM

Residual clay; equivalent of 5 NTU = 8.33 mg/L

Sludge composition with alum

	mM	mg/L	
$3CaSO_4$	0.202	27.47	soluble
$2Al(OH)_3$	0.135	10.51	solids
Clay		41.67	solids
	Total solids produced	**52.18**	
		2.09	**mg/L-NTU removed**

$2FeCl_3 + 3Ca(HCO_3)_2 \leftrightarrow 2Fe(OH)_3 + 3CwaCl_2 + 6CO_2$

Ferric chloride (20 mg/L) = 0.12 mM

Sludge composition with ferric chloride

	mM	mg/L	
$1.5CaCl_2$	0.18462	20.49	soluble
$Fe(OH)_3$	0.12308	13.17	solids
Clay		41.67	solids
	Total solids produced	**54.84**	
		2.19	**mg/L-NTU removed**

PROBLEM 5.5.4

Jar tests were conducted with slightly turbid, synthetic water at alum doses ranging from 0 to 120 mg/L. Estimate the pH at different alum doses based on the solution to Problem 5.5.1. Jar tests results with a real water sample show a pH drop from 7.7 to 4.8 over an alum dose range of 0 to 110 mg/L. Can the experimental results be explained?

Solution

Based on the results from Problem 5.5.1 for alum, we know that 6 mM of $[H^+]$ are generated per mM of alum added. The following table of pH values was generated for alum doses ranging from 0 to 120 mg/L.

Alum dose, mg/L	Alum dose, mM	$[H^+]$ generated, M	pH final
0	0.000	1.00×10^{-7}	7
10	0.017	1.0×10^{-4}	4.00
20	0.034	2.0×10^{-4}	3.69
30	0.051	3.0×10^{-4}	3.52
40	0.067	4.0×10^{-4}	3.39
50	0.084	5.1×10^{-4}	3.30
60	0.101	6.1×10^{-4}	3.22
70	0.118	7.1×10^{-4}	3.15
80	0.135	8.1×10^{-4}	3.09
90	0.152	9.1×10^{-4}	3.04
100	0.168	1.0×10^{-3}	3.00
110	0.185	1.1×10^{-3}	2.95
120	0.202	1.2×10^{-3}	2.92

Actual jar test results show a pH drop from 7.7 to 4.8 over an alum dose range of 0 to 110 mg/L.

If the solution had no alkalinity in it, the pH would have dropped to 3. However, due to the alkalinity added to the solution, the pH drop was much less.

5.6 SEDIMENTATION OR CLARIFICATION

The term 'Type 2 settling' is used for the settling of flocculating particles in a dilute suspension. Flocculant particles change shape, size and specific gravity and are unlike discrete particles which have constant shape, size and specific gravity. Examples of discrete particles are sand, and clay, while flocculant particles include proteins, bacteria, fibers, natural organic matter, i.e., most organic colloidal particles. Stokes' and Newton's equations describe the settling of discrete particles and cannot be used for flocculating particles.

5.6.1 Flocculant Settling Analysis (Type 2)

Analysis of the settling behavior of flocculating suspensions requires actual settling tests. A settling test for flocculating particles is similar in many respects to that for a discrete particle suspension and is described below (Peavy, Rowe, and Tchobanoglous 1985):

- A dilute homogeneous suspension of flocculating particles is poured into a column with multiple ports.
- Samples are removed from every port at periodic time intervals.
- Suspended-solids concentrations in the samples are determined.
- Percent removal is calculated for each sample and is given as

$$X_{ij} = (1 - C_{ij}/C_o)*100 \qquad 5.6.1$$

where X_{ij} is the mass fraction in percent that is removed at *i*th depth and *j*th time interval.

- Percent removal is plotted on a graph where depth in the column is on the y-axis and time on the x-axis.
- Curves of equal percent removals are drawn by interpolating between the measured points.

Examples of such settling curves can be found in most textbooks that include Type 2 settling.

5.6.2 Design of Clariflocculator

Clariflocculators are similar to circular settling tanks as shown in Figure 5.14. There is an inlet pipe in the center of a circular tank from where water (after coagulation) emerges at the top of the basin. Unlike the circular settling tank in Section 5.4, there is a circular baffle in clariflocculators which separates the flocculation zone from the settling zone. A central shaft with motor has two major mechanisms attached to it: a sludge scraper for removing solids from the bottom of the tank and flocculator paddles for gentle mixing of the coagulated water. Water enters the flocculation zone moving downwards and then upwards into the settling zone. Floc formed in the flocculation zone is removed in the settling zone and the sludge scrapers collect them in a sludge hopper. Dimensions of a clariflocculator are similar to those of a circular settling tank and were provided in Section 5.4.

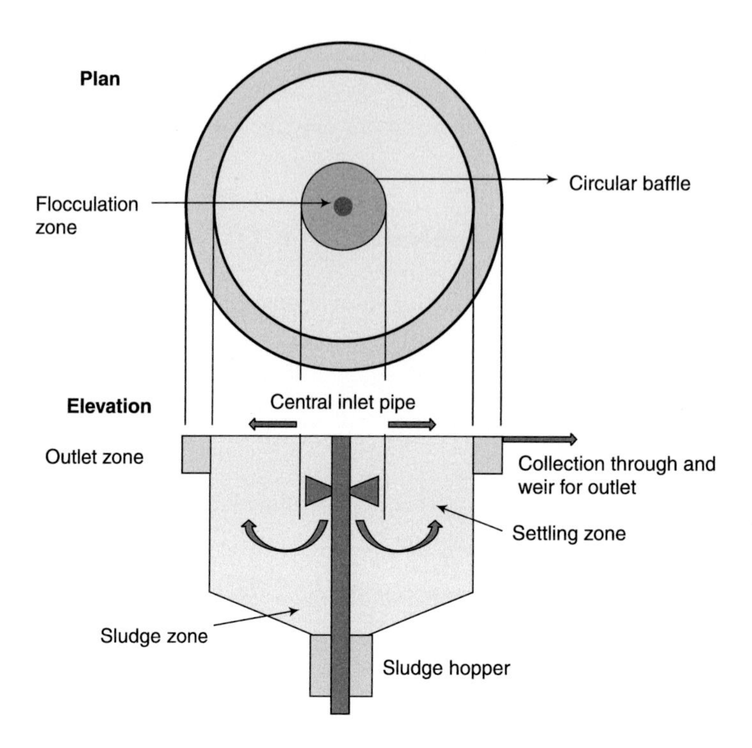

Figure 5.14a Schematic of a clariflocculator.

Figure 5.14b Photo of a clariflocculator in a water treatment plant (see color plate).

PROBLEM 5.6.1

A water treatment plant for treating 10 MLD of water has a clariflocculator for turbidity removal. Maximum turbidity in the raw water is 25 NTU. The optimum alum dose for complete turbidity removal was determined in jar tests and found to be 40 mg/L. TSS concentration is assumed to be

1 mg/L per NTU. Calculate the maximum amount of solids that can be produced per day. Assume sludge from the clariflocculator has 5 percent solids. Determine the amount of sludge produced per day.

Solution

Q, flow rate = 10000000 L/d

Maximum turbidity = 25 NTU = 25 mg/L

Alum to be added = 400000000 mg/d = 400 kg/d

Turbidity removed (assume 100 percent removal) = 250000000 mg/d = 250 kg/d

Assuming alum and turbidity both are included in the solids removed, total solids produced/d = (400 +250) kg/d = 650 kg/d

Assuming 5% solids concentration, sludge produced/d = 13000 kg/d

For water with density of approximately 1000 kg/m^3, volume of sludge produced = 13 m^3/d

PROBLEM 5.6.2

A clariflocculator is to be designed for a water treatment plant of capacity 10 MLD. The design surface overflow rate is 40 m^3/m^2-d, and detention time is 3 h.

Solution

Q, flow rate = 10^7 L/d = 10^4 m^3/d

Surface Overflow Rate (SOR) = 40 m^3/m^2-d

Surface area to be provided = $2.5*10^2$ m^2

Therefore, radius of clariflocculator = 8.92 m, rounded off to the next higher integer, i.e., 9 m.

Q = $4.17*10^5$ L/h

To determine depth of clariflocculator = $1.25*10^6$ L = $1.25*10^3$ m^3

Depth = 5.0 m

Check for weir overflow rate:

Perimeter = 56.55 m

Weir Overflow Rate (WOR) = 176.8 m^3/m-d; OK since the acceptable range for WOR = 140 to 3000 m^3/m-d

PROBLEM 5.6.3

Collect a sample of wastewater, backwash water or highly turbid river water. If these samples are not available, a synthetic solution can be made by mixing approximately 1000 mg/L of silt, sludge or clay in water. Measure turbidity in NTU and total suspended solids (TSS) for different dilutions of the same water sample and determine the relation between the two parameters.

Solution

A sample of backwash water was collected from the local water treatment plant, analyzed for turbidity (NTU) and TSS (mg/L) at different dilutions and the data are shown in the following table. A relationship between the two parameters was determined and the graph is provided below. Based on the regression equation for this water sample, the relation between turbidity and TSS holds only for turbidity > 41 NTU since x = 41 NTU when y = 0.

Sample no	Turbidity, NTU	TSS, mg/L
1	475	363
2	374	228
3	169.9	90
4	89.3	56
5	61.2	14

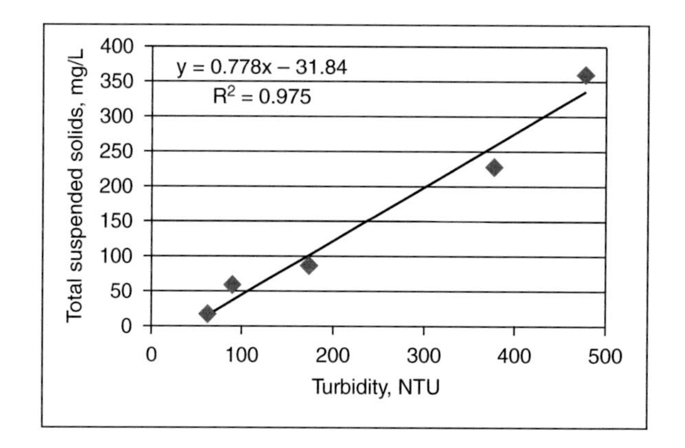

5.7 SOFTENING: CHEMICAL

Hardness of water is due to the presence of two divalent cations: calcium (Ca) and magnesium (Mg). Other cations like Fe, Mn and Sr are also found in natural water sources and may be associated with anions like carbonates, sulphates, nitrates, chlorides, and silicates. Hardness prevents lather formation with soap, results in scum formation, and makes the cleaning process difficult. Ca and Mg tend to precipitate at high temperatures forming hard-scales on the sides of boilers, on tubes in distillation units or any heating elements associated with water leading to loss of energy and poor efficiency or even failure of these units.

Removal of hardness-causing cations is essential for improving water quality and can be done by either the lime-soda ash method or by ion exchange.

Choice of the process depends on various factors such as:

- Quality of raw water
- Quality desired in treated water
- Competing ions
- Alkalinity
- Contaminant concentration
- Affinity of resin or alumina for contaminant versus competing ions

Softening by chemical means is the most common method for removing hardness from water, since it is cheaper than ion exchange process, and therefore, can be carried out on a large-scale. Further, pre-treatment or turbidity removal is not required prior to chemical softening unlike ion exchange. However, it should be noted that ion exchange is a more efficient albeit more expensive process.

Lime – soda ash softening: Softening by chemical means is done mainly by using the lime-soda ash method. In the lime-soda ash process, quick lime (CaO) is added to hard water to convert highly soluble salts of calcium and magnesium to insoluble carbonates and hydroxides, respectively, which are then removed by precipitation.

Table 5.5 Equilibrium of solid and dissolved species of common ions (Lowenthal and Marais 1982)

Mineral	Solubility, mg/L as $CaCO_3$ at 0 °C
Calcium carbonate	15
Calcium bicarbonate	1,61,000
Calcium chloride	3,36,000
Calcium sulphate	1,290
Calcium hydroxide	2,390
Magnesium bicarbonate	37,100
Magnesium carbonate	101
Magnesium chloride	3,62,000
Magnesium hydroxide	17
Magnesium sulphate	1,70,000
Sodium bicarbonate	38,700
Sodium carbonate	61,400
Sodium chloride	2,25,000
Sodium hydroxide	3,70,000
Sodium sulphate	33,600

Reactions that can occur on addition of lime and/or soda ash are (Hammer and Hammer 2008; Peavy, Rowe, and Tchobanoglous 1985):

$$CaO + CO_2 \rightarrow CaCO_3$$

$$Ca(HCO_3^-)_2 + CaO + H_2O \rightarrow 2CaCO_3(s) + 2H_2O$$

$$Mg(HCO_3^-)_2 + CaO + H_2O \rightarrow CaCO_3(s) + Mg^{2+} + 2H_2O + CO_3^{2-}$$

$$Mg^{2+} + CO_3^{2-} + CaO + H_2O \rightarrow CaCO_3(s) + Mg(OH)_2(s)$$

$$Mg^{2+} + \{SO_4^{2-}, 2Cl^-, 2NO_3^-\} + CaO + H_2O \rightarrow Ca^{2+} + \{SO_4^{2-}, 2Cl^-, 2NO_3^-\} + Mg(OH)_2(s)$$

Non-carbonate hardness in water can be removed by using soda ash as shown below:

$$Ca^{2+} + \{SO_4^{2-}, 2Cl^-, 2NO_3^-\} + Na_2CO_3 \rightarrow CaCO_{3(s)} + 2Na^+ + (\{SO_4^{2-}, 2Cl^-, 2NO_3^-\})$$

Since precipitation of calcium carbonate and magnesium hydroxide is a pH dependent process, optimum pH has to be maintained for effective precipitation. A pH range of 9-9.5 is required for calcium carbonate to precipitate. For magnesium hydroxide, pH $\geq$ 11 has to be maintained. It can be achieved by adding excess lime as shown:

$$CaO + H_2O \rightarrow Ca^{2+} + 2OH^-$$

Caustic Soda (NaOH): Hardness can also be removed by using caustic soda (sodium hydroxide) as shown below:

$$CO_2 + 2NaOH \rightarrow 2Na^+ + CO_3^{2-} + H_2O$$

$$Ca^{2+} + 2(HCO_3)^- + 2NaOH \rightarrow CaCO_3(s) + 2Na^+ \; CO_3^{2-} + 2H_2O$$

$$Mg^{2+} + 2(HCO_3)^- + 2NaOH \rightarrow Mg(OH)_2(s) + 2Na^+ + CO_3^{2-} + 2H_2O$$

$$Mg^{2+} + SO_4^{2-} + 2NaOH \rightarrow Mg(OH)_2(s) + 2Na^+ + SO_4^{2-}$$

All calcium carbonate and magnesium hydroxide cannot be removed from water due to solubility limits (shown in Table 5.5) and process inefficiencies. Under normal conditions, 40 mg/L of $CaCO_3$ and 10 mg/L of $Mg(OH)_2$ are assumed to remain in water. To prevent deposits or flakes of calcium carbonate from floating in solution, these remaining precipitates have to be converted back to soluble sodium bicarbonate. This can be achieved by neutralizing the alkaline water using either sulphuric acid or recarbonating with carbon dioxide.

$$2CaCO_3 + H_2SO_4 \rightarrow 2Ca^{2+} + 2(HCO_3)^- + SO_4^{2-}$$

$$CaCO_3 + CO_2 + H_2O \rightarrow Ca^{+2} + 2(HCO_3)^-$$

A very important point in calculating chemical requirements in the lime-soda ash process is the use of milliequivalents/liter (meq/L) versus moles/L. Bar diagrams are always drawn on a meq/L basis since they are a visual method of showing charge balance of cations and anions in a solution. Most textbooks use meq/L for calculating chemical requirements, but this can be confusing based on the stoichiometry. Therefore, it is better to stick to the general convention of calculating stoichiometric requirements for chemicals in mmol/L (mM) and then converting them to mg/L or meq/L as needed. All reactions written here are on a molar basis.

PROBLEM 5.7.1

A water sample has been analyzed with the following constituents in meq/L: CO_2 = 1; Ca^{2+} = 3; Mg^{2+} = 1; Na^+ = 2; HCO_3^- = 1; SO_4^{2-} = 4.5. It is to be softened to the minimum possible calcium hardness by the lime-soda ash process. Magnesium removal is not necessary. A flow of 200 MLD is provided to a city. Calculate the daily chemical requirement and mass of solids produced. Assume lime used is 75 percent pure and soda ash is 60 percent pure. Draw bar diagrams for the raw and finished water.

Solution

Step 1: Create a table with the concentrations of each constituent in meq/L, millimoles/L (mM) and mg/L as shown below.

Ion	conc., meq/L	Eq. wt., mg/meq	mg/L	MW	mM
CO_2	1	22	22	44	0.5
Ca^{2+}	3	20	60	40	1.5
Mg^{2+}	1	12.2	12.2	24.4	0.5
Na^+	2	23	46	23	2
HCO_3^-	1.5	61	91.5	61	1.5
SO_4^{2-}	4.5	48	216	96	2.25

Step 2: Write the reactions that are applicable as shown below –

For removal of *0.5 mM or 1 meq/L of* CO_2, 1 meq/L of lime or 0.5 mM of lime is required to form calcium carbonate which will precipitate.

$$0.5\,CaO + 0.5\,CO_2 \rightarrow 0.5\,CaCO_3(s)$$

Calcium is generally associated with carbonates and bicarbonates in water and calcium bicarbonate is highly soluble. It can be removed by adding lime which will result in the formation of calcium carbonate precipitates. Since there are no carbonates in this water and only *1.5 meq/L (or mM) of bicarbonate*, the corresponding reaction is shown below:

$$0.75\,Ca^{2+} + 1.5\,HCO_3^- + 0.75\,CaO + 0.75\,H_2O \rightarrow 1.5\,CaCO_3(s) + 1.5\,H_2O$$

For removing the remaining Ca from solution, it is assumed to be associated with sulfate and can be removed by addition of soda ash (sodium carbonate) as shown below.

$$0.75Ca^{2+} + 0.75SO_4^{2-} + 0.75Na_2CO_3 \rightarrow 0.75CaCO_3(s) + 1.5Na^+ + 0.75SO_4^{2-}$$

Recarbonation, which is addition of carbon dioxide to water, is done to redissolve any precipitates of calcium carbonate that remain after settling of the bulk precipitates. Assuming that 40 mg/L (0.8 meq/L or 0.4 mM) of calcium carbonate remains in the softened water of which 15 mg/L (0.3 meq/L or 0.15 mM) is in dissolved form, and the remaining 25 mg/L (0.5 meq/L or 0.25 mM) of calcium carbonate needs to be converted to soluble calcium bicarbonate.

$$0.25CaCO_3 + 0.25CO_2 + 0.25H_2O \rightarrow 0.25Ca(HCO_3)_2$$

Step 3: Calculate the total chemical requirements as shown below based on the preceding reactions

Lime = 0.5 + 1.5/2 + 1.5/2 = 2.0 mM

CO_2 = 0.25 mM

The daily chemical requirements for a flow of 200 MLD

Lime = [200*10^6 L/d*2.0 mmoles/L*56 mg/mmoles]/0.75 = 29.87 tons/d

Carbon dioxide = [200*10^6 L/d*0.25 mmoles/L*44 mg/mmoles] = 2.20 tons/day

Mass of solids produced, i.e., precipitates of $CaCO_3$ are calculated as shown below.

Solids produced as $CaCO_3$ = [0.5 + 3.0] – 0.8 = 2.7 meq/L

The mass of calcium carbonate produced = [200*10^6 L/d*2.7 meq/L*100 mg/mmoles] = 54 tons/day

If we assume that impurities in lime will also precipitate then additional solids generated are = 29.87*0.25 = 7.5 tons/day

Total mass of solids produced per day = 54 + 7.5 = 61.5 tons/d.

(b) Bar diagrams for the raw and finished water are:

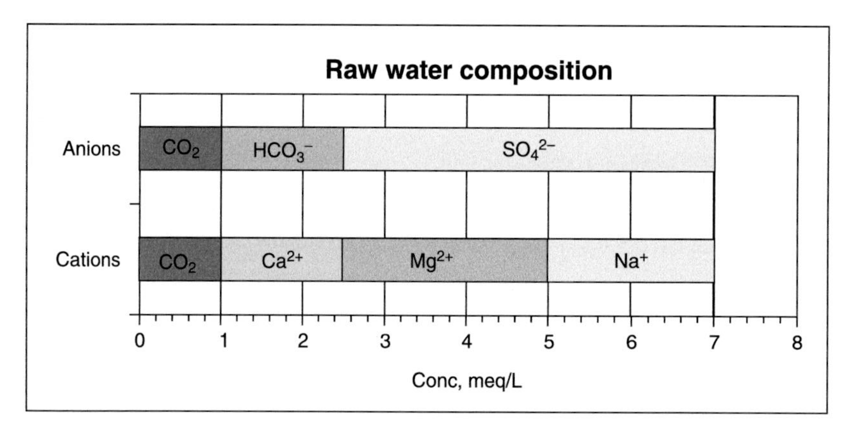

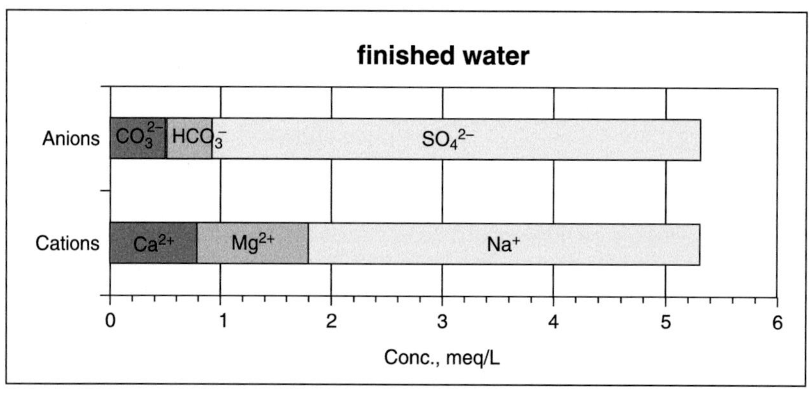

PROBLEM 5.7.2

A water treatment plant processes 50,000 m^3/d of water with the following ionic concentrations in meq/L: carbon dioxide = 0.5; calcium = 4; magnesium = 3; sodium = 2; bicarbonate = 2.5; sulphate = 3.5; chloride = 3.3.

a. Determine the quantities of chemicals (kg/d) required to soften this water to the minimum possible hardness by two-stage lime-soda ash softening.
b. Draw a bar diagram of the finished water and calculate the amount of dry solids generated as sludge.

Solution

Step 1: Create a table with the concentrations of each constituent in meq/L, moles/L and mg/L as shown below.

Ion	Eq. wt., mg/meq	Concentration, meq/L	Concentration, mM
CO_2	22	0.5	0.25
Ca^{2+}	20	4	2
Mg^{2+}	12.2	3.00	1.50
Na^+	23	2	2
HCO_3^-	61	2.50	2.5
SO_4^{2-}	48	3.50	1.75
Cl^-	35.5	3.30	3.3

Step 2: Write the reactions that are applicable as shown below:

1 Removal of 0.25 mM of carbon dioxide:

$$0.25CO_2 + 0.25Ca(OH)_2 \rightarrow 0.25CaCO_3\downarrow + 0.25H_2O$$

2 For removal of 2 mM of calcium:

Reaction with 2.5 mM of bicarbonate will remove only 1.25 mM of calcium

$$1.25Ca(HCO_3)_2 + 1.25Ca(OH)_2 \rightarrow 2.5CaCO_3\downarrow + 2.5H_2O$$

Remaining 0.75 mM of calcium is now associated with sulphate (next anion after bicarbonate):

$$0.75CaSO_4 + 0.75Na_2CO_3 \rightarrow 0.75CaCO_3\downarrow + 0.75Na_2SO_4$$

3. For removal of 1.5 mM of magnesium, there are several reactions and excess lime (1.25 meq/L or 0.625 mM) is to be added to increase the pH to >11:

In the reaction where magnesium (1.5 mM) is associated with the remaining sulphate (1.75 – 0.75 = 1.0 mM), only 1 mM (lesser amount determines Mg removal) can be removed.

$$1.0MgSO_4 + 1.0Ca(OH)_2 \rightarrow 1.0Mg(OH)_2\downarrow + 1.0CaSO_4$$

Calcium from added lime is removed using soda ash:

$$1.0CaSO_4 + 1.0Na_2CO_3 \rightarrow 1.0CaCO_3\downarrow + 1.0Na_2SO_4$$

Removal of remaining Mg (0.5 mM) in association with the remaining anion, chloride

$0.5MgCl_2 + 0.5Ca(OH)_2 \rightarrow 0.5Mg(OH)_2\downarrow + 0.5CaCl_2$

$0.5CaCl_2 + 0.5Na_2CO_3 \rightarrow 1.0CaCO_3\downarrow + 1.0H_2O$

4 Recarbonation of residual precipitates of calcium carbonate (as explained in previous problem) by carbon dioxide:

$0.25CaCO_3 + 0.25CO_2 \rightarrow 0.25Ca(HCO_3)_2 + 0.25H_2O$

A summary table with each chemical in each of the above reactions is provided below.

Step 3: Calculate total chemical requirements as shown below based on above reactions:

Lime = 0.25 + 1.25 + 1.0 + 0.5 (excess) = 3.625 mM = 3.625*56 = 203 mg/L

Soda ash = 2.25 mM = 1 mmoles/L *106 mg/mmoles = 238.5 mg/L

Carbon dioxide = 1.125 mM = 1.125 mmoles/L *44 mg/mmoles = 49.5 mg/L.

	Conc., mM	Lime, mM	Soda, mM	CO_2, mM
CO_2	0.25	0.25	0	0.25
$Ca(HCO_3)_2$	1.25	1.25	0	
$CaSO_4$	0.75		0.75	
$MgSO_4$	1	1	1	
$MgCl_2$	0.5	0.5	0.5	
excess		0.625		0.625
dissolution of $CaCO_3$				0.25
	mol/L	3.625	2.25	1.125
	mg/L	203	238.5	49.5

Chemical requirements for 50,000 m^3/d (50 MLD):

Lime = 203 mg/L*50*10^6 L/d = 10.15 tons/d

Soda ash = 238.5 mg/L *50*10^6 L/d = 11.925 tons/d

Carbon dioxide = 49.5 mg/L*50*10^6 L/d = 2.475 tons/d.

5.8 SOFTENING: ION EXCHANGE

Ion exchange is the displacement of one ion by another. The displaced ion is originally part of a solid matrix while the displacing ion is originally in solution. The insoluble part of the exchange material is the 'host' and is generally sold in the form of resinous beads of diameter ranging from 0.3 to 1.2 mm and a specific gravity ranging from 1.1 to 1.4.[52] The resin beads are filled into a vertical steel pressure vessel which functions as a pressurized porous bed reactor. Ion exchange resins are either cation exchangers with sodium or potassium as the displaced ion, or anion exchangers with hydroxide as the displaced ion.

52 http://msdssearch.dow.com/PublishedLiteratureDOWCOM/dh_0032/0901b803800326ca.pdf.

Ion exchange was first used for water softening in 1905[53] and has several applications in water and wastewater treatment. Some of the major applications include hardness removal, TDS removal from treated water or wastewater, removal of specific pollutants like arsenic, other heavy metals, fluoride, different forms of nitrogen, and nutrients like phosphate and sulphate. Commercially available ion exchange resins show similar relative affinities for various ions as shown below (Clifford, Sorg, and Ghurye 2011).

Relative affinities (with respect to Na^+) of strong acid cation resins:

$$Ra^{2+}\ (13) > Ba^{2+}\ (5.8) > Pb^{2+}\ (5.0) > Sr^{2+}\ (4.8) > Cu^{2+}\ (2.6) > Ca^{2+}\ (1.9)$$
$$> Zn^{2+}\ (1.8) > Fe^{2+}\ (1.7) > Mg^{2+}, K^{+}\ (1.67) > Mn^{2+}\ (1.6) > NH_4^{+}\ (1.3)$$
$$> Na^{+}\ (1.0) > H^{+}\ (0.67)$$

Relative affinities (with respect to Cl^-) of strong base anion resins:

$$SO_4^{2-}\ (9.1) > HAsO_4^{2-}\ (4.5) > HSO_4^{-}\ (4.1) > NO_3^{-}\ (3.2) > Br^{-}\ (2.3)$$
$$> SeO_3^{2-}\ (1.3) > HSO_3^{-}\ (1.2) > NO_2^{-}\ (1.1) > Cl^{-}\ (1.0)$$
$$> HCO_3^{-}\ (0.27) > F^{-}\ (0.07)$$

These sequences are important for determining the appropriate resin depending on the objective of the treatment process, i.e., removal of hardness, arsenic, nitrate, fluoride or any other contaminant.

Ion exchange is a highly efficient process with efficiencies >99.99 percent in distilled water solutions of specific ions. However in field conditions, if high concentrations of TOC, SS and TDS exist, the efficiency of this process is very poor or the life of the resin bed is reduced drastically. The porous bed reactors that are generally used get choked under these conditions, and the resin is 'blinded' to the ions that need to be exchanged. Therefore, pre-treatment is essential for raw waters of poor quality.

In ion exchange processes for water softening, calcium and magnesium ions in the feed water are adsorbed to the resin beads while non-hardness causing cations such as sodium or potassium are displaced from the beads. This exchange takes place at the solid surface of the beads. Synthetic resins were developed first and soon natural zeolites like Greensand (naturally occurring aluminium silicate) became popular ion exchange materials. Nowadays, synthetic resins are used, and are coated with the desirable cation exchange material depending on the application. These resins have an advantage over natural zeolite due to their larger number of exchange sites. Sodium-based natural zeolites have an exchange capacity of 15 to 100 kg/m^3 (Peavy, Rowe, and Tchobanoglous 1985). Also, they can be regenerated easily.

When hard water comes in contact with the resins, the following reaction occurs (Qasim, Motley and Zhu 2000):

$$Ca/Mg - \{anions\} + Na_2R \rightarrow \{Ca/Mg\} - [R] + 2Na - \{anions\}$$

53 https://www.suezwatertechnologies.com/handbook/ext_treatment/ch_8_ionexchange.jsp.

R represents the resin to which a cation such as Na is attached and exchanged with hardness causing cations like Ca and Mg. Over time the resin exchange capacity gets exhausted, and hardness appears in the treated water. The resins can be regenerated using a sodium chloride solution.

$\{Ca/Mg\}[R] + 2NaCl \rightarrow \{Ca/Mg\}Cl_2 + Na_2R$

The efficiency and capacity of the process depends on the following factors:

- Type of solid medium
- Type of exchange material used for coating
- Quantity of regeneration material
- Regeneration time
- Quality of water (before and after treatment).

Ion exchange processes are more efficient than conventional lime soda processes in providing better quality water and less sludge production. The process is simpler, i.e., easier to control than the lime-soda process. However, a major pre-requisite is that the feed water should be of good quality which means it should be free from turbidity. Turbidity is due to colloids which can clog the resin bed and reduce the process efficiency. In general, ion-exchange has high pre-treatment requirements (it is rarely or never used as a 'stand-alone' process).

Examples of natural zeolites that can be used for ion exchange are shown in the figure below (Wikipedia 2007). The microscopic structure of zeolite crystals is shown on the right and demonstrates the large surface area to volume ratio that these adsorbents have. The large pore spaces and surface area result in high cation exchange capacity of these adsorbents.

Figure 5.15 Macroscopic and microscopic structures of natural zeolites (Source: Wikipedia) (see color plate).

Ion exchange columns or reactors are generally run in downflow mode. There are four major steps in an ion exchange process:[54]

1. **Exhaustion:** Ca and Mg from water attach themselves to the cation sites on the resin beads while Na or K ions from the beads are released into the water. This exchange continues until

54 https://www.wqpmag.com/resin-regeneration-fundamentals.

all sites on the beads are 'occupied' by Ca or Mg. At this point, the bed is considered to be exhausted. In practice, exhaustion of bed capacity is marked by an increase in concentration of Ca or Mg in the treated water.

2. **Backwashing:** Similar to rapid sand filters, water is pumped through the reactor in upflow mode along with air at high velocity to dislodge particles and resin fines clogging the porous bed. These particles are removed during the backwashing step. Bed expansion during backwashing is about 50 percent and the backwash cycle lasts for 10 to 15 minutes.
3. **Regeneration:** The resin beads which are completely 'occupied' by Ca and Mg can be regenerated by using 8 to 12 percent solution of common salt (NaCl) for a sodium cation exchange resin. The salt solution serves as eluent or regenerant and sodium occupies the active sites on the resin beads during this step while Ca and Mg that were adsorbed on the resin are released into the brine solution. A regeneration time of 30 minutes is desirable and is measured from the moment the regenerant is introduced into the reactor until the beginning of the next step, the slow rinse step. Effluent from the regeneration cycle contains hardness from the softening process as well as sodium chloride used during regeneration. Post-regeneration, fresh soft water should be used to flush out the excess sodium chloride.
4. **Slow and fast rinse:** The slow rinse is done with treated (softened) water at a low flow rate to ensure that the brine is not flushed out of the reactor too quickly. The final step is the fast rinse with treated water for removing any traces of the regenerant in the reactor. It is done at the same flow rate as the service flow rate.

Typical design parameters for water softening using ion exchange are:

- Typical bed depths: 1 to 2 m
- Flow rates: 0.2 to 0.4 m^3/m^2-min
- Total exchange capacity: 50,000 to 80,000 g/m^3 as $CaCO_3$

PROBLEM 5.8.1

The same water sample as in Problem 5.7.1 is to be softened to the minimum calcium hardness by an ion-exchange process. Magnesium removal is not necessary.

a. A flow of 20 MLD of relatively high quality groundwater is to be supplied to a town. Ion exchange resin which has an adsorptive capacity of 80 kg of $CaCO_3/m^3$ of resin is to be used at a flow rate of 0.5 m^3/m^2-min. Regeneration is accomplished using 150 kg of sodium/m^3 of resin in 15 percent solution. Determine the volume of medium required and the physical arrangement for continuous operation in fixed bed.

b. Also determine the chemical requirements for regeneration.

Solution

Total Ca hardness = 4 meq/L = 200 mg/L as $CaCO_3$

Minimum hardness to be achieved = 40 mg/L as $CaCO_3$

Therefore, flow that can be bypassed assuming ion exchange provides close to 100 percent efficiency = 0.2 fraction

Flow to be treated = 0.8 fraction = 16000 m^3/d = 11.11 m^3/min

Hardness to be removed = 200 mg/L*16*10^6 L/d = 3,200 kg/d

Volume of medium required for 1 day's flow = 3200 kg/d/(80 kg/m^3 medium) = 40 m^3/d

Surface area of tank = 11.11 m^3/min/ (0.5 m^3/m^2-min) = 22.22 m^2

If tank has diameter = 2.2 m

Cross-sectional area of tank = 3.8 m^2

No. of tanks = 22.22/3.8 = 5.84 units

Using 6 units for further design and three additional units for cleaning and maintenance, Therefore, a total of 9 units are to be used.

Height of tank = volume of tank/surface area of tank = 40/22.22 = 1.8 m

Total volume of resin for 9 tanks = 61.56 m^3

Chemical requirements for regeneration:

Volume of one unit = 1.8*3.8 = 6.84 m^3/unit

Salt requirement = 150*6.84 = 1026.36 kg/unit

Regeneration of 6 units/d,

NaCl required = 6*1026.36 = 6158.16 kg/d of NaCl

15 percent solution of NaCl is equal to 0.15 kg/L assuming 1 L weighs approximately 1 kg. Therefore, 6158.15 kg NaCl/d/(0.15 kg/L) = 41054 L/d of a 15 percent salt solution has to be provided.

PROBLEM 5.8.2

An ion exchange process is to be designed for a flow of 1 MLD and provides treated water with hardness of 5 mg/L as $CaCO_3$. The feed water concentration is 500 mg/L as $CaCO_3$ and is to be reduced to ≤ 300 mg/L. Split treatment is the most economical way of treating this water. Determine the flow to be passed through the ion exchange column.

Solution

A schematic of the split flow treatment process is provided below:

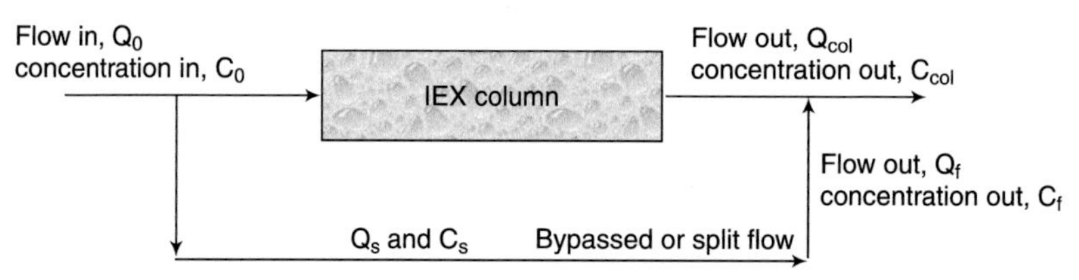

Flow in, $Q_0 = 1.00 \times 10^6$ L/d

Hardness in feed water, $C_0 = 500$ mg/L

Hardness in treated water, $C_f = 300$ mg/L

Hardness in flow out of column, $C_{col} = 50$ mg/L

$$Q_0 = Q_f = Q_s + Q_{col}$$

Where split flow (by-passed) is Q_s and flow through column is Q_{col}

Mass in = $Q_0*C_0 = 5.0 \times 10^8$ mg/d

1. Mass out = $Q_f*C_f = Q_{col}*C_{col} + Q_s*C_s = 3.0 \times 10^8$ mg/d
2. $Q_{col} = Q_0 - Q_s$

Substituting for Q_{col} in (1) and solving for Q_s

$$Q_f*C_f = (Q_0 - Q_s)*C_{col} + Q_s*C_s$$

$$300*10^6 = 50*(10^6 - Q_s) + 500*Q_s$$

$$Q_s = 5.56 \times 10^5 \text{ L/d}$$

$$Q_{col} = 4.44 \times 10^5 \text{ L/d}$$

5.9 GRANULAR MEDIA FILTRATION

The objective of filtration is the removal of suspended particles and floc that cannot be removed by settling. It is an important unit in conventional water treatment and is used in advanced wastewater treatment for polishing the effluent. In conventional water treatment plants, filtration follows coagulation, flocculation and settling. However, if the feed water quality is relatively high and the floc formed after coagulation and flocculation is small in size and low in concentration, direct filtration can be used without settling (Cleasby 1990).

There are three general classes of filtration (Metcalf and Eddy 2003):

1. **Granular medium filtration:** Filter beds packed with media are used to create porous bed filters and are the most common type of filtration processes used in water treatment plants. The most common filtration process involves a stationary granular medium filter bed which removes solids from water and allows filtered water to pass through. The media used can be sand, anthracite, garnet or even granular activated carbon (GAC) which can be used for adsorption and filtration (Cleasby 1990). Fundamental properties of these materials are summarized in Table 5.6. These filters can be operated in different flow modes: upflow, biflow, pressure, vacuum or gravity. Gravity flow is the most common flow mode in water and wastewater treatment where the weight of the water column above the filter provides the driving force. Examples include e.g., slow and rapid sand filters. Within granular medium filtration processes, there are two types of processes:

Depth filtration: If solids penetrate into the depth of the porous bed and are removed within the pores of the granular media, it is termed depth filtration. Rapid sand filters are the most common type of depth filters.

Cake filtration: If solids are removed on the entering face of the granular material, forming a layer of fine material (cake) at the surface of the filter, it is termed cake filtration. The most common examples of cake filtration are slow sand filters, and precoat filters (diatomaceous earth filters).

Table 5.6 Fundamental properties of common filter media (Cleasby 1990)

Filter media	Grain size, d_{60} (mm)	Specific gravity	Loose bed porosity, Percent	Sphericity
Silica sand	0.76	2.65	0.42 to 0.47	0.7 to 0.8
Anthracite	2.2	1.45 to 1.73	0.56 to 0.60	0.46 to 0.6
Garnet	-	3.6 to 4.2	0.45 to 0.55	0.6
GAC	-	1.3 to 1.5	0.5	0.75

2. **Surface filtration:** Straining or mechanical sieving through different types of filters is termed surface filtration. Examples include coarse filtration using paper or cloth; laboratory analysis for TSS is done using cellulose, cellulose acetate or glass fiber filters (GFF) while diatoms are used as fine filter media.
3. **Membrane filtration:** Porous synthetic membranes of different pore sizes are used to remove suspended solids and are covered in Section 5.10.

5.9.1 Types of Granular Media Filters

Several different types of granular media filters can be designed and are listed here based on the factor that is varied:

- Bed depth: conventional, shallow or deep
- Filter media: stratified or unstratified
- Type of media: mono-, dual- or multi-
- Type of operation: downflow or upflow
- Method of solids management: surface or internal storage
- Driving force: gravity or pressure

Further, the most common types of filters used in drinking water treatment are described here and can be designed based on the choice of media, size, size distribution and density of the filter medium. The most common filters used in water treatment are slow and rapid sand filters. Granular media filtration is used only as an advanced treatment process for wastewater.

Slow sand filters

Slow sand filtration, also known as biological filtration, is one of the oldest processes in use for drinking water treatment. It was first used in 1804 and incorporated into London's public water supply systems in Chelsea in 1829 (Huisman and Wood 1974). These filters are of fine sand of effective size 0.2 mm. The small grain size means that almost all suspended particles are removed at the surface of the filter. A mat of biological organisms generally develops at the water-sand interface and aids the filtration process forming what is called the *schmutzdecke* (German word for slime) layer. Therefore, filters with very large surface area are required to provide sufficient quantities of treated water. Cleaning is periodically done by draining the filter and manually or mechanically removing the top few centimeters of sand along with the accumulated solids and biological mat. Each filter may run for months depending on the quality of the feed water. These filters require large land area and are capital-intensive but require less operation and maintenance than rapid sand filters. They are also not efficient in treating highly turbid waters as the surface plugs quickly, and are best for low to moderate levels of turbidity. They are ideal for feed waters with <10 mg/L of TSS and can handle up to 50 mg/L TSS over longer periods of time.

Slow sand filters were found to be effective in removing turbidity to <1 NTU with stream turbidity >3 NTU along with Fe and Mn removal (Riesenberg et al. 1995). Slow sand filters can achieve 2-log removal of *Giardia* and 1-log removal of virus. It can provide 3-log *Giardia* and 2-log virus removal when combined with disinfection.

Design elements of a slow sand filter: They are constructed as open-top boxes that are either fully or partially below the ground surface or they can be constructed as elevated tanks or filters to provide the required head for gravity flow depending on the elevation of the feed water supply (Riesenberg et al. 1995). It is important to create a roof or shed for these filters to prevent algal growth on the surface. The head loss is very high due to the small media size and results in low flow rates, ranging from 0.1-0.4 m^3/h (Huisman and Wood 1974). Contact time in a slow sand filter is more than 2 hours while the depth of the filter bed ranges from 0.6 to 1.2 m. A constant head (height above the filter bed) is maintained above the filter using a feed water reservoir and under no conditions should the water level be allowed to fall below the surface of the filter bed.

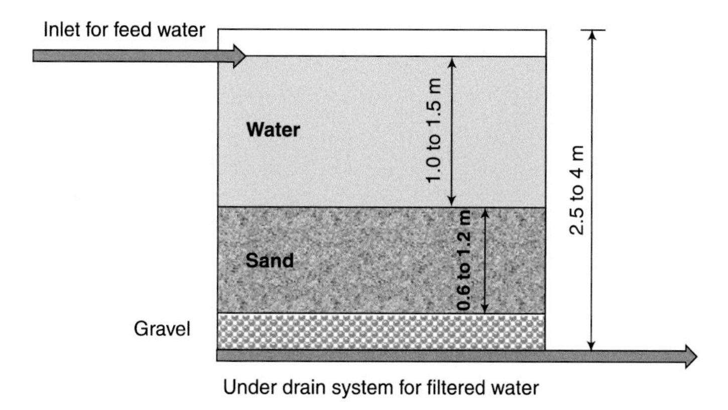

Figure 5.16 Schematic of a slow sand filter, based on Huisman and Wood (1974).

Rapid sand filter

Rapid sand filters include a bed of silica sand ranging from 0.6 – 0.75 m in depth with an effective size of 0.45 – 0.55 mm. A uniformity coefficient of 1.65 is commonly specified. Here the biological mat is absent but due to the higher filtration rates, these filters require more frequent cleaning. General filtration rates range from 2.5 to 5 m/h. Cleaning is generally done by backwashing, which results in stratification of the media with the smallest grain size at the top and increase in grain size with increase in depth of bed. This stratification results in inefficient use of the storage capacity of the filter bed. If the gradation of the filter is reversed then the entire media would function efficiently with larger pores at the top. Then the result would be longer filter runs, less head loss and high filtration rates. Dual-media filters are based on this principle and involve reverse gradation to some extent (Peavy, Rowe, and Tchobanoglous 1985).

Dual-media filter

These are usually made with layers of silica sand at the bottom and anthracite coal at the top since sand is heavier than coal (specific gravity of sand is 2.65 versus coal with 1.5) (Peavy et al. 1985). The depth of sand may range from 0.15 – 0.4 m with a coal bed depth of 0.3 – 0.6 m. The large pores in the coal layer remove large particles and floc while the sand layer removes smaller particles. The advantage of dual-media filters is greater storage capacity of the filter medium.

Mixed media filter

An ideal filter bed would be of media graded evenly with large particles at the top to small particles at the bottom. This can be done by using three or more types of media with selected size, density and uniformity coefficients. A typical installation might consist of a 0.75 m bed with 60 percent anthracite (layer 1 – top), 30 percent silica sand (layer 2 – middle) and 10 percent garnet sand (layer 3 – bottom), with specific gravities of 1.6, 2.6, and 4.2, respectively. Effective sizes ranging from a maximum of 1 mm for anthracite to a minimum of 0.15 mm for the garnet, with selected uniformity coefficients, will produce intermixing and appropriate pore size gradation (Peavy et al. 1985).

Operation of Depth Filters

When water is passed through a filter bed, the filter media removes solids from water by entrapping them in its pores. As a result, the filtered water has less turbidity than the influent water due to removal of floc and SS. During the filtration process, the pores get clogged and the total available pore volume decreases. In other words, as more and more solids are stored in the filter media, available pore spaces decrease, resulting in increased head loss and reduced filtration rates. The capacity of the filter media refers to the maximum amount of solids that a filter media can store before reaching critical head loss. Filtration efficiency refers to the quality of filtration, i.e., the purity of water after filtration. Thus, small pore openings increase filtration efficiency but the filtration rate is low. Large pore openings provide good filtration rate at lower filtration efficiency. In order to clean the pores, i.e., to remove the solids stored in the filter media, the filter bed can either be manually or mechanically cleaned or the filter bed can be backwashed or the filter media can be replaced. Cleaning of the filters is done as soon as terminal head loss is reached or when the quality of the treated water does not meet standards.

Replacing large quantities of filter media, and manual or mechanical cleaning are difficult and expensive options. Backwashing involves passing of water and air at high velocity in the reverse direction (counter-current) and is an easier and more efficient method than the other options. Filter media vary in diameter within a selected size range. The effect of varying size becomes important during backwashing as smaller grains are lifted farther and settle slowly compared to larger grains. This results in smaller pore openings at the top which is an inefficient arrangement for the filtration process. Density of the medium also affects expansion of the filter bed during backwashing. Lighter material naturally settles above denser material of the same size.

Design of depth filters requires computation of number and size of filter units. The surface area of the filters can be calculated based on peak filtration rates and peak plant flow rates. Depth and porosity of the filters are defined by the nature of the media used and whether the filter is a single-media or multi-media filter.

5.9.2 Design Considerations

Head loss for a clean filter media can be calculated using several equations and the most common one is the Carman–Kozeny equation shown below for a single media filter (Metcalf and Eddy 2003). This equation is derived from the Darcy–Weisbach equation for flow through circular pipes by assuming that flow through a porous medium is analogous to flow through several capillary tubes or parallel cylinders.

$$h_f = \frac{f' L(1-e)v_s^2}{e^3 g d_p} \tag{5.9.1}$$

$$f' = \frac{150(1-e)}{Re} + 1.75 \tag{5.9.2}$$

For a multi-media filter, the equation is modified as

$$h_f = \frac{L(1-e)v_s^2}{e^3 g} \sum \frac{f'_{ij} X_{ij}}{d_{ij}} \tag{5.9.3}$$

h_f = head loss through a bed of uniform particles of diameter, d_p, m

L = depth of filter bed, m

e = porosity of bed,

X_{ij} = weight fraction of particles between two adjacent sieve sizes

d_{ij} = average particle size based on the two adjacent sieve sizes

f'_{ij} = friction factor for two adjacent sieve sizes

v_s = filtering velocity or superficial velocity = Q/surface area of filter bed, m/s

g = gravitational acceleration, m/s^2

During a filter run, solids are 'stored' in the pores of the filter bed resulting in a decrease in porosity and an increase in head loss. Eventually, filtering velocity is no longer constant and the filter run has to be stopped for cleaning. In some cases, solids stored in the pores 'breakthrough' into the treated water so that water quality requirements are no longer met.

Backwashing of filter bed

Decrease in filtering velocity, terminal head loss or inadequate quality of filtered water are used to indicate the end of a filter run (Metcalf and Eddy 1991; Peavy, Rowe, and Tchobanoglous 1985). At this point, backwashing is necessary and a stream of water and air at high velocities in upward flow are sent through the filter bed to fluidize it. Fluidization is essential for mobilizing the solids that were stored in the pores of the fixed bed and collisions between the media particles result in 'freeing' of solids that adhere to the media particles. During fluidization, the porosity and depth of the filter bed change to e_{fb} and L_{fb}, respectively.

Head loss through a fluidized bed is calculated as the weight (gravitational force) of the bed as shown below (Crittenden et al. 2012; Peavy, Rowe, and Tchobanoglous 1985):

$$F_g = m^*g = (\rho_p - \rho_w)(1 - e)A^*L^*g/\rho_w \qquad 5.9.4$$

Where

F_g = gravitational force which is also the weight of the bed, N

ρ_p and ρ_w = specific gravity of particles and water, respectively

m = mass of the entire bed, kg

A = cross-sectional area of the filter bed, m^2

The pressure drop across the bed, $\Delta P = F_g/A$, N/m^2 and the head loss is

$$h_f = \frac{F_g}{Ag\rho_w} = \frac{L(1-e)(\rho_p - \rho_w)}{\rho_w} \qquad 5.9.5$$

Since the weight of the fixed bed is equal to the weight of the fluidized bed,

$$F_g = m^*g = (\rho_p - \rho_w)(1 - e)^*ALg/\rho_w = (\rho_p - \rho_w)(1 - e')AL'g/\rho_w$$

The above equation can be simplified to

$$L' = L\frac{(1-e)}{(1-e')} \qquad 5.9.6$$

The porosity of the fluidized bed is related to the backwash velocity and the settling velocity of the particles by the following equation:

$$e_{fb} = \left(\frac{v_b}{v_s}\right)^{0.22} \qquad 5.9.7$$

PROBLEM 5.9.1

Water is to be filtered at 20 °C through a uniform bed of garnet at a filtering velocity of 3 m/h. The garnet beads are 1 mm diameter and can be assumed to be perfect spheres. The specific gravity of garnet is 4.0. The depth of the bed is 0.8 m and porosity is 0.47. Determine the head loss through the bed.

Solution

Filtering velocity (v) = 3 m/hr = 8.33×10^{-4} m/s

Diameter of bead (d_p) = 1 mm

Specific Gravity of the particle, SG_p= 4

Depth (D) = 0.8 m

Porosity (e) = 0.47

Porosity after backwash (e') = 0.75

Density of water at (ρ_w) = 998.2 kg/m^3

Dynamic Viscosity (μ) = 0.001002 kg/m.s

$g = 9.81$ m/s^2

Step 1: Calculate Reynolds Number

$Re = 0.83$

Step 2: Calculate Friction Factor

$f' = 97.51$

Step 3: Head Loss

$h_f = 0.0282$ m

Head loss = 28 mm

PROBLEM 5.9.2

In Problem 5.9.1, during backwash the filter expands to a porosity of 0.75. Determine the required backwash velocity and resulting expanded depth of the filter.

Solution

Step 1 Calculate terminal velocity for the particles in the filter assuming laminar flow and using Stokes' Law (equation 5.4.9)

v_t (assuming laminar flow) = 1.63 m/s

Step 2 Check Reynold's number using equation

$\Phi = 1$ assuming particles are perfect spheres

Re = 1622.62 implies transitional flow. Therefore, v_t calculated in step 1 is incorrect.

Step 3 Check for Coefficient of Discharge using equation 5.4.5 for transitional flow.

$C_d = 0.43$

Based on Newton's law and equation 5.4.3, $v_t = 0.30$ m/s

Check for Re = 301.2 which in turn results in $v_t = 0.26$ m/s

Since v_t in each iteration is different, continue iterating until there is no significant change in v_t as shown in the following table.

Iteration	Reynolds' Number	Coefficient of Discharge	v_t^2	v_t
	1622.6223	0.429266		0.302344
1	301.19724	0.92543	0.066223	0.257338
2	256.36255	0.620985	0.06319	0.251376
3	250.4228	0.625414	0.062742	0.250484
4	249.5344	0.626093	0.062674	0.250349
5	249.39918	0.626196	0.062664	0.250328
6	249.37854	0.626212	0.062662	0.250325

Final $V_t = 0.25$ m/s

$e' = 0.75$

$L = 0.8$ m

$e = 0.47$

Now, backwash velocity, $v_B = 0.0677$ m/s

Also, L' (depth of expanded or fluidized bed) = 1.7 m

PROBLEM 5.9.3

A filter of 0.8 m depth is composed of garnet grains of specific gravity 4.0 and shape factor 0.8. The porosity of the bed is 0.4 and results of the sieve analysis are shown below. Determine head loss through the bed. Make any reasonable assumptions to complete the problem.

Mass retained, %	1	8	20	30	25	8	4	3	1	100
Average particle size, mm	0.9	0.8	0.7	0.6	0.5	0.4	0.3	0.2	0.1	

Solution

Step-wise calculations for determining head loss are shown below:

Step 1 Calculate Reynolds number in terms of d_{ij}

At 20 °C, density of water, ρ_w =	998.2	kg/m^3
μ =	0.001	kg/m-s
Assuming filtering velocity, v = 3 m/h	3	m/h
	0.0008333	m/s
Reynolds number, Re = $\varnothing \rho_w v d_{ij}/\mu$	665.46667	dij

Step 2 Calculate f'_{ij} for each particle size

$$f'_{ij} = 150 \times (1 - e)/Re + 1.75$$

Step 3 Determine $\Sigma f'_{ij} X_{ij}/d_{ij}$ as shown in the following table:

d_{ij} (× 10^{-3})	Xij	f'_{ij}	$f'_{ij}(X_{ij}/d_{ij})$
			1/m
0.9	0.01	152.02	1,689.117
0.8	0.08	170.80	17,080.430
0.7	0.2	194.95	55,701.403
0.6	0.3	227.16	1,13,577.865
0.5	0.25	272.24	1,36,118.438
0.4	0.08	339.86	67,971.719
0.3	0.04	452.56	60,341.528
0.2	0.03	677.97	1,01,695.079
0.1	0.01	1,354.18	1,35,418.438
		Σ	6,89,594.017

Step 4 Determine head loss

$$h = L^*(1 - e)^*v^2/[(g^*e^3) \times \Sigma f'_{ij} X_{ij}/d_{ij}] = 0.366 \text{ m}$$

PROBLEM 5.9.4

A 5 MLD plant is to be built for treating surface water. Since it is a semi-urban community, a possible treatment option is slow sand filtration. Determine the area required for slow sand filters. The following values may be assumed for the design.

Filtration velocity ranges from 0.1 to 0.2 m/h, so use 0.15 m/h

Depth of filter bed = 1 m

Height of water above filter ranges from 0.7 to 1.5 m, so use 1 m

Solution

For a flow rate of 5 MLD, i.e., 208.3 m^3/h.

Area required assuming flow rate is 0.15 m/h = 1389 m^2

Provide additional area of 20 percent for standby filters during operation and maintenance

Therefore, total area required = 1667, i.e., 1700 m^2.

5.10 MEMBRANE FILTRATION

Membrane processes have become popular in both water and wastewater treatment, and are defined as processes in which a membrane is used to permeate high-quality water while rejecting passage of dissolved and suspended solids. In drinking water treatment, membrane processes are used most often for demineralization (also called desalination) and in several cases for particulate and/or NOM removal. They are used for tertiary or higher level treatment of wastewater for the removal of microbes, and dissolved and suspended particles. In almost all cases, some degree of pre-treatment is required.

Membrane filtration is used to filter suspended solids including colloidal particles and dissolved solids up to a size of 0.1 nm in water or wastewater treatment. The influent to the membrane process (conventional symbols are shown in the schematic below) is called the feed water. The liquid that passes through the membrane is called permeate and the liquid rejected is called concentrate or retentate or rejectate. The rate of flux is defined as the rate of flow of permeate through the membrane.

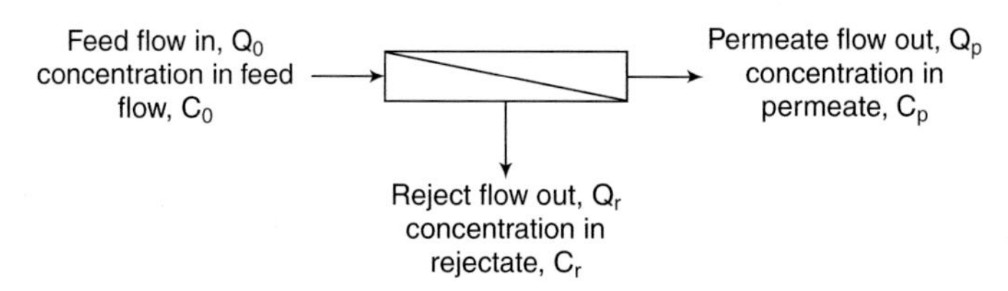

Figure 5.17 Schematic of a single module membrane process (the diagonal represents the membrane).

It is important to note that membrane processes are generally 'high-end' processes, i.e., the quality of feed water is often high and these processes are used to achieve further (greater) contaminant removal.

Membranes have a thin skin (0.20-0.25 μm thickness) which is the working surface of the membrane. This layer rests on a thicker, more porous layer (100 μm thickness) which acts as supporting material. Membranes can be made of various materials including ceramic, cellulose, polypropylene, cellulose acetate and thin-film composites (TFC). They are available in the form of sheets, fibers or in tubular form. Pores in membranes are categorized as macropores (>50 nm), mesopores (2-50 nm) and micropores (<2 nm) (Metcalf and Eddy 2003).

5.10.1 Types of Membrane Filters

Membrane processes are of two types: pressure-driven or electric current-driven. Pressure driven processes include microfiltration (MF), ultrafiltration (UF), nanofiltration (NF) and reverse osmosis (RO). Electric-current driven processes include electrodialysis (ED) and electrodialysis reversal (EDR). Salient features of different membranes are summarized in Table 5.7.

Table 5.7 Characteristics of different membrane processes

Membrane process	Particle size removal or pore size, micron	Principle	Operating pressure	Removal
Microfiltration (MF)	0.1 to 10	Size exclusion	100 to 400 kPa (15 to 60 psig)	Suspended particles
Ultrafiltration (UF)	0.005 to 0.1	Size exclusion	200 to 700 kPa (30 to 100 psig)	Sand, silt, clay, Giardia and Cryptosporidium, algae, NOM, bacteria, virus and colloidal particles, >10,000 MW, heavy metals
Nanofiltration (NF)	0.001 to 0.005	Diffusion of solvent (water) and size exclusion	600 to 1000 kPa (90 to 150 psig)	NOM, bacteria and virus removal, alkalinity, heavy metals
Reverse osmosis (RO)	< 1 nm	Diffusion of solvent (water)	2000 to 7000 kPa (300 to 1000 psig)	salts and low MW organics, heavy metals
Electrodialysis (ED) and EDR	< 1 nm	Donnan exclusion (ion exchange or exclusion)	electricity driven	salts

Desalination capacity has increased exponentially all over the world, since its beginning in the 1950s, and includes various membrane and distillation technologies. Reverse osmosis (RO) accounts for 44 percent of the installed desalination capacity world-wide followed by multi-stage flash distillation (MSF) with 40 percent, electrodialysis (6 percent), vapor compression (5 percent) and multiple effect distillation (MED) at 3 percent (Greenlee, Lawler, Freeman, Marrot, and Moulin 2009). Distillation was the method of choice in the Gulf countries prior to the use of membrane technologies. In non-Gulf countries, reverse osmosis is the method of choice for desalination.

Applications of different types of membranes

- Microfiltration (MF) is often used as a substitute for depth filtration to achieve lower turbidity and TSS concentrations in the treated water. It can be used for NOM removal as well.
- Membrane bioreactors (MBRs) are being used for biological treatment of wastewater.
- Ultrafiltration (UF) is being used to produce high-purity process rinse water.

- Nanofiltration (NF) is being used to remove hardness and NOM and for desalination of brackish water (Greenlee, Lawler, Freeman, Marrot, and Moulin 2009).
- Reverse osmosis (RO) is used for desalination of brackish water and is the only membrane technology that can be used for desalination of seawater.
- ED and EDR are used for desalination of brackish water.
- UF, NF and RO have been successfully used for removing heavy metals like Co, Ni, Zn, Cd, Cr and Cu from wastewater (Kurniawan et al. 2006).
- Other applications of membrane processes include separation of different components in various body fluids (an example is electrodialysis for those with poorly functioning kidneys) and laboratory analytical methods for separation of different constituents of the samples (an example is separation of suspended solids from dissolved solids by membrane filtration).

5.10.2 Osmosis and Reverse Osmosis

An osmotic membrane allows diffusion of water only. In osmosis, when a semi-permeable osmotic membrane separates two solutions with different salt concentrations, water diffuses from the solution of lower salt concentration (higher water concentration; higher chemical potential) to the one of higher salt concentration (lower water concentration; lower chemical potential). The flow continues until the chemical potential difference is balanced by the pressure difference (called osmotic pressure) as shown in Figure 5.18. Routine examples of osmosis include the swelling of raisins in water. Bacterial cells are generally harvested in isotonic solutions so that they do not lyse due to a sudden influx of water into the cell by osmosis.

When a pressure in excess of osmotic pressure is applied on a cell which has higher salt concentration, water will flow in the reverse direction. This phenomenon is called reverse osmosis.

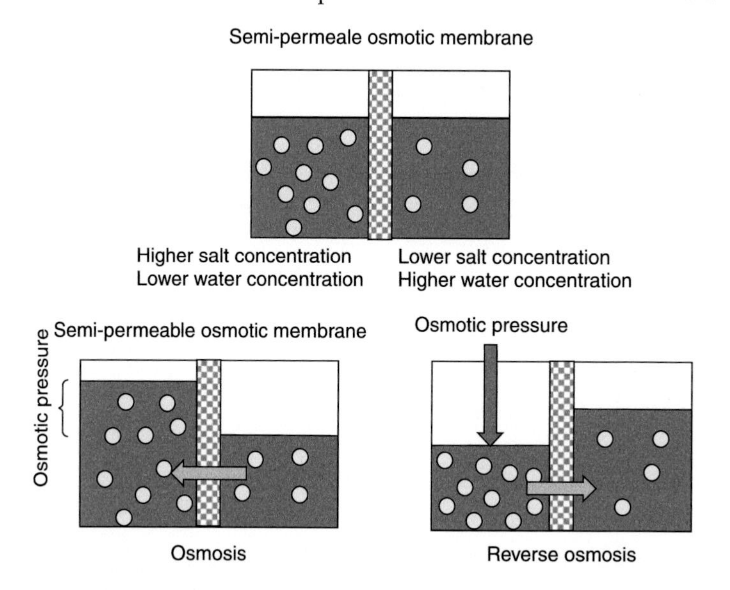

Figure 5.18 Schematic of the osmotic and reverse osmosis process.

Van't Hoff Law

Van't Hoff published his paper relating osmotic pressure to the concentration of ions in dilute solutions based on the ideal gas law in 1887.[55] He was awarded the first Nobel prize in chemistry in 1901. The osmotic pressure of any solution is directly proportionate to the concentration of ions in solution and is expressed mathematically as shown below:

$$\Pi = \phi i C^{*} R^{*} T \quad 5.10.1$$

Π = osmotic pressure, kPa (1 atm = 101.3 kPa)

C = concentration of solute, mol/m^3

R = universal gas constant

= 8.31451 N-m/g-mol-degree K

= 8.31451 Joules/g-mol-degree K

= 0.0820578 L-atm/g-mol-degree K

T = absolute temperature of the solution, degree K

Van't Hoff (VH) factor, $i = 1 + \alpha(n - 1)$ where α = degree of dissociation of solute in a specific solvent, n = number of moles of ions or molecules for a specific solvent assuming 100 percent dissociation,[56]

Φ = correction factor for non-ideal behavior (osmotic coefficient).

The Van't Hoff law implies that solutions of salts with the same molar concentrations have the same osmotic pressures. Osmotic pressure for a mixture of salts is additive.

The RO membrane was first made in the laboratory in 1959 in UCLA by Sidney Loeb and Srinivasa Sourirajan.[57] In 1965, the first commercial RO plant was set up in Coalinga, CA, USA. Reverse osmosis (RO) is used for desalinating water with high TDS in arid and semi-arid regions of the world. The world's largest seawater-to-drinking water treatment plant is based on RO and has a capacity of 6,24,000 m^3/d. Located in Sorek, south of Tel Aviv, Israel, it became fully operational in 2013.[58] Recovery rates in seawater reverse osmosis (SWRO) plants range from 45 to 55 percent (Elimelech and Phillip 2011).

5.10.3 Electrodialysis and Electrodialysis Reversal

Use of electrochemical separation processes for desalinating water, i.e., removal of salts or TDS, began in the 1930s while the first desalination plant with ED was set up in 1969 (Hendricks 2011). Mineral salts and other ionic species are transported through ion selective membranes from one solution to another under the driving force of a direct current (DC) electrical potential. Membranes of high-capacity, highly cross-linked ion exchange resins allow passage of ions but not water. Pre-treatment of water is required before using this process.

---

55 https://web.lemoyne.edu/giunta/classicalcs/vanthoffnote.html.

56 https://www.chemteam.info/Solutions/vantHoff-factor-and-degree-of-dissoc.html.

57 https://www.freedrinkingwater.com/reverse-osmosis/knowledge-base/history-of-reverse-osmosis-filtration.htm.

58 https://en.wikipedia.org/wiki/Reverse_osmosis_plant.

An electrodialysis unit consists of cation permeable and anion permeable membranes placed alternately in stacks and separated by spacers. A set of adjacent components, consisting of a diluting compartment spacer, an anion membrane, a concentrating compartment spacer, and a cation membrane, is called a cell pair. Membranes are always arranged in pairs, and the number of membranes in any unit will always be even. If there are m number of membranes, then there will be $m/2$ number of deionizing compartments in every ED or EDR unit. The distance between any two adjacent membranes is a few mm or less.

An electric potential is applied between two electrodes at the two ends of the ED unit. It causes cations to move towards the negative electrode and anions towards the positive electrode. Due to the arrangement of membranes, alternate cells of dilute and concentrated salts are formed (Metcalf and Eddy 2003). Electrodialysis stacks can contain as many as 600 cell pairs. Energy requirement is proportionate to concentration of salts.

A major problem with ED is membrane fouling due to deposition of salts and NOM. Reversal of polarity leads to resuspension of the salts and NOM that are deposited on the membrane surfaces. This process is called electrodialysis reversal (EDR). In EDR, the alternate cells of dilute and concentrated salts are switched when the polarity of the electrodes is reversed. EDR results in longer membrane life and therefore, lower costs.

ED and EDR can be used for demineralization (removal of ions) only; they cannot be used for removal of NOM and SS; both are responsible for membrane fouling.

From Faraday's law, the current required for a stack of membranes is given by

$$I = \frac{FQN\eta}{nE_c} \qquad 5.10.2$$

Therefore, the number of gram equivalents removed per unit time is:

Gram eq/unit time = $QN\eta$, where

Q = flow rate (L/s), N = normality of solution (g-eq/L),

η = coulombic efficiency of the process; F = Faraday's constant, 96,485 Coulombs/mol or Joules/volt-g-equivalent

n = number of cells in the stack, E_c = current efficiency

If the resistance of the unit, R, is known, the power required is given by

$$P = I^2R \qquad 5.10.3$$

5.10.4 Design of Membrane Filters

Membrane filters are usually run as modules in series or parallel. The feed solution is pressurized and circulated through the module using a pump. The pressure of the concentrate (also called rejectate or retentate) is adjusted using a valve, while the permeate is withdrawn at atmospheric pressure. As solids accumulate on the membranes, the membrane flux decreases and after a predetermined point, backwashing or cleaning of the membranes is done.

Membrane processes can be categorized based on their flow mode and schematics of these flow configurations are shown in Figure 5.19.

Dead-end (direct flow) or cross-flow

Flow configuration is termed *direct flow* or *dead end flow* when feed water is passed through the membrane perpendicular to the surface of the membrane. The plate and frame model shown in Figure 5.19 is an example of a direct flow configuration. A common laboratory example is the use of membrane filters at the tip of a syringe to filter small quantities of sample for various analytical purposes. Direct flow configuration results in the formation of a 'cake layer' on the surface of the membrane. This additional layer leads to a significant increase in trans-membrane pressure and requires more frequent cleaning than cross-flow membranes. Direct flow configuration is better used for concentrated suspensions such as municipal wastewater and wastewater from some types of food processing units.

When the feed water flow is tangential (parallel) to the membrane it is termed *cross-flow*. Cross-flow configuration results in shear forces at the surface of the membrane which prevent formation of the cake layer. The result is greater run times for the same feed water and flow rate while energy requirement may be greater compared to direct flow configuration. Most full-scale membrane filtration plants operate in cross-flow mode. Common cross-flow membrane filters are large membrane tubes, spiral-wound membranes or hollow fiber membranes. Spiral-wound membranes and hollow fiber membranes are the most common membranes in water treatment.

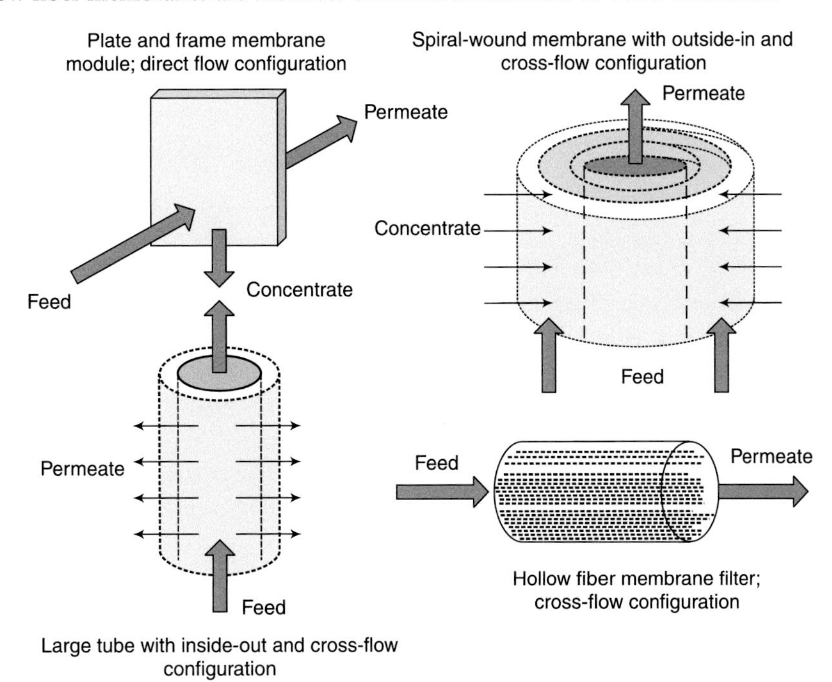

Figure 5.19 Different types of membrane filters and their flow configurations.

Inside-out or outside-in

Feed water can be sent through a membrane module in two different ways: inside-out and outside-in as shown in Figure 5.19.

Inside-out configuration is where the feed enters under pressure into a central perforated pipe which has a membrane around it. Permeate is forced out of the tube through the membrane and is collected from the outside of the tube. The outflow from the central tube is the concentrate.

Outside-in configuration is used in spiral-wound membranes where the feed water is fed into a tube under pressure and the permeate passes through the membranes into the central collection tube. Layers of membrane wound over the central perforated tube have spacers between them for the permeate to flow into the central tube.

5.10.5 Operation of Membrane Filters

Membrane processes can be operated at constant flux (of water) or constant trans-membrane pressure or variable flux and trans-membrane pressure (Metcalf and Eddy 2003).

For cross-flow modes, the trans-membrane pressure gradient is given by

$$P_{tm} = \frac{P_f + P_c}{2} - P_p$$

where P_f = pressure of feed stream, P_c = pressure of concentrate stream, P_p = pressure of permeate stream.

The pressure drop across the membrane module is given by

$$P = P_f - P_p$$

For the direct feed mode, the trans-membrane pressure gradient is given by

$$P_{tm} = P_f - P_p$$

The total permeate flow (kg/s) is given by

$$Q_p = F_w A$$

Where F_w = trasmembrane water flux rate, kg/m^{2}s, A = membrane area, m^2

The recovery rate, r is defined as

$$\text{r, Percent} = \frac{Q_p}{Q_f} \times 100 \qquad 5.10.4$$

where Q_p and Q_f are flow and feed stream flow respectively in kg/s.

The salt rejection rate, R is defined as

$$\text{R, \%} = \frac{C_f - C_p}{C_f} \times 100 \qquad 5.10.5$$

Where C_f and C_p are the concentrations of the solutes in the feed stream and permeate, respectively in units of mass/volume. Other measures of the salt rejection factor, S include $S = Q_r C_r / Q_0 C_0$ or C_r / C_0 (Sincero and Sincero 1996; Qasim, Motley, and Zhu 2000).

The mass balance equations are

$$Q_f = Q_p + Q_c \text{ and}$$

$$Q_f C_f = Q_p C_p + Q_c C_c$$

where C_f, C_c and C_p are the solute concentrations in feed stream, concentrate and permeate.

Different modes of operation can be achieved by varying flux and trans-membrane pressure (TMP) simultaneously or by keeping any one of them constant.

The flux of water through the membrane is given by

$$F_w = k_w(\Delta P_a - \Delta\Pi) = \frac{Q_p}{A} \qquad 5.10.6$$

Where K_w = water mass transfer coefficient, s/m

ΔP_a = average imposed pressure gradient = $\dfrac{P_f + P_c}{2} - P_p$

$\Delta\Pi$ = osmotic pressure gradient = $\dfrac{\Pi_f + \Pi_c}{2} - \Pi_p$

Q_p = permeate stream flow, kg/s and A = membrane area, m^2

Solute flux through the membrane is given by

$$F_i = k_i \Delta C_i = \frac{Q_p C_p}{A} \qquad 5.10.7$$

Where k_i = solute mass transfer coefficient, and

ΔC_i = change in solute concentration

The efficiency of a membrane process is dependent on several factors, and efficiencies in excess of 90 percent are common in water and wastewater treatment. Some of the factors that affect the performance of a membrane process can be divided into three major categories.

Feed water characteristics such as pH, ionic strength, hardness and concentration of solutes to be removed such as NOM, salts, and heavy metals influence the efficiency of the membrane process (Bellona, Drewes, Xu, and Amy 2004; Sincero and Sincero 1996). Degree of dissociation of many compounds is a function of pH and therefore, their removal by membrane processes is influenced by pH. The presence of solutes other than the contaminant of concern in the feed water will also affect removal of the contaminant due to synergistic or antagonistic effects. For example, TDS removal efficiency is affected by the presence of NOM in the feed water.

Solute parameters such as molecular weight (MW) and solute molecule size (length and width), pK_a – acid dissociation constant, octanol-water partitioning coefficient (K_{ow}) and diffusion coefficient of the solute through the membrane also influence performance of a membrane process. MW or size of the solute molecule which is a result of its steric configuration such as branched or chain molecules will impact their removal through a membrane.

Membrane characteristics such as the pore size of the membrane, zeta potential, contact angle and roughness of membrane have a direct impact on membrane efficiency. Pore sizes are often described in terms of molecular weight (MW) cut-off, based on the passage of standard protein solutions of defined MW. Zeta potential describes the surface charge of a membrane and is a function of pH. The contact angle is a measure of the hydrophobicity or hydrophilicity of the membrane.

Major problems in membrane filtration are the high energy requirements, high capital costs for membrane processes, membrane fouling and large amounts of wastewater generated. In conventional water treatment, 5 to 10 percent of the influent water (feed water) is discharged as wastewater while in membrane processes, about 30 to 60 percent of the influent is lost as wastewater. The wastewater generated does not contain chemicals and therefore, can be disposed without treatment into sewers and wells. However, it cannot be reused for irrigation due to the high concentration of salts.

5.10.6 Membrane Fouling

Membrane fouling refers to the accumulation of feed water constituents on the membrane, formation of chemical precipitates on the membrane or damage to the membrane due to reactive chemicals. Membrane fouling occurs in three different ways (Metcalf and Eddy 2003):

1. During an operational run, solids larger than the pore size of the membrane form a cake or gel layer on the surface of the membrane,
2. Solids that are equal to the pore size of the membrane block the pores, or
3. Solids that are smaller than the pore size of the membrane can adhere to the pore surface thereby narrowing the pores leading to greater trans-membrane pressure.

Membrane fouling can be controlled in several ways: (a) by improving feed water quality, i.e., pre-treating the feed water to remove NOM and suspended solids, (b) backwashing with air and water, and (c) using dilute acid or other chemical solutions for cleaning the membranes. Nanofiltration and reverse osmosis require high quality feeds to ensure that the membrane lasts long and provides adequate treatment efficiency. Pre-treatment for membrane filtration units can be done by any of the following methods:

1. Multimedia filtration and ultrafiltration to remove colloids,
2. Disinfection using chlorine, ozone or UV to kill bacteria,
3. Removal of Ca, Mg, Fe and Mn to limit scaling, and
4. Ensuring absence of oxygen to prevent oxidation and deposition of precipitates of Fe and Mn.

PROBLEM 5.10.1

Determine the osmotic pressure for the following solutions: $CaCl_2$ of 1000 mg/L or 2000 mg/L; NaCl of 1000 mg/L, 5000 mg/L or 35,000 mg/L; glucose of 1000 mg/L or sucrose – a disaccharide of glucose of 1000 mg/L. Assume osmotic coefficient is 0.95 in all cases and the dissociation constant for the salts is 1. What is the osmotic pressure of a mixture of 2000 mg/L of $CaCl_2$ and 1000 mg/L of NaCl? Assume a temperature of 25 °C and R = 0.0820578 atm-L/degree K-mol. The molecular formula for glucose is $C_6H_{12}O_6$ while that for sucrose is $C_{12}H_{22}O_{11}$.

Solution

Step 1: Create a table for each solution and its concentration. Convert mg/L to moles/L.

Step 2: Since the degree of dissociation (α) of all salts in this problem is 100 percent which means $\alpha = 1$, the VH factor (i)= n where n = number of ions or molecules of the solute/mole of solute. Therefore, NaCl has 2 moles of ions/mole, $CaCl_2$ has 3 moles of ions/mole and sucrose and glucose each have 1 mole ion or compound/mole. Since $\alpha = 1$, n = i.

Step 3: Using equation 5.10.1, determine osmotic pressure in atm or kPa.

	C, mg/L	C, mol/L	VH factor, i	Osmotic pressure, π	
				atm	kPa
$CaCl_2$	1,000	9.01×10^{-3}	3	0.63	63.65
	2,000	1.80×10^{-2}	3	1.26	127.30
NaCl	1,000	1.71×10^{-2}	2	0.80	80.51
	5,000	8.55×10^{-2}	2	4.00	402.57
	35,000	5.98×10^{-1}	2	27.99	2817.98
Sucrose	1,000	2.92×10^{-3}	1	0.07	6.89
Glucose	1,000	5.56×10^{-3}	1	0.13	13.08

PROBLEM 5.10.2

What is the osmotic pressure of a solution containing 1000 mg/L of $CaCl_2$ which dissociates 80 percent only? Assume osmotic coefficient is 0.9.

Solution

Since α = degree of dissociation = 80 percent or 0.8; n = number of ions for $CaCl_2$ = 3, the VH factor, i = 1 + 0.8(3 – 1) = 2.6

Concentration of $CaCl_2$ = 1000 mg/L which is 9.01 mM. For this solution, osmotic pressure, $\pi = 0.9*2.6*9.01*10^{-3}*0.0820578*298 = 0.515$ atm

PROBLEM 5.10.3

What is the osmotic pressure of a solution containing 2000 mg/L of $CaCl_2$ and 1000 mg/L of NaCl? Assume complete dissociation of both salts as in Problem 5.10.1.

Solution

Osmotic pressure of solution = osmotic pressure of its components

From table in Problem 5.10.1, $CaCl_2$ (1.26) + NaCl (0.8) = 2.06 atm.

PROBLEM 5.10.4

For a reverse osmosis system as shown in Figure 5.19 with an input concentration C_0 of 1000 mg/L TDS, C_p = 200 mg/L and C_r = 3000 mg/L, determine the salt rejection factor and the recovery factor.

Solution

$$Q_0 = Q_p + Q_r \quad 1$$

$$C_0{*}Q_0 = C_r{*}Q_r + C_p{*}Q_p \quad 2$$

Based on concentration values, [2] can be modified to

$$1000{*}Q_0 = 3000{*}Q_r + 200{*}Q_p \quad 3$$

or $$Q_0 = 3Q_r + 0.2Q_p \quad 4$$

Subtracting [4] from [1] results in

$$2Q_r = 0.8Q_p \text{ or } Q_r/Q_p = 0.4 \text{ or } Q_r = 0.4{*}Q_p$$

$$S = Q_rC_r / Q_0C_0 = 3Q_r / Q_0 = 3{*}0.4Q_p / 1.4Q_p = 0.86$$

$$R = Q_p/(Q_p + Q_r) = Q_p/1.4\,Q_p = 0.71 \text{ based on mass balance}$$

Therefore, R = 0.71 implies that 29% of the feed water is lost as retentate or wastewater and the salt rejection factor is 0.86 which implies that RO efficiency in terms of TDS removal is 86%.

PROBLEM 5.10.5

An RO system has a salt rejection factor S (Q_rC_r/ Q_0C_0) = 0.8 and a recovery factor R (Q_p/Q_0) of 0.6. A feed water concentration C_0 of 1000 mg/L, is to be treated and the feed flow rate is 3000 L/d. What are the concentrations of permeate and rejected water (brine or rejectate)?

Solution

$Q_0 = Q_r + Q_p = 3000$ L/d

$C_0Q_0 = Q_rC_r + Q_pC_p$

$S = Q_rC_r/Q_0C_0 = 0.8$

$R = Q_p/Q_0 = 0.6$; therefore $Q_p = 0.6*3000 =$ **1800** L/d

$Q_r = Q_0 - Q_p =$ **1200** L/d

C_r can be obtained from S; $Q_r/Q_0 = 0.4$; therefore, $C_r =$ **2000** mg/L

C_p can be obtained from mass balance; 333.33 mg/L.

PROBLEM 5.10.6

For a feed water with TDS concentration of 2500 mg/L and a flow rate of 2000 m^3/d, design an RO system with the following data: recovery rate = 70 percent, salt rejection rate = 95 percent, flux rate 0.8 m^3/m^2-d and a design pressure of 4000 kN/m^2. Determine membrane area, permeate flow rate and concentration, and power consumption.

Solution

Flow rate = $Q_0 = 2000$ m^3/d

Flux = Q_0/A or $Q_p/A = 0.8$ $m^3/d\text{-}m^2$

Concentration in permeate = $C_p = C_0(1\text{-}S)/R = 178.57$ mg/L

$S = Q_rC_r/Q_0C_0$; $R = Q_p/Q_0$

$Q_p = 1400$ m^3/d

$Q_r = 600$ m^3/d

Concentration in rejectate, $C_r = 7916.7$ mg/L

Based on Q_0, Area 1 = 2500.00 m^2

Based on Q_p, Area 2 = 1750.0 m^2

Taking the larger membrane area, based on Q_0 not Q_p

Power consumption = design pressure * Q_0 = 92.59 kN-m/s = 92.59 kW

PROBLEM 5.10.7

A saline groundwater has NaCl conc of 1000 mg/L. An RO process is to be used and the RO membrane has a water flux rate coefficient, k_w of 1.0×10^{-6} s/m while the solute flux rate coefficient is $k_i = 2 \times 10^{-6}$ s/m. The product water should meet IS standards for potable water and 1 MLD of water is to be supplied to the community. The net operating pressure is 3000 kPa and the recovery rate is 80 percent. Determine membrane area, rejection rate and concentrate of rejectate.

Solution

IS standards for desirable TDS level in potable water = 500 mg/L

Recovery rate = Q_p/Q_0 = 0.8

$Q_p = 0.8*Q_0$ = 1 MLD = 11.57 L/s = 0.0116 m^3/s

$Q_0 = Q_p/R$ = 1.25 MLD

$Q_r = Q_0 - Q_p$ = 0.25 MLD

Membrane area required

Water flux = F_w, kg/m^2-s based on $F_w = k_w$*net pressure drop (or net operating pressure) = 0.003 kg/m^2-s

Area of membranes = Q_p (0.01 m^3/s*1000 kg/m^3)/F_w = 3858.0 m^2

Concentration of rejectate for the given conditions = 3000 g/m^3 or mg/L

Rejection rate = 50 percent

Study Outline

Flow equalization

The objective of flow equalization is to reduce (dampen) fluctuations in volumetric flow and concentrations of contaminants in water or wastewater so as to get consistent plant performance. Flow equalization is essential in water and wastewater treatment plants since flow is never constant and is subject to seasonal and diurnal variations, fluctuations in water demand or generation of wastewater. These tanks help to provide uniform flow throughout the day in water supply systems and also reduce the volume of tanks required, and costs of pumping. Tank volume for water supply systems includes balancing storage, breakdown storage, and fire demand.

Aeration

The transfer of air constituents from gas phase to liquid phase is termed aeration. The objective of aeration is removal of offensive compounds like hydrogen sulphide, carbon dioxide, volatile organic compounds (VOCs), Fe, Mn, and As from water and addition of oxygen to wastewater for removal of BOD under aerobic conditions.

The driving force for aeration is the concentration gradient which is the difference in concentration of a contaminant in the fluid and its saturation concentration under the same conditions. Since temperature affects the saturation concentration of a compound, variations in temperature will also lead to change in mass transfer of the compound. Temperature correction and design for appropriate temperature conditions is necessary.

Different types of aerators such as fountains, tray towers, cascade aerators, diffusers, mechanical mixers, and waterfalls can be used for aeration of water or wastewater depending on site conditions.

Screening

A screen is a device with openings used for removing large objectionable solids from influent water or wastewater.

Screens can be vertical or inclined and they can be cleaned manually or mechanically.

In practice, bar spacing of 20 to 40 mm is commonly used in wastewater treatment plants.

Head loss through the screens can be calculated using Bernoulli's equation for vertical screens and Kirschmer's equation along with Bernoulli's equation for inclined screens.

Type 1 settling

Particles dispersed in any fluid constitute a suspension, in contrast to dissolved compounds which constitute a solution. Suspensions can be either dilute or concentrated. Particles in suspension can be discrete or flocculent. Settling of discrete particles in dilute suspensions is termed Type 1 settling.

Discrete settling and Stokes' law

Newton's law can be derived for discrete particles in dilute suspensions. The three major forces acting on a particle are: gravity, buoyancy, and drag forces. The terminal settling velocity of discrete particles depends on the flow regime: laminar, transitional or turbulent. Stokes' law is a special case of Newton's law and is applicable to spherical particles under laminar flow conditions. Settling column analysis (with a single port) is done to determine the removal efficiency of a dilute suspension of discrete particles of different sizes and is the basis for the design of settling tanks.

Design of settling tanks

Settling tanks can be rectangular or circular and the most important design parameter is the surface overflow rate (SOR). SOR is also the settling velocity of particles that will be removed 100 percent, while those particles with velocities less than SOR will be removed in proportion to their velocities.

Coagulation–flocculation

Coagulation, flocculation, and settling are used in water treatment with the primary objective of turbidity removal. Turbidity is caused by colloidal particles which tend to be stable under normal environmental conditions and is measured in terms of nephelometic turbidity units (NTU).

Coagulation is the *chemical conditioning* of particles. In this process, particles are destabilized and their physico-chemical characteristics are modified so as to allow aggregation and floc formation to occur. Flocculation is the *physical conditioning* of particles. In this process, gentle mixing of the destabilized particle suspension is done to accelerate inter-particle contact, which in turn promotes aggregation and settling of the aggregated particles or 'floc'.

The coagulation–flocculation process facilitates removal of colloidal particles as well as dissolved compounds like natural organic matter (NOM), heavy metals, color-, taste- and odor-causing dissolved compounds, and synthetic organic compounds (SOCs) like pesticides.

Stability of particles

Colloids are particles that are stable since they do not settle under gravity and remain suspended in fluid media. Their stability is attributed to electrostatic stabilization and steric stabilization.

Coagulation theory

There are four major mechanisms by which particles are coagulated: double layer compression, adsorption and charge neutralization, sweep floc formation, and inter-particle bridging.

The Schulze–Hardy rule states that destabilization of colloids by an indifferent electrolyte is due to counter-ions and coagulation effectiveness increases with ion charge.

Rapid mixing of coagulants is essential for achieving uniform dispersion of coagulant species in the particle suspension. Some of the important coagulants used in water treatment include alum (hydrated aluminum sulphate salts), ferric salts like ferric chloride and ferric sulphate, activated silica, and organic polymers.

Flocculation

The objective of flocculation is to enhance inter-particle contact and promote aggregation of the destabilized colloids generated during coagulation. Three physical processes that lead to inter-particle collisions are: perikinetic flocculation or Brownian diffusion, orthokinetic flocculation or fluid shear, and differential settling. G^*t is an important dimensionless design parameter used in flocculator design and is related to the power input required for mixing.

Settling or sedimentation or clarification

The objective of clarification or sedimentation is removal of suspended colloidal particles or floc. The term 'Type 2 settling' is used for the settling of flocculating particles in a dilute suspension. Flocculant particles change shape, size and specific gravity and are unlike discrete particles which have constant shape, size, and specific gravity.

Settling column analysis (with multiple ports) is done to determine the removal efficiency of a dilute suspension of flocculant particles of different sizes and is the basis for the design of clariflocculators.

Softening – chemical

The objective of chemical softening is removal of hardness-causing cations: calcium and magnesium. Hardness in water prevents lather formation with soap, results in scum formation, and makes the cleaning process difficult. Calcium and magnesium salts tend to precipitate at high temperatures forming hard scales on the sides of boilers, on tubes in distillation units or any heating elements associated with water, leading to loss of energy and poor efficiency or even failure of these units.

Lime is added to remove hardness associated with carbonates while soda ash (sodium carbonate) is added to remove non-carbonate hardness. Calcium carbonate and magnesium hydroxide are removed in the form of precipitates thereby removing the hardness causing cations.

Softening – ion exchange

Another method for removing hardness is ion exchange where natural or synthetic zeolites are used for removing Ca and Mg. This method is more efficient and also more expensive than lime-soda ash softening. Since the removal efficiency of ion exchange is generally >90 percent and in excess of what is required for meeting potable water quality standards, split treatment is commonly used. In split treatment, a fraction of the feed water bypasses the ion exchange reactor and is mixed with the treated water to meet drinking water standards.

Filtration

The objective of filtration is the removal of suspended particles and floc that cannot be removed by settling. In conventional water treatment plants, filtration follows coagulation, flocculation and settling. However, if the feed water quality is relatively high and the floc formed during coagulation and flocculation is of small size and low concentration, direct filtration can be used without settling.

There are three general classes of filtration:

a. Granular medium filtration: Filter beds packed with media are used to create porous bed filters through which water is passed. Gravity flow is the most common flow mode in water and wastewater treatment where the weight of the water column above the filter provides the driving force; e.g., slow and rapid sand filters. Within granular medium filtration processes, there are two types of processes:

> ***Depth filtration*:** If solids penetrate into the depth of the porous bed and are removed within pores of the granular media, it is termed depth filtration. Rapid sand filters are the most common type of depth filters.
>
> ***Cake filtration*:** If solids are removed on the entering face of the granular material, forming a layer of fine material (cake) at the surface of the filter, it is termed cake filtration. The most common examples of cake filtration are slow sand filters, and precoat filters (diatomaceous earth filters).

b. Surface filtration: Straining or mechanical sieving through different types of filters is termed surface filtration. Examples include coarse filtration using paper or cloth; laboratory analysis for TSS is done using cellulose, cellulose acetate or glass fiber filters (GFF) while diatoms are used as fine filter media.

c. Membrane filtration: Porous synthetic membranes of different pore sizes are used to remove suspended solids and are covered in some detail in Section 5.10.

Design considerations for granular media filters

Head loss for a clean filter media can be calculated using several equations and the most common one is the Carman–Kozeny equation for a single media filter. The equation is easily modified for multi-media filters. Backwashing velocity can be determined based on the extent to which the bed is fluidized and the terminal settling velocity of the filter media.

Membrane filtration

The objective of membrane processes is to filter suspended solids including colloidal particles and dissolved solids up to a size of 0.1 nm in water or wastewater treatment.

The rate of flux is defined as the rate of flow of permeate through the membrane. In general, pre-treatment of the feed water is required before any membrane process.

Types of membrane processes

Membrane processes are of two types: pressure-driven or electric current-driven. Pressure driven processes include microfiltration (MF), ultrafiltration (UF), nanofiltration (NF), and reverse osmosis (RO). Electric-current driven processes include electrodialysis (ED) and electrodialysis reversal (EDR).

Membrane processes, especially RO, are used mainly for desalination of water while microfiltration and ultrafiltration can be used to remove suspended solids.

Osmosis and reverse osmosis

Osmosis is the diffusion of water through a semi-permeable membrane and the driving force is the concentration gradient of water. In reverse osmosis, pressure greater than the osmotic pressure of the feed water is applied to force water to diffuse through the osmotic membrane against the concentration gradient. The resulting rejection of suspended and dissolved solids is used to purify water. Recovery rates in RO processes range from 45 to 55 percent for seawater reverse osmosis plants.

Van't Hoff law is used to determine the osmotic pressure of a solution with known constituents.

Major issues in membrane filtration are the large amounts of water lost as rejectate or brine, membrane fouling which reduces membrane life and performance and the need for pre-treatment for increasing feed water quality to increase membrane life and performance.

Electrodialysis (ED) and electrodialysis reversal (EDR)

The objective of ED and EDR is removal of salts or TDS from brackish water, i.e., desalination of water. In ED, mineral salts and other ionic species are transported through ion selective membranes from one solution to another under the driving force of a direct current (DC) electrical potential. Polarity reversal helps to increase the life of the membranes by re-suspending NOM and salts that are deposited on the membrane surface. This process is EDR.

Design of electrodialysis units is based on Faraday's law. Power consumption can be calculated based on the current applied and the resistance of the electrodialysis unit.

Design and operation of membrane filters

Membrane filters are usually run as modules in series or parallel. Flow configuration can be dead end (or direct flow) or cross-flow. The latter is more popular since the formation of the cake layer on the membrane surface is less and a longer run time is possible with cross-flow as compared to direct flow.

Another method of categorizing flow configuration is, 'outside-in' versus 'inside-out'. These configurations are based on the input of feed to the central tube supporting the membrane. If the feed enters outside the tube so that permeate is generated inside the central tube, it is termed outside-in. On the other hand, if the feed enters inside the tube supporting the membrane and permeate is generated outside the central tube, then the configuration is termed inside-out.

Two important parameters in membrane filtration are recovery rate which is the ratio of permeate or product water generated to feed water treated, and salt rejection rate which has been defined in many different ways.

Membrane fouling refers to the accumulation of feed water constituents on the membrane, formation of chemical precipitates on the membrane or damage to the membrane due to reactive chemicals. Fouling can be controlled using pre-treatment of feed water, chemical treatment of membranes, and backflushing of membranes.

Major problems in membrane filtration are the high energy requirements, membrane fouling and large amounts of wastewater generated. The wastewater generated cannot be reused for irrigation due to the high concentration of salts.

Study Questions

1. What are the benefits of flow equalization or balancing storage?
2. Describe different methods of aeration with applications.
3. What is the impact of temperature on process design and should one design for lower or higher water temperatures?
4. Why is screening essential in both surface water and wastewater treatment plants?
5. How can screens be cleaned?
6. What is Newton's law and how can Stokes' law be derived from it?
7. Describe the double layer model and how it is used to describe the behavior of stable particles.
8. How does Type 1 settling differ from Type 2 settling?
9. How does a clariflocculator differ from a circular settling tank?
10. Why is surface overflow rate a design parameter for settling tanks, but weir overflow rate is not a design parameter?
11. When is water 'hard' and why is it a problem? What can be done to remove hardness from water?
12. What is granular media filtration? Describe with examples.
13. What is backwashing? When and why is it used?
14. Compare and contrast slow sand and rapid sand filtration.
15. Compare and contrast chemical softening versus ion exchange for softening.

16. What pre-treatment processes are necessary prior to using ion exchange processes?
17. List in order of pore size (largest to smallest), all the pressure-driven membrane filtration processes that can be used in water and wastewater treatment.
18. In reverse osmosis systems, if the salt rejection factor is increased, what is the impact on the volume of water wasted and the operational costs of the process?
19. Name at least four pollutant categories that can be removed by membrane processes.
20. RO membranes can generally reject particles that are greater than 1 nm. Which of the following particles will pass through the membrane and which ones will be rejected: bacteria, virus, proteins, heavy metals, sugar, and salt?

CHAPTER 6

Biological Processes for Water and Wastewater Treatment

Learning Objectives

- *Estimate the quantity and quality of wastewater generated in a community*
- *Analyze variations in wastewater quantities, flow and quality and determine treatment requirements*
- *Explain the advantages and disadvantages of attached versus suspended biological growth processes*
- *Use available models and equations to design attached or suspended growth biological processes*
- *Analyze and design clarifiers for hindered and compression settling (for thickened sludge)*
- *Evaluate the extent and type of sludge management options available and applicable for a given water or wastewater*
- *Compare and contrast anaerobic versus aerobic digestion*
- *Analyze potential for energy recovery from wastewater*
- *Evaluate and design appropriate processes for sludge dewatering and disposal*
- *Describe conditions under which Type 3 and 4 settling takes place and how it can be modeled*
- *Design sludge thickeners for hindered and/or compression settling*
- *Explain the conditions under which anaerobic sludge digestion is an appropriate treatment process, i.e., pre-conditions*
- *List the different types of anaerobic digesters that can be used*
- *Explain the conditions under which aerobic digestion of sludge is more appropriate than anaerobic digestion*
- *Describe the need for sludge dewatering prior to disposal*

- *Describe the need for disinfection in water or wastewater treatment and design the process*
- *Evaluate the merits and demerits of different types of disinfectants*

The main objective of wastewater treatment is to ensure that water resources to which wastewater is returned are not contaminated for downstream users and for future uses. Reasons for treating wastewater can be divided into two categories:

a. *Public health concerns*: Untreated wastewater can cause long-term contamination not only of water but also of soil, sediments and the biota that grows in contact with the contaminated water. Contaminated wastewater may contain pathogenic organisms and serve as a habitat for disease vectors like mosquitoes. Further, contamination of any parts of the food web can lead to bioaccumulation of contaminants in the tissue of organisms leading to various health problems such as cancers, reproductive and developmental defects.

b. *Environmental concerns*: There is growing awareness of the importance of maintaining and preserving ecosystems. Any activities that are detrimental to the health of these ecosystems, including discharge of untreated wastewater, are not acceptable. There are several other reasons for ensuring that wastewater discharged to land, water or sea is adequately treated. These reasons include maintaining adequate water quality for commercial activities such as aquaculture, aesthetic and recreational purposes, and for ensuring that civic pride and property values are maintained or improved.

Wastewater treatment is often neglected in developing countries due to reasons of high cost, lack of uninterrupted power supply, and ignorance about the importance of wastewater treatment.

6.1 WASTEWATER CHARACTERISTICS

Wastewater can be classified as industrial or municipal depending on its source. Only municipal wastewater (sewage) is discussed in this text. Collection of municipal wastewater, including sewage and rainwater, was practiced in several cities of the ancient civilizations, i.e., Mesopotamian, Indus valley, Egyptian and Greek (Lofrano and Brown 2010). The sophistication of the ancient civilizations was followed by the Dark Ages, and it is only in the last 200 years that urban centers have incorporated water and wastewater systems into their planning and design again.

Milestones in the history of sanitation and wastewater treatment (Lofrano and Brown 2010)
3500 to 2500 BC – Mesopotamia; drains from homes and latrines leading to cesspits were found.

Before 2500 BC – A highly sophisticated drainage system with wastewater treatment (perhaps, sedimentation) was found in Mohenjo-daro; toilets and sewers were present in every house and covered drains were present in the middle of streets (Webster 1962).

There is some evidence that flush toilets may have existed in the Indus Valley and Greek civilizations.

Before 2100 BC – City of Herakopolis in Egypt had bathrooms, sewers, and toilets seats made of limestone.

Before 1800 BC – Dholavira which was part of the Harappan civilization had water reservoirs, stormwater drains, septic tanks, cess pits (Agarwal 2009).

600 BC – Drainage structure in ancient Rome called the Cloaca Maxima was built to collect rainwater, sewage, and water from swamps and discharge it into the River Tiber (Delleur 2003).

300 BC to 500 AD – Public toilets, sewers and collection mains for conveying wastewater and stormwater were present in ancient Greece. Wastewater was collected and used for irrigating fields.

1189 AD – Cesspits for collection of wastewater were constructed in London and sludge from them was conveyed to fields for land application (sold to farmers as fertilizer).

1370 AD – A vaulted stonewall sewer was built in a section of Paris (Delleur 2003).

Beginning of the fourteenth century – Recognition that River Thames was extremely polluted.

1596 – Invention of the modern flush toilet by Sir John Harington.

1858 to 1865 – The London sewer system was constructed.

1859 to 1875 – Sewer system constructed in Kolkata, India.

1890 – Sewer system constructed in Chennai, India.

As noted above, wastewater was historically discharged directly over land (and was sometimes used for irrigation) or into surface water bodies like ponds, lakes, rivers or oceans. As populations in urban centers increased, there was a simultaneous increase in the amount of water consumed per person due to increase in living standards. These two factors: increase in population and increase in per capita water consumption led to an increase in the amount of wastewater generated. When large quantities of wastewater are discharged into surface water bodies without treatment, the result is exhaustion of the self-purification capacities of surface water bodies. The consequences are damage to ecosystems, lack of potability of such water supplies and pollution of groundwater resources and soils. Polluted water also results in noxious and unhealthy conditions that serve as a breeding ground for disease-vectors like mosquitoes and flies which spread diseases like malaria, dengue, cholera and typhoid.

Further, industrial and other activities within urban centers result in the addition of contaminants that were not historically present in sewage such as toxic heavy metals, and synthetic organic compounds like detergents and pesticides. These relatively 'new' contaminants have contributed to an increase in treatment requirements since conventional secondary wastewater treatment is not capable of removing them.

The design of conventional municipal wastewater treatment plants is based on the characteristics of the wastewater and the concentrations of its various constituents. Wastewater quality is defined on the basis of different physical, chemical and biological parameters. Physical parameters of importance are total solids [TS], total suspended solids [TSS] and volatile suspended solids [VSS]. Chemical parameters include measures of oxygen demand (indirectly the organic content) like BOD and COD, alkalinity, pH, sulphur, nitrogen, phosphorous, and dissolved gases like dissolved oxygen, carbon dioxide, ammonia and hydrogen sulphide. Biological parameters include microbes such as virus, bacteria, protozoa, fungi and algae including pathogenic microbes and higher organisms.

Discharge standards for different wastewater parameters are prescribed by regulatory agencies like Central Pollution Control Board (CPCB) and can be obtained from their website.

6.1.1 Quantity Generated

The quantity of wastewater generated is expressed in terms of flow rates, which is the quantity of wastewater discharged or treated per day. Its unit can be m^3/day, cumec (m^3/s) or MLD (million liters per day). Similar to the design of water treatment plants, design of wastewater treatment plants is based on estimates of current and future populations that are to be served. Wastewater generation can fluctuate on a daily as well as annual or seasonal basis.

Wastewater collection and treatment systems can be of two types: combined systems where municipal wastewater is combined with stormwater collection systems, and separate wastewater systems where a city has two separate systems for collection of stormwater and wastewater.

The amount of sewage generated in a separate sewer system (ignoring inputs from precipitation) = Water used in the public water supply system + water used by private sources – water losses due to leakage or diversion or consumptive uses like gardening, cooking and drinking – water lost by infiltration into the subsurface.

In combined sewer systems, precipitation is a major input in the total flows to the sewerage system.

Daily Variations

Wastewater quantities reaching the treatment plant vary throughout the day. In general, the amount of wastewater generated at source can vary from 60 percent to more than 100 percent of the water consumed (Metcalf and Eddy 1991). In areas where a significant portion of the water supplied is for consumptive uses like gardening, the amount of wastewater generated as a percentage of water used is relatively low and varies from 50 to 80 percent. In areas with high population density, the consumptive use of water is much less in comparison to non-consumptive uses, and wastewater generated can be assumed to be equal to water used (Henry and Heinke 1996). Due to use of private borewells in residential and industrial areas, stormwater infiltration into sewer systems, and water from roof tops that enters the sewer systems, the amount of wastewater generated can be higher than the water supplied by the municipal authorities (Metcalf and Eddy 1991).

Data for water supply and wastewater generation in Indian cities with more than 1,00,000 population show that water is used at an average rate of 179 L/capita-day and on average 156 L/capita-day is generated as sewage (CPCB 2009). This amounts to 87.3 percent of the water used ending up as wastewater in large cities in India. Maximum water usage was reported for Chandigarh at 540 L/capita-d and was not included in the above average. Per capita water usage ranged from 80 L/capita-d in Tamil Nadu to 310 L/capita-d in Maharashtra.

Wastewater flow generated follows the same trend as water consumed, i.e., it is bimodally distributed. Flow is at its minimum, early in the morning, rises to a maximum during the morning hours from 8 AM to 12 noon, decreases in the afternoon and again increases in the evening as shown in Figure 6.1. In general in tropical climates, the first peak of the day is higher than the second peak. In colder regions, the second peak in the day may well be the higher peak.

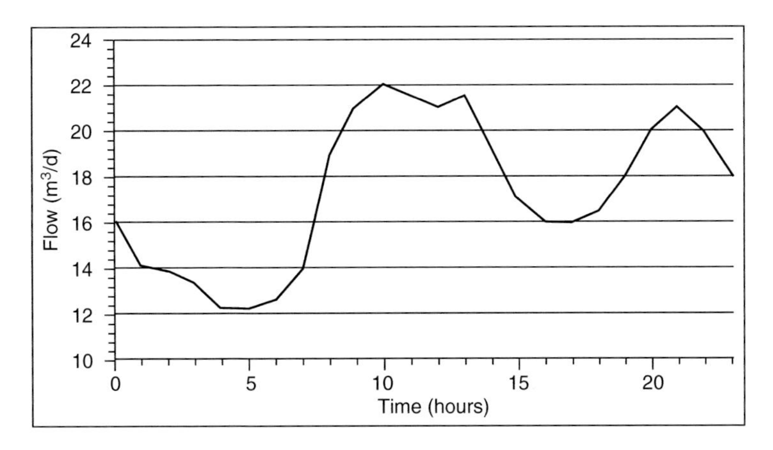

Figure 6.1 Variations in wastewater flow generated or treated during the day.

Seasonal Variations

Flow rates also vary as the seasons change. For example, in a combined collection system, flow rates are higher in the monsoon season as compared to the summer due to additional stormwater flows along with wastewater.

Other than the above flow rates, the following ratios are used as a check in the design of some treatment units and collection channels.

$$\text{Peaking Factor} = \frac{\text{maximum hourly or daily flow}}{\text{average daily flow}}$$ and it ranges from 1.5 to 4.0 (Metcalf and Eddy 2003)

$$\text{Minimum Factor} = \frac{\text{minimum flow}}{\text{average flow}}$$

Ratio of peak flow to average flow can also be estimated by empirical formulae such as Harmon formula (Henry and Heinke 1996):

$$\frac{\text{Peak flow rate}}{\text{Average flow rate}} = 1 + \frac{14}{4 + \sqrt{P}} \qquad 6.1.1$$

Where P = population in thousands.

6.1.2 Quality

Basic water quality parameters that are generally used to characterize municipal wastewater are BOD, COD, TS, TSS, TDS, pH, temperature, turbidity, TKN and phosphate. Average values expected in untreated municipal wastewater are (Masters 1998; Metcalf and Eddy 2003):

- BOD_5 = 100 to 350 mg/L
- COD = 250 to 1000 mg/L
- Solids: total solids = TSS + TDS; 390 to 1230 mg/L
  - ♦ Total suspended solids (TSS) = 100 to 350 mg/L
  - ♦ Total dissolved solids (TDS) = 200 to 1000 mg/L
- TKN: all nitrogen species = 20 to 80 mg/L
- Total phosphorus = 5 to 20 mg/L

Data from 152 sewage treatment plants in India show that the average ± standard deviation values for the treatment plants were: inlet BOD_5 was 110 ± 82.73 mg/L, and inlet COD was 301 ± 212 mg/L (CPCB 2013). Maximum value reported for BOD_5 was 433 mg/L and for COD was 925 mg/L. Average COD/ BOD_5 ratio for all plants (operational or not) was 2.74 ± 2.56 at the inlet and 2.96 ± 1.63 at the outlet. The average COD/ BOD_5 ratio for 73 of the sewage treatment plants that were fully operational was 3.0 ± 1.04 at the inlet and increased to 4.85 ± 3.51 at the plant outlet. These values indicate the increase in non-biodegradable content of the wastewater after biological treatment due to preferential removal of the biodegradable fraction.

Solids present in municipal wastewater can be characterized in several ways as shown in Figure 6.2. As mentioned in Section 3.1.4, total solids (TS) can be divided into two fractions: fixed solids (FS) and volatile solids (VS) after burning total solids at 550 °C. Fixed solids are ash or inorganic materials that do not burn while volatile solids are mainly organic material that is lost after burning at 550 °C. Solids can be further categorized as dissolved or suspended, settleable or non-settleable solids and biodegradable or non-biodegradable as shown in the following figures. Estimated fractions of total solids in each category are shown in Figures 6.2 a and b, and are useful for computing various fractions used in the design of biological processes.

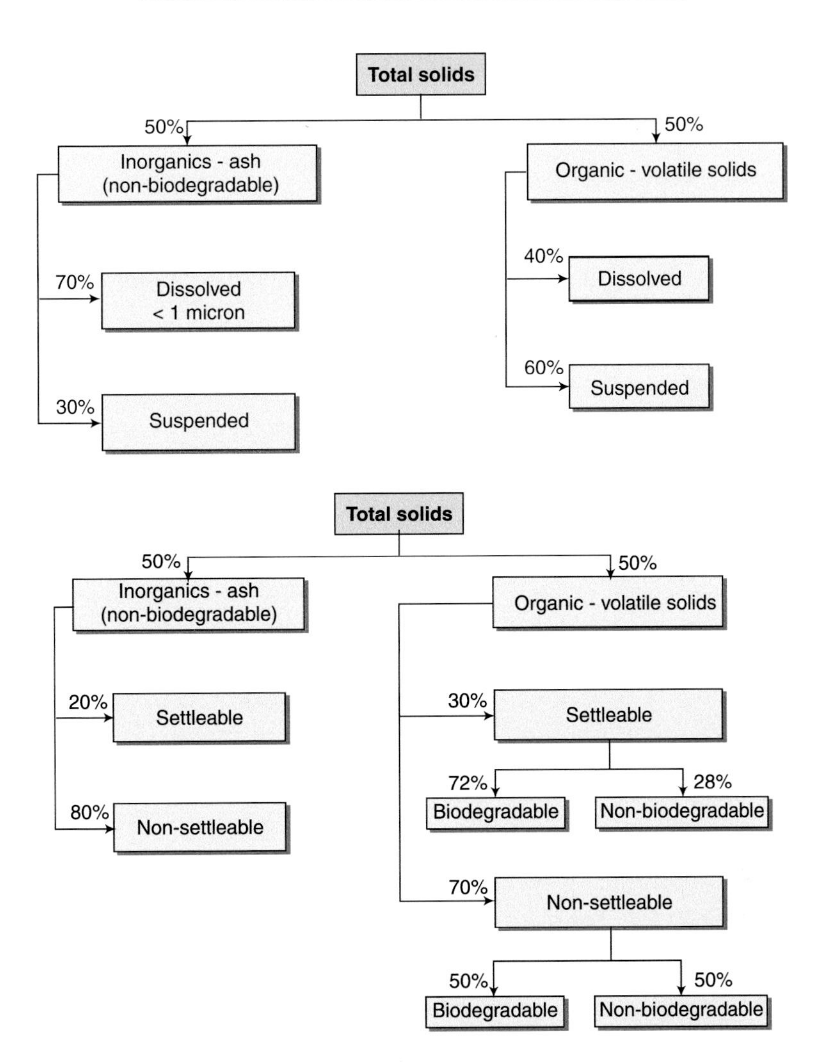

Figure 6.2 Characterization of solids in municipal wastewater, based on Bouwer (1991).

In biological treatment processes, the concentration of total suspended solids (TSS) in the reactor is an important design parameter and is denoted by X with units of mg/L (Bouwer 1991). The different fractions of TSS that are relevant for biological processes are shown below.

$$X = X_{in} + X_v \qquad 6.1.1$$

$$X_v = X_a + X_i + X_d + X_r \qquad 6.1.2$$

where X_{in} = inorganic fraction of TSS

X_v = organic or volatile fraction of TSS, i.e., VSS which is assumed to be mostly biomass

X_a = active microorganisms in VSS

X_i = inert biomass or dead microorganisms in VSS

X_d = biodegradable VSS

X_r = refractory VSS

The biodegradable fraction, f_d of active biomass, X_a is assumed to be 0.8 while the fraction of inert biomass ranges from 0.7 to 0.9.

PROBLEM 6.1.1

A community is supplied 10 MLD of water for a population of 25,000 residents. Determine the amount of wastewater generated at present and what will be generated after 25 years at an annual exponential population growth rate of 3 percent? What peaking factors should be used in the design of wastewater treatment plant facilities?

Solution

It is safest to follow a conservative design approach and assume that the amount of wastewater generated = amount of water used/supplied, i.e., 100 percent of the water supplied ends up as wastewater.

Therefore, wastewater generation at present = 10 MLD = 10 MLD

Population at present, P_0 = 25,000 persons

Based on urban exponential growth, we can use the equation: $P = P_0 * \exp(kt)$ where k = 3 percent or 0.03 and t = 25 years,

Population after 25 years = 52,925 persons

Per capita water usage at present = 10^7/25,000 = 400 L/capita-d

Estimated total water used 25 years from now (assuming per capita usage remains the same) = 21.17 MLD

Estimated wastewater generated 25 years from now = 21.17 MLD

Therefore, average daily flow rate for wastewater treatment plant design = 21.17 MLD

The plant hydraulic capacity is designed for maximum flow rates while treatment processes are designed for a factor of the average daily flow rate (Table 2.8, Chapter 2). A peaking factor of 3 to 4 for plant hydraulic capacity and a factor of 2 for treatment processes is assumed for designing wastewater treatment facilities (Mihelcic and Zimmerman 2010).

PROBLEM 6.1.2

Data from Sewage Treatment Plants for major cities in India were available in a CPCB report (CPCB 2005). The three major sewage parameters for raw sewage are noted in the following table for some cities. Answer the following questions based on the data provided.

	BOD_5 mg/L	COD mg/L	TSS mg/L
Bhilai, Chattisgarh	66	672	180
Chandigarh	227	548	311
Singanapore, Surat	62	601	128
Panipat	955	2187	326
Bangalore	203	567	302

a. If the average BOD_5 in municipal sewage is 100-350 mg/L, what would account for a BOD_5 value of 62 (Surat) and 66 mg/L (Bhilai)?

b. Similarly, what would account for a BOD_5 value of 955 mg/L (Panipat)?

c. If the average COD: BOD_5 ratio is 3, what would account for a ratio > 5?

d. What are the implications of discharging raw sewage with a BOD_5: COD ratio that is close to 1, and at the other extreme when the ratio is approximately 0.1?

Solution

All questions are best answered on the basis of the COD/BOD_5 ratio.

(a) The concentration of BOD_5 in Bhilai and Singanapore is 62 and 66 mg/L, respectively, while the COD/BOD_5 ratios are 10.2 and 9.7, respectively. As compared to the COD values in both locations, the BOD_5 values are extremely low and suggest that the organic material in the wastewater is either toxic or resistant to biodegradation. These values suggest high influx of industrial wastewaters. Both cities are well-known industrial hubs.

(b) The sewage for Panipat shows a very high BOD_5 value but a close to average COD/BOD_5 ratio of 2.29. It is likely that the extremely high values are due to industrial wastewater discharges which are not too different from municipal sewage in terms of their biodegradability.

(c) If the average COD/BOD_5 ratio is 3, then a ratio >5 implies that the organic content of the wastewater is toxic or resistant to biodegradation, i.e., only about 20 percent of the total organic matter is biodegradable.

(d) If BOD_5: COD ratio is close to 1, it implies that all of the organic material in the wastewater is biodegradable. If the ratio is <0.1, it implies that the wastewater organic content is not biodegradable to any significant extent.

PROBLEM 6.1.3

Based on Figure 6.2, determine the fraction of suspended solids and the fraction of biodegradable solids for the typical distribution of total solids shown.

Solution

To determine suspended solids (TSS) in wastewater, the fraction present in inorganic solids and the fraction present in organic solids are added together, i.e., 0.3*0.5 + 0.6*0.5 = 0.45

Therefore, the fraction of suspended solids (TSS) in total solids is 45 percent.

Biodegradable solids in wastewater are part of organic solids only and are not present in inorganic solids. The fraction of biodegradable solids present in volatile solids can be determined from Figure 6.2 as: 0.72*0.3 + 0.5*0.7 = 0.566 or 56.6 percent of VS.

Since volatile solids are 50 percent of TS, the biodegradable fraction of TS is 0.283 or 28.3 percent.

6.2 SECONDARY TREATMENT: SUSPENDED GROWTH PROCESSES

The objective of biological treatment is removal of BOD from wastewater. Biological processes can be categorized based on the state of the biomass. If the biomass is in suspension in the fluid, the process is a suspended growth process while if the biomass is attached to a solid substratum, the process is an attached growth process. Aerobic suspended growth processes are covered in this section. Examples of suspended growth processes include the activated sludge process (ASP), sequencing batch reactors (SBR), aerated lagoons (AL), oxidation ponds (OP), waste stabilization ponds (WSP) and anaerobic digesters (AD).

Activated sludge process, sequencing batch reactors, aerated lagoons, oxidation ponds, and waste stabilization ponds are covered in this section while anaerobic digesters are covered in Section 6.6.1.

6.2.1 Activated Sludge Process

Activated sludge process (ASP) is so named because it involves the production of an active mass of microbes which stabilize waste under aerobic conditions. It is a suspended growth process, i.e., the microorganisms are in suspension. Conventional ASP is a two-stage process with an aeration tank followed by a secondary clarifier. Aerobic microorganisms oxidize biodegradable organic matter (a fraction of the total organic contaminants present in wastewater, measured as ultimate BOD or BODu) in the aeration tank. The biomass generated in the aeration tank is removed by settling in a subsequent clarification tank known as the secondary clarifier.

The biodegradable organic matter is known as the substrate (or food for microbes) and the end products of this process are gases like carbon dioxide, hydrogen sulphide and ammonia along with new biomass as shown in the following reaction:

Biodegradable organics + aerobic microbes → CO_2 + other gases [NH_3, H_2S] + new biomass + H_2O

This biomass is called activated sludge since it comprises living microbes that keep the biodegradation process going. In the aeration tank, contact time is provided for mixing of the influent with suspended biomass solids (microbes) and aeration of the influent as shown in Figure 6.3.

Biomass from the aeration tank is sent to a secondary clarifier where the solids settle to the bottom. The clarified effluent at the top of the clarifier is removed and discharged after disinfection to a receiving water body or over land. A fraction of the solids removed from the bottom of the clarifier is taken for sludge treatment (digestion and dewatering) prior to disposal in a landfill, incinerator or for use as compost. The remaining solids are recycled back to the head of the aeration tank to maintain a constant biomass concentration in the tank.

The mixture of solids resulting from combining the recycled sludge with influent wastewater in the aeration tank is termed MLSS (mixed liquor suspended solids). Biomass solids are commonly measured in terms of VSS (volatile suspended solids) and the concentration of biomass in the aeration tank is measured in terms of MLVSS.

$$\text{MLSS} = \text{MLVSS} + \text{other SS}$$

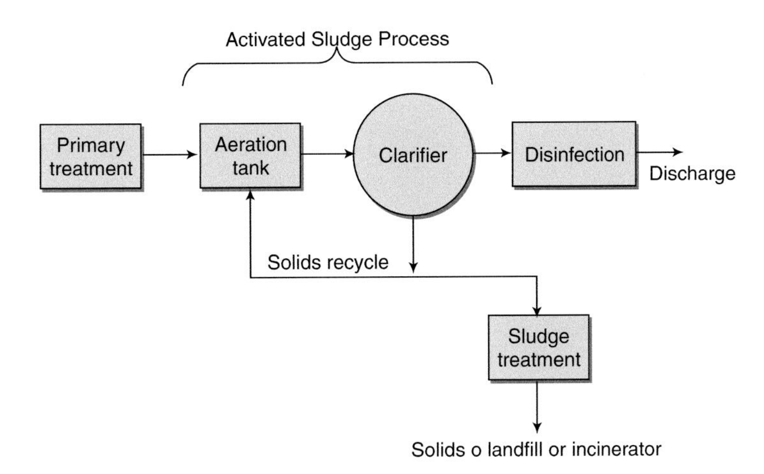

Figure 6.3 Schematic of an activated sludge process (ASP).

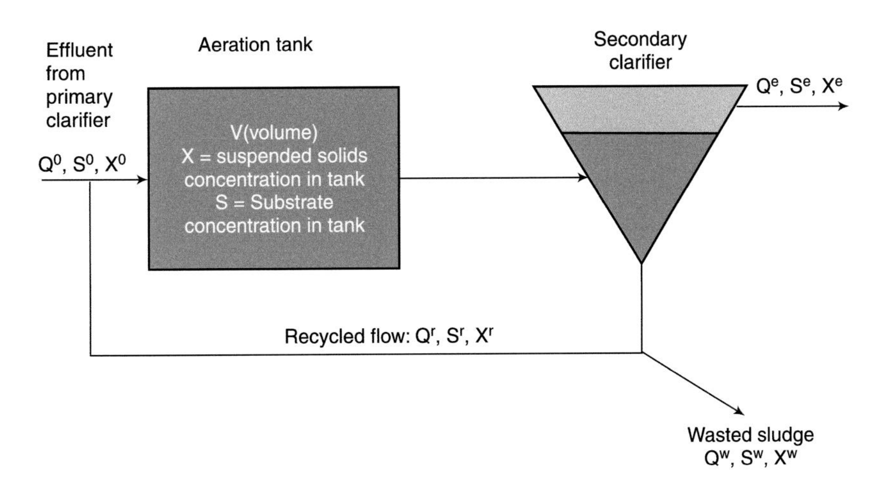

Figure 6.4 Definitional sketch of an ASP process.

Figure 6.4 is a schematic of an ASP for doing a complete mass balance on any of the three variables: flow rate, Q; concentration of TSS (or MLSS), X and concentration of substrate, S. A superscript on Q, X or S denotes the location of the variable and subscripts denote the nature of the variable. Therefore, X^0 implies TSS in the influent, X^e implies TSS concentration in the effluent, X^r is the TSS concentration in the underflow or recycle line, and X^w is the concentration of TSS in the wasted sludge (Bouwer 1991).

For an aeration tank of volume V, a mass balance on active biomass X_a can be written based on Figure 6.4.

Accumulation = Input – Output + net growth of biomass

$$\text{Accumulation} = Q^0 X_a^0 - (Q^e X_a^e + Q^w X_a^w) + V\frac{dX_a}{dt} \qquad 6.2.1$$

At steady-state, accumulation = 0. It is further assumed that X_a^0 is negligible since bacteria in sewage are mostly anaerobes and we assume that no active aerobic microbial cells are present in the influent. Therefore, equation 6.2.1 can be modified as:

$$0 = -(Q^e X_a^e + Q^w X_a^w) + V\frac{dX_a}{dt} \qquad 6.2.2$$

and based on Monod kinetics (described in Chapter 4, Section 4.4)

$$V\frac{dX_a}{dt} = V\left[\frac{YkSX_a}{K_s + S} - bX_a\right] \qquad 6.2.3$$

Substituting 6.2.3 in 6.2.2, we get:

$$VX_a\left[\frac{YkS}{K_s + S} - b\right] = (Q^e X_a^e + Q^w X_a^w) \qquad 6.2.4$$

Hydraulic detention time, $\theta = V/(Q^0 + Q^r)$ 6.2.5

Mean cell retention time (MCRT) is defined as $\theta_c = \dfrac{VX_a}{(Q^e X_a^e + Q^w X_a^w)}$

Therefore, assuming S^e = S since the aeration tank is assumed to be a CSTR, equation 6.2.4 can be modified as

$$\frac{1}{\theta_c} = \left[\frac{YkS}{K_s + S} - b\right] \qquad 6.2.6$$

$$S^e = \frac{K_s(1 + b\theta_c)}{\theta_c(Yk - b) - 1} \qquad 6.2.7$$

Microbes in ASP

Bacteria present in sewage are mostly anaerobic while those used in ASP are aerobic bacteria. Yield (Y) and kinetic rate constants for different bioprocesses are summarized in Table 6.1. The kinetic parameters are k = maximum rate of substrate utilization per unit mass of microbes, b = endogenous decay coefficient, and Ks = half-velocity constant or the substrate concentration at half the maximum growth rate.

Table 6.1 Kinetic rate constants for various biological processes (Bouwer 1991)

Biological process	θ_c^{lim}, days	Y, g cells/g COD	k, g COD/ g cells-d	b, 1/d	Ks, mg COD/L
Aerobic treatment					
Organic removal	0.1 to 0.5	0.45	5 to 20	0.05 to 0.2	25 to 200
Nitrification	1.8	0.3 g cells/ g NH_4-N	2 g NH_4-N/ g cells-d	0.05	0.5 to 2 mg/L NH_4-N
Anaerobic treatment					
Denitrification	0.2	0.35	14	0.05	20
Sulphate reduction	1.1	0.1	9.3	0.03	400
Methane fermentation					
Fats	3.8 to 10	0.031	4 to 8.4	0.03	20 to 2,000
Proteins	3.8 to 10	0.081	4 to 8.4	0.03	20 to 2,000
Carbohydrates	3.8 to 10	0.23	4 to 8.4	0.03	20 to 2,000
Sewage sludge	3.8 to 10	0.081	4 to 8.4	0.03	20 to 2,000

Important operating parameters for ASP and other aerobic suspended growth processes are summarized in Table 6.2.

Table 6.2 Operating parameters for aerobic suspended growth processes, (Metcalf and Eddy 2003)

Process type	MCRT, d	HRT, h	F/M, kg BOD_5/kg MLVSS-d	Volumetric organic loading, kg BOD_5/m^3-d	MLSS, mg/L
Conventional ASP (CSTR)	3 to 15	3 to 5	0.2 to 0.6	0.3 to 1.6	1,500 to 4,000
Extended aeration	20 to 40	20 to 30	0.04 to 0.10	0.1 to 0.3	2,000 to 5,000
Oxidation ditch	15 to 30	15 to 30	0.04 to 0.1	0.1 to 0.3	3,000 to 5,000
Sequencing batch reactor	None	15 to 40	0.04 to 0.1	0.1 to 0.3	1,500 to 5,000
High-rate aeration	0.5 to 2	1.5 to 3	1.5 to 2.0	1.2 to 2.4	200 to 1,000

ASP processes can be designed with a single aeration tank or a series of aeration tanks. If only one aeration tank is used, it can be modeled as a CSTR. If a series of aeration tanks is used, the process approximates a PFR in terms of its performance. In all cases, a portion of the sludge from the clarifier is recycled back to the aeration tank to ensure that an adequate biomass concentration is maintained for biodegradation of the substrate. The remaining or excess sludge is sent to an anaerobic digester for further degradation and volume reduction. BOD removals >90 percent have been achieved in several ASP-based wastewater treatment plants in India (CPCB 2013).

Sludge from secondary clarifiers has about 5 percent solids and a range of 3 to 8 percent solids.

Design procedure for ASP

A detailed schematic for an ASP is shown in **Figure 6.4.** Steps in the design process are as follows and their use is illustrated in Problem 6.2.1:

1. Determine mean cell detention time, θ_c^d in days
2. Determine the effluent concentration, S^e in mg BODu/L
3. Determine aeration tank volume, m^3
4. Determine quantity of sludge produced in kg VS/day
5. Determine quantity of air required in kg O_2/day

PROBLEM 6.2.1

Design a CSTR-type ASP for a flow of 10 MLD with a single aeration tank and a single clarifier. Influent BOD_5 to the treatment plant is 300 mg/L and effluent from the primary clarifier has 200 mg/L BOD_5. The secondary treated wastewater should have less than 10 mg/L BOD_5 in the effluent.

Solution

For a flow rate, Q = 10^7 L/d = 10^4 m^3/d

Influent to aeration tank is the effluent from the primary clarifier and ultimate BOD which is BODu = 200/0.68 = 294 mg/L BODu. Use 300 mg/L of BODu for design.

The following assumptions are made for design:

The aeration tank is completely mixed since it is a CSTR-type ASP.

Y = 0.5 g bacteria/g BODu,

Cell decay constant, b = 0.05 1/d,

k = 20 g BODu/g bacteria-day, and

K_s = 50 mg/L BOD_u

MLSS concentration in the aeration tank = 3000 mg/L with a ratio of MLVSS: MLSS of 0.8

Therefore, MLVSS (assumed to be bacteria), X_v = 2400 mg/L

For design purposes: X_i, X_r and X_d in the influent are 50, 200 and 300 mg/L, respectively.

Return sludge or sludge underflow concentration from the secondary clarifier = 10,000 mg/L of SS.

1. Determine design mean cell detention time, θ_c^d

$$\theta_c^{\lim} = (Yk - b)^{-1} = (0.5*20 - 0.05)^{-1} = 0.100 \text{ days}$$

Using a safety factor of 70, $\theta_c^d = 70* \theta_c^{\lim} = 7$ days. If this is not sufficient, the safety factor can be increased to 100 or more.
Acceptable range for θ_c^d = 3 to 15 days. Therefore, 7 days is OK

2. Determine the effluent concentration, S^e in mg BODu/L

$$S^e = \frac{Ks\,(1 + b\theta_c^d)}{\theta_c^d\,(Yk - b) - 1} = \frac{50\,(1 + 0.05*7)}{7\,(0.5*20 - 0.05) - 1} = 0.98 \text{ mg BODu/L}$$

which is less than the design objective and therefore, OK.

3. Determine hydraulic detention time[59]

$$\theta = \frac{\theta_c^d}{X_v}\left[X_r^0 + \frac{S^e}{S^0} X_d^0 + \frac{Y(S^0 - S^e)}{1 + b\theta_c^d}(1 + (1 - f_d)\,b\theta_c^d \right]$$

$$\theta = \frac{7}{2500}\left[200 + \frac{0.98}{300} 300 + \frac{0.5(300 - 0.98)}{1 + 0.05*7}(1 + (1 - 0.8)0.05*7 \right] = 0.895 \text{ days}$$

$= 21.5$ hours

4. Determine aeration tank volume, m^3

Volume of tank = $Q*\theta = 0.895*10^4 = 8946$ m^3 which can be rounded off to 9000 m^3.

5. Determine quantity of sludge produced in kg VSS/day and the different fractions of sludge solids

$$\text{Sludge produced} = \frac{VX_v}{\theta_c^d} = 3067.2 \text{ kg VSS/d}$$

X_r = refractory SS in sludge remain the same as in influent = 200 mg/L

$$X_d = \text{biodegradable SS} = \left[\frac{S^e}{S^0} X_d^0 \right] = 0.98 \text{ mg/L}$$

$$X_a = \text{active microorganisms in TSS} = \left[\frac{Y(S^0 - S^e)}{1 + b\theta_c^d} \right] = 110.74 \text{ mg/L}$$

$$X_i = \text{inert biomass or dead microorganisms} = \left[\frac{Y(S^0 - S^e)}{1 + b\theta_c^d}(1 - f_d)\,b\theta_c^d \right] = 7.75 \text{ mg/L}$$

$X_v = X_a + X_i + X_d + X_r = 200 + 0.98 + 110.74 + 7.75 = 319.47$ mg/L

59 A detailed derivation for this equation can be found in Rittmann and McCarty (2012).

Refractory portion of solids produced = $(X_r + X_i)/(X_v)$ = (200 + 7.75)/319.47 = 0.65 or 65 percent.

6. Determine quantity of air required in kg O_2/day

BOD removed per day = $(S^0 - S^e)*Q = (300 - 0.98)*10^7 = 2.99*10^9$ mg/d = 2990 kg O_2/d

Air contains 23.15 percent oxygen by weight. Therefore, air supply needed per day = 12,916 kg/d.

Extended aeration process: These systems are modifications of ASPs where the cells are maintained in endogenous phase. The advantage of doing so is that most of the sludge is mineralized. This minimizes sludge volume and further treatment and handling requirements. However, the hydraulic detention time is much greater in extended aeration as compared to a conventional ASP and based on STP data, the efficiency may not be as high as an ASP (CPCB 2013) (Metcalf and Eddy 2003).

Extended aeration is often used for small treatment facilities (up to 7.5 MLD) where land is readily available. Power requirements may be higher or lower than conventional ASP. It is ideal for small flow, modular applications that require low maintenance such as residential subdivisions. It is also used for larger treatment plants in the form of oxidation ditches.

Other modifications of the conventional ASP include aerated lagoons, waste stabilization ponds and oxidation ditches. BOD removals >90 percent have been achieved with these processes in several wastewater treatment plants (CPCB 2013).

PROBLEM 6.2.2

Design a conventional ASP having an inlet BOD_u of 300 mg/l and average flow rate of 10 MLD which has to be treated so that effluent BOD_u should be 10 mg/L. Use the same assumptions as in Problem 6.2.1.

Solution

The ASP can be designed on the basis of the required BODu in the effluent instead of a safety factor (SF) as in the previous problem.

The following assumptions have been made for design purposes:

MLSS in aeration tank = 3000 mg/l and MLVSS/MLSS = 0.8

Mean cell residence time θ_c = 7 days [Between 3 to 15 days]

Y (yield coefficient) = 0.50 kg cells (VSS)/ kg substrate (BODu) consumed, b = 0.05 1/d

$$BOD_5 = BODu*0.68$$

BODu in influent, S^0 = 300 mg/L

BODu required in effluent, S^e = 10 mg/L

1. Determine treatment efficiency based of BOD_5 removal

$$\eta = (BOD_i - BOD_f)*100 / BOD_i = (300-10) * 100 / 300 = 96.7 \text{ percent}$$

2. Determine reactor volume, V

$$X_v * V = \frac{\text{Design average flow rate } (Q) * \theta_c * Y * (S^0 - S^e)}{(1 + b * \theta_c)}$$

$$X_v V = 10^7 * 7 * 0.5 * (300 - 10) / (1 + 0.05 * 7] = 7518.5 * 10^6 = 7519 * 10^3$$

If X_v which is the MLVSS concentration = 0.8*3000 then the volume of aeration tank is

$$V = 3133 \text{ m}^3$$

3. Determine Hydraulic Retention time, HRT

$$HRT = V/Q = 3133 * 24 / 10^4 = 7.52 \text{ hours.}$$

4. Calculation of Food to Microorganism ratio, F/M

$$\text{F/M ratio} = Q * S^0 / V * X_v$$

$$= 10^4 * 300 / 3133 * 0.8 * 3000$$

$$= 0.4 \text{ mg BOD/mg MLVSS-d}$$

Since range is 0.2 to 0.6 kg BOD/ kg VSS-day, hence OK.

5. Calculation for volumetric organic loading

Volumetric organic loading = $Q * S^0 / V = 10^4 * 300 * 10^{-3} / 3133$

$= 0.96$ kg BOD/m^3-day

Range: 0.3 to 1.6 kg BOD/m^3-day; Hence OK.

6. Sludge produced

$$Y' = Y/(1 + b * \theta_c) = 0.5/(1 + 0.05 * 7) = 0.37 \text{ kg VSS/kg BOD removal}$$

Sludge produced (kg VSS/day) = $Y' * Q * (S^0 - S^e) = 0.37 * 10^7 * (300 - 10) = 1073$ kg VSS/ day

If MLVSS/MLSS = 0.8, then MLSS produced is 1341.25 kg TSS/day.

7. Determine volume of sludge to be wasted, Q_w

Assume that VSS in influent and effluent are insignificant in concentration, and return sludge (or sludge underflow from clarifier) has a concentration of 10,000 mg/L.

Since $\theta_c = V/Q^w$, we can solve for $Q^w = 3133/7 = 447.6\ m^3/day$

$$Qin = Qout = Q^e + Q^w = 10{,}000\ m^3/day$$

Therefore, $Q^e = 9552.4\ m^3/day$

8. Recirculation ratio, R

$$R = Q^r/Q$$

Based on a mass balance around the clarifier underflow:

$$(Qin + Q^r)*X_v = Q^r X^r$$

$$0.8*3000\ (Q + Q^r) = 10{,}000*\ Q^r$$

Solving for Q^r, we get $Q^r = 3158$ m/d and R = 0.316

This is within the typical range of R values of 0.25 to 0.5 and is acceptable.

9. Determine quantity of air required in kg O_2/day

BOD removed per day = $(S^0 - S^e)*Q = (300 - 10)*10^7 = 2.9*10^9$ mg/d = 2900 kg O_2/d

Air contains 23.15 percent oxygen by weight. Therefore, air supply needed per day = 12527 kg/d.

Assuming 8 percent oxygen transfer efficiency and factor of safety of 2, air required = 12527*2 /0.08 = 43316.5 m^3/day.

PROBLEM 6.2.3

A conventional activated sludge process has the following characteristics:

Flow rate = 18000 m^3/d, Influent BOD_5 = 150 mg/L, Effluent BOD_5 = 15 mg/L, Cell coefficient, Y = 0.55 kg biomass/kg BOD_5 utilized, Endogenous decay coefficient, b = 0.04 1/d, aeration tank volume = 6000 m^3, mean cell residence time = 8 days

a. Find the average biomass concentration in the aeration tank and the food/microorganism ratio for the system,
b. If the system has a recycle ratio of 0.4, determine the underflow solids (MLSS) concentration and the mass and volume of solids that must be wasted each day.

Solution

Flow rate, Q = 1.80 x 10^4 m^3/d

Influent BOD_5 = 150 mg/L

Effluent BOD_5 = 15 mg/L

Cell yield, Y = 0.55 kg biomass/kg BOD_5 utilized

Mean cell residence time, θ_c = 8.0 d

$b = 0.04\ d^{-1}$

Aeration tank volume, V = 6000 m^3

Hydraulic detention time, θ = 6000/18000 = 0.33 day = 8 hours

Average biomass concentration in tank

$X = \theta_c *Y*(S^0 - S)/(\theta*(1 + b* \theta_c)) = 8*0.55*(150 - 15)/(0.33*(1 + 0.04*8))$

$= 1363.6\ mg/L = 1.36\ kg/m^3$

F/M ratio = $Q*S^0/V*X_v$ = 18000*150/(6000*1363.6) = 0.33

Mass of solids wasted = $Q^w*X^u = VX/\theta_c$ = 6000*1.36/8 = 1020 kg/d

For Q^r/Q = 0.4 and $Q^r = Q*X - Q^w*X^u/(X^u - X)$

Solving for X^u results in X^u = 4631 mg/L

Volume of solids wasted, Q^w = 220 m^3/d

PROBLEM 6.2.4

Under what conditions can extended aeration be used in an ASP?

Solution

Extended aeration is a type of ASP for treating either raw wastewater or primary treated effluent (after screening and grit removal, generally without primary settling). Very long hydraulic detention times are provided, ranging from 12 to 36 h to incorporate endogenous phase of bacterial growth. This results in less sludge generation.

6.2.2 Sequential Batch Reactors (SBR)

Unlike an ASP which is a continuous flow system, SBRs are run in batch mode with complete mixing of the contents of the reactor. A typical operating sequence for SBRs includes the following steps (Metcalf and Eddy 1991)

1. Fill tank (25 percent of one cycle time)
2. Aerate: Allow for reaction time; provide air or O_2 (35 percent of one cycle time)
3. Allow settling of biomass and its removal (20 percent of one cycle time)
4. Decant clarified effluent (15 percent of one cycle time)
5. Idle time where sludge is wasted while maintaining a requisite amount for the next cycle (5 percent of one cycle time).

SBRs can be run in parallel for a pseudo-continuous operating mode, where one reactor is filled while another reactor is in the last stage of its cycle (Karia and Christian 2006). Very high BOD removals (98.3 percent) were achieved with SBRs in one municipal wastewater treatment (CPCB 2013). In this plant, 4 SBR units with a total capacity of 6875 m^3 and 220 mechanical aerators of 3HP each were used. The first two steps (fill and aerate) lasted for 108 min, next two steps (settling and sludge removal) were done for 50 min, clarified effluent was decanted for 58 min resulting in a total time for one cycle of 216 min.

Higher BOD removal efficiencies can be achieved with SBRs compared to CSTR processes like ASP, since batch processes are inherently more efficient than CSTRs as described in Chapter 4.

6.2.3 Aerated Lagoons

Aerated lagoons are similar to extended aeration ASPs with an important difference. These are continuous flow-through systems which can be operated without sludge recycle. Oxygen or air is provided by surface or diffused aerators. The mean cell residence times provided in these lagoons is very high compared to ASPs and can range from 3 to 6 days without recycle and 10 to 30 days with recycle for municipal wastewater (Karia and Christian 2006).

Depth of the lagoons range from 2 to 5 m while hydraulic detention times range from 2 to 10 days for flow-through systems, 3-20 days for facultative lagoons and 0.7 to 2 days for extended aeration type systems (Karia and Christian 2006).

6.3 SECONDARY TREATMENT: FIXED FILM PROCESSES

In fixed-film or attached growth processes, microorganisms that degrade the influent organic matter are attached to an inert medium or substratum such as plastic or rock forming a biofilm (layer of microbial cells). Compared to activated sludge processes, these processes require less energy, are easier to control and have better sludge thickening properties. Examples of attached growth processes include trickling filters, biofilters, rotating biological contactors, fluidized bed reactors, and roughing filters. Some of these processes are described here. In all cases, a secondary clarifier follows the main unit for removal of biomass generated in the wastewater.

6.3.1 Trickling Filters (TF)

Trickling filters are attached growth aerobic biological processes where the media is not submerged in water. The reactor is packed with plastic or rock materials and wastewater is distributed over the packing continuously using a rotary distributor. Large void spaces in the reactor allow air to circulate freely and naturally (natural draft) or air is provided mechanically (forced draught). Bed diameters can be as large as 60 m. Bed depths with rocks are generally 3 m while they can be much greater with plastic resulting in tall structures called biological towers or biotowers.

Other components of the filter are an underdrain system and a containing structure. The underdrain system provides circulation of air and is designed for collection of trickling filter effluent. The effluent contains VSS due to sloughing of the biomass which has to be removed in a secondary clarifier in the form of sludge. A portion of the effluent is sent back to the feed flow for dilution, maintaining adequate biomass in the influent and for keeping the biological slimy layer moist. Recycling of the treated wastewater (effluent) enables higher efficiency by providing greater biomass for BOD removal.

Major advantages of trickling filters in comparison to ASP are higher removal efficiency and less land area requirement. There are several problems associated with the operation and maintenance of trickling filters such as washing out of the biomass in monsoons and drying out in summer; extreme sensitivity to fluctuations in quantity and quality of influent; and inability to withstand toxic or shock loads.

Biological growth in trickling filters includes a biofilm formed with bacteria, algae, fungi, and protozoa. Macro fauna are found in filters in the form of worms, snails, filter flies and insect larvae and their removal is a major operational problem. Several methods have been tried for addressing these issues and are described by Daigger and Boltz (2011).

A schematic explaining mass transfer of contaminants from the bulk liquid into the biofilm is shown in Figure 6.5. The biofilm consists of facultative bacteria, fungi, algae and protozoa. With the growth of microorganisms, the slime layer thickness increases. Aerobic microorganisms in the outer portion of the biofilm degrade the substrate and consume oxygen present in the bulk fluid during the degradation process. As the biofilm grows and thickens (becomes deeper), substrate is unable to penetrate through the entire depth of the biofilm. In the layer at the bottom, where no substrate or oxygen is available, bacteria undergo endogenous respiration. This results in loss of the slime layer and bacteria lose their ability to cling to the surface (substratum). The biofilm is then washed away in a process called sloughing and a fresh slime layer starts to grow on the substratum.

Important considerations for designing trickling filters are the nature of the filter media, hydraulic and organic loading rates, depth of the filter and the nature of the effluent.

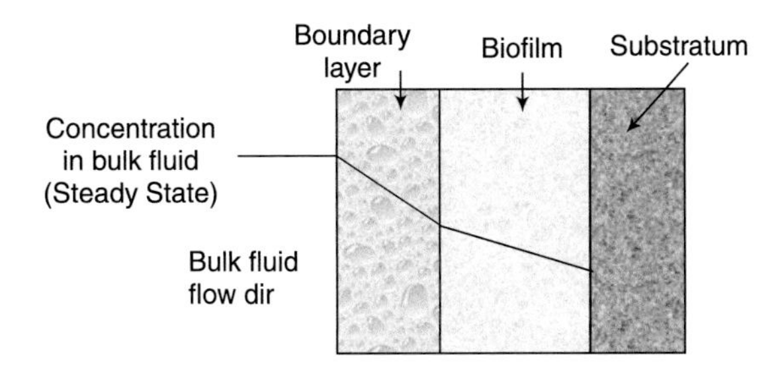

Figure 6.5 Mass transfer of contaminants or oxygen in an attached growth process.

Filter Media: The most common trickling filter media are rocks and slag while plastic media (cross-flow or vertical flow) are more modern and have been popular for a long time now. Ideally, the media should have high specific surface area (m^2 of media surface/m^3 of filter volume), and high durability. It should also be cheap and should have high porosity to prevent clogging and ensure air circulation. The properties of the filter media are summarized in Table 6.3. Rock material with crevices are not preferred since water can collect in the crevices and the daily cycle of heating and cooling results in cracking of the material and its release into the treated effluent. Smooth river rocks are preferred if they are locally available. Rock media filters are shallower in depth (1 to 6 m) compared to plastic media filters (>12 m) (Daigger and Boltz 2011).

Table 6.3 Properties of common trickling filter media (Daigger and Boltz 2011)

Media type	Nominal size, mm	Bulk density, kg/m^3	Specific surface area, m^2/m^3	Void space, percent
River rock	24 to 76	1400	62	50
Slag	76 to 128	1600	46	60
Plastic – cross flow	610 x 610 x 1220	24 to 45	100 and 223	95
Plastic – vertical flow	610 x 610 x 1220	24 to 45	102 and 131	95
Random	185 diameter and 51 height	27	98	95

***Dosing Rate*:** It is defined as the volume of liquid discharged per unit area of the filter surface in every pass of the distributor with units of m^3/m^2-pass or m/pass. Higher dosing rates are known to give better flushing of solids, thinner biofilms and less odor. They are achieved by lowering the distributor speed. Operational and flushing dosing rates for different hydraulic and organic loading rates are given by Daigger and Boltz (2011).

***Distribution Systems*:** A distributor has at least two arms mounted on a pivot in the center of the filter, which revolves in a horizontal plane. There are nozzles in the arms through which influent is discharged over the filter bed. The distribution system should be easy to clean, able to handle flowrate fluctuations and be corrosion-resistant.

The rotational speed of a distributor, n in rpm is given by (Metcalf and Eddy 2003):

$$n = \frac{(1+R)(Q)\left(\frac{1000\text{ mm}}{\text{m}}\right)}{(A)\,(DR)\left(\frac{60\text{ min}}{\text{h}}\right)} \qquad 6.3.1$$

Where Q = influent applied hydraulic loading rate, m^3/m^2-h, R = recycle ratio, A = number of arms, DR = dosing rate, mm/pass.

***Underdrains*:** They convey the filtered wastewater and solids to the sedimentation tank. Blocks of vitrified clay or fibreglass grating are laid on a reinforced concrete subfloor. Therefore, the underdrains must be able to support the packing, slime growth and influent wastewater.

***Airflow*:** Proper airflow is necessary to provide efficient treatment and to prevent odors. In natural draft, the airflow is driven by the temperature difference between ambient air and the air inside the pores or void spaces.

The pressure head due to temperature difference, called draft, is given by

$$D_{\text{air}} = 353\left(\frac{1}{T_c} - \frac{1}{T_h}\right)Z \qquad 6.3.2$$

where

D_{air} = natural air draft, mm of water, Z = height of filter, m

T_c = cold temperature, K and T_h = hot temperature, K.

The average pore air temperature is given by

$$T_m = \frac{T_2 - T_1}{\ln\left(\frac{T_2}{T_1}\right)} \qquad 6.3.3$$

where

T_2 = colder temperature, K and

T_1 = warmer temperature, K

Assuming oxygen transfer efficiency of 5 percent, the required oxygen is given by

$$R_o = 20\left[0.80e^{-9L_B} + 1.2e^{-0.17L_B}\right](PF),$$ for BOD removal only

$$R_o = 40\left[0.80e^{-9L_B} + 1.2e^{-0.17L_B} + 4.6N_{ox}/BOD\right](PF),$$ for BOD removal and nitrification

Where, R_o = oxygen supply, kg O_2/kg BOD applied,

L_B = BOD loading to filter, kg BOD/m^3.d,

N_{ox}/BOD = ratio of influent nitrogen oxidized to influent BOD,

PF = peaking factor, maximum to average load.

The air application rate at 20 °C and 1.0 atm is given by

$$AR_{20} = \frac{3.58R_oQS_o}{1000\text{g/kg} \times 1440\text{ min/d}} \qquad 6.3.4$$

where

AR_{20} = airflow rate at 20 °C and 1.0 atm, m^3/min, Q =wastewater flowrate, m^3/d

S_0 = primary effluent BOD, g/m^3.

The airflow rate is corrected for temperature and pressure according to

$$AR_{T_A} = AR_{20}\left(\frac{273.15 + T_A}{273.15}\right)\left(\frac{760}{P_o}\right) \qquad 6.3.5$$

Where, AR_{TA} = airflow rate at ambient air temperature, T_A = ambient air temperature, °C and

P_o = pressure at treatment plant site, mm Hg.

For each degree Celsius above 20 °C, the airflow rate is corrected by 1 percent as

$$AR_{T>20°C} = AR_T\left(1 + \frac{T_A - 20}{100}\right) \qquad 6.3.6$$

The pressure drop through the packing is given by

$$\Delta P = N_p\left(\frac{v^2}{2g}\right) \qquad 6.3.7$$

where

N_p = tower resistance as head loss relating to airflow,

v = superficial air velocity

$$N_p = 10.33De^{(1.36\times10^{-5})\left(\frac{L}{A}\right)} \qquad 6.3.8$$

where

D = packing depth, m,

L = liquid loading rate, kg/h,

A = tower cross-sectional area, m^2.

Settling Tanks

They produce a clarified effluent, which has lower suspended solids content than ASP settling tanks. All the sludge from these tanks is sent for processing or returned to primary clarifiers.

Types of Trickling Filters

Based on hydraulic or organic loading rates, trickling filters have been classified as low-rate, intermediate-rate, high-rate and roughing filters. Low-rate TF can be used for BOD removal and nitrification while the remaining are used for BOD removal only. A comparison of the process parameters for each of these filters is provided in Table 6.4.

Table 6.4 Process parameters for different types of trickling filters (Metcalf and Eddy 2003)

Type of TF	Low or standard rate	Intermediate rate	High rate	High rate	Roughing
Media	Rock	Rock	Rock	Plastic	Rock/plastic
Hydraulic loading, m^3/m^2-d	1 to 4	4 to 10	10 to 40	10 to 75	40 to 200

Organic loading, kg BOD/m^3-d	0.07 to 0.22	0.24 to 0.48	0.4 to 2.4	0.6 to 3.2	>1.5
Recirculation ratio, Q^r/Q^0	0	0 to 1	1 to 2	1 to 2	0 to 2
Depth, m	1.8 to 2.4	1.8 to 2.4	1.8 to 2.4	3 to 12.2	0.9 to 6
BOD removal efficiency, percent	80 to 90	50 to 80	50 to 90	60 to 90	40 to 70

Low-Rate Filters

They are simple devices which use rock materials as packing and produce effluents of reasonable quality for different influent strengths. Nitrification is possible along with carbonaceous BOD removal.

Intermediate-Rate and High-Rate Filters

They are usually continuous-flow circular filters and use rock or plastic packing. In these filters, recirculation of the effluent takes place. This facilitates higher organic loading, gives better slime layer thickness and less odor problems.

Roughing Filters

These filters are high-rate type filters which can have organic loading >1.5 kg/m^3-d and hydraulic loading up to 200 m^3/m^2-d (Metcalf and Eddy 2003). They generally use plastic packing and consume less energy for BOD removal as compared to ASP.

Several models are available for design of trickling filters and the most common ones are the NRC model published in 1946 and the Eckenfelder model published in 1961 (Metcalf and Eddy 2003). Only the NRC and Eckenfelder's models are described here.

NRC equation for 1st-stage TF (1946)

$$E_1 = \frac{100}{1 + 0.4432\sqrt{\dfrac{W1}{VF}}} \qquad 6.3.9$$

For the 2nd-stage TF, removal efficiency is determined by

$$E_2 = \frac{100}{1 + \dfrac{0.4432}{1 - E_2}\sqrt{\dfrac{W2}{VF}}} \qquad 6.3.10$$

where

E_1 and E_2 = BOD removal efficiency in 1st-stage TF and 2nd-stage TF, respectively with recirculation at 20 °C, percent

W_1 and W_2 = BOD loading to 1st-stage and 2nd-stage filter, kg/d

V = volume of filter packing, m^3

F = recirculation factor = $\dfrac{1+R}{\left(1+\dfrac{R}{10}\right)^2}$ where R = recirculation ratio

Eckenfelder Model (1961)

Eckenfelder's model has been modified by several others since it was first published (Metcalf and Eddy 2003).

Eckenfelder's equation: $S^e/S^0 = \exp(-kD/q^n)$ where

D = depth of filter media, m

q = hydraulic loading rate of wastewater, L/m^2-s

n = coefficient depending on media characteristics, value of 0.5 is used in general.

Treatability constant, k varies for different types of wastewater, concentrations and temperature and type of filter media. It should be determined in pilot-plant studies for a specific wastewater and filter media. The k value has to be corrected for temperature and scaling up for different depths and influent BOD_5 concentrations as shown in equations 6.3.11 and 6.3.12.

Temperature correction for $k_T = k_{20} * \theta^{(T-20)}$ where $\theta = 1.035$ 6.3.11

$$k_2 = k_1 * (D_1/D_2)^{0.5} * (S_1/S_2)^{0.5} \quad 6.3.12$$

Where D_1, S_1 and k_1, are applicable for standard conditions of 20 °C, 6.1 m height of trickling filter, and influent BOD concentration of 150 mg/L, respectively. D_2, S_2 and k_2 values are applicable for site-specific conditions. The k_1 value for domestic wastewater is 0.21 $(L/s)^{0.5}/m^2$ and k = 0.06, for plastic media.

PROBLEM 6.3.1

Design a trickling filter for a flow of 1 MLD. The influent BOD_5 is 150 mg/L and the effluent BOD_5 required is 10 mg/L. Use Eckenfelder's equation. Assume a treatability constant, k_{20} for plastic of 0.21 $(L/s)^{0.5}/m^2$, hydraulic loading rate of 10 m^3/m^2-d and filter media constant, n as 0.5. Design for a minimum temperature of 10 °C.

Solution

Using Eckenfelder's equation: $S^e/S^0 = \exp(-kD/q^n)$

Flow rate, Q = 1 MLD = 11.574 L/s

Temperature correction for 10 °C, $k_{10} = k_{20}*\theta^{(T-20)} = 0.21*(1.035)^{\wedge}(10\text{-}20) = 0.149$

Since k_{10} is for a 20 ft tower or 6.1 m height of filter, solve for q

$$Ln(S^e/S^0) = -k_{10}*D/q^n$$

$$Ln(10/150) = -0.149*6.1/q^{0.5}$$

Hydraulic loading rate (HLR), q = 0.113 L/m^2–s = 9.73 m^3/m^2–d

Surface area of filter, A = Q/ HLR = 102.8 m^2

Area of one filter with 10 m diameter = 78.54 m^2

Number of filters to be provided for a total area of 103 m^2, and diameter of 10 m, i.e., 1.3 or 2 filters and an additional filter is to be provided as standby capacity. Therefore, at least 3 filters are to be provided.

PROBLEM 6.3.2

A municipal wastewater with an average flow of 10 MLD and a BOD_5 of 200 mg/L is to be treated in a bioreactor with plastic modular medium. The treatability constant for the system is 0.21 $(L/s)^{0.5}/m^2$ at 20 °C. The maximum temperature expected is 30 °C and minimum temperature expected is 5 °C. For a recycle ratio of 2:1 and 7 m depth of bed, determine the area of the tower required to produce 10 mg/L of BOD_5 in the effluent.

Solution

Flow rate, Q = 10,000 m^3/d = 6.94 m^3/min

S^0, influent BOD_5 = 200 mg/L

S^e, effluent BOD_5 = 10.0 mg/L

Treatability constant, k = 0.21 $(L/s)^{0.5}/m^2$ at 20 °C

Taking a conservative approach, we design for the minimum temperature of 5 °C, by correcting the treatability constant k

Temperature correction for 5 °C, $k_5 = k_{20}*\theta^{(T-20)} = 0.21*(1.035)^{\wedge}(5-20) = 0.125$

Correction for different depth and influent concentration

$$k_2 = k_1*(D_1/D_2)^{0.5}*(S_1/S_2)^{0.5}$$

$$k_2 = 0.125*(6.1/7)^{0.5}*(150/200)^{0.5} = 0.101$$

Recycle ratio, R = Q_r/Q = 2.0 which means Qr = 2Q

Concentration S_a, BOD_5 of mix of recycled and raw influent is obtained by doing a mass balance around the inlet to the biotower:

$$200*Q + 10*Q^r = 220*Q$$

Dividing by the flow at inlet to tower = $Q + Q^r = 3Q$

Therefore, 220*Q/3Q = 73.3 mg/L

n, medium coefficient = 0.5

Using the equation for recirculated flow, the loading rate q can be determined:

$$S^e/S^a = \exp(-kD/q^n)/((1 + R)-R*\exp(-kD/q^n))$$

$$\exp(-kD/q^n) = (S^e/S^a)*(1 + R)/[1 + R*S^e/S^a]$$

$$S^e/S^a = 10/200 = 0.05, \text{ therefore } \exp(-kD/q^n) = (0.05*3)/(0.05*2 + 1) = 0.136$$

Hydraulic loading rate, q = 0.369 L/m^2-s = 31.9 m^3/m^2-d

Surface area of filter required = 10000/31.9 = 313.4 m^2 or 320 m^2

Provide at least 25 percent excess capacity, therefore, total surface area = 400 m^2

If each tower has a diameter of 8 m, surface area of each biotower = 50.27 m

Number of towers to be provided (having diameter of 8 m) = 7.95, so 8 towers.

6.3.2 Rotating Biological Contactors (RBC)

Rotating biological contactors (RBCs) were first installed in Germany in 1959 and thereafter in the USA and Canada.[60] An RBC is followed by a secondary clarifier and serves a similar purpose as the aeration tank in ASP. It has a tank with a central shaft on which several closely spaced circular disks made of high density polypropylene, polystyrene or polyvinyl chloride are mounted as shown in Figure 6.6. Corrugations on the disks increase the available surface area for biofilm formation. Wastewater flows into the tank at one end resulting in submergence of 35 to 40 percent of the disk area. The disks rotate at speeds of 1 - 2 rpm and are attached to mechanical drive units. Due to continuous rotation, the biofilm formed on the surface of the disk is alternately submerged and then exposed to air resulting in aerobic growth of biomass and degradation of wastewater BOD.

RBCs are available in preformed standard units of 3.5 m diameter and 7.5 m length with a surface area of 9300 m^2 and a side water depth of 1.5 m for 40 percent submergence of disk (Metcalf and Eddy 2003). Earlier installations in the US had expanded polystyrene disks with thickness of 1.25 cm and a spacing of 3.3 cm (USEPA 1984). In 1972, polyethylene was used for disks which increased the available surface area due to its lighter weight (more disks) and corrugated surfaces. Generally, the tanks with shafts are covered with fiberglass reinforced plastic covers and the entire unit is enclosed to reduce algal growth, process heat loss and UV exposure which can damage the plastic disks. The treated wastewater comes out of the RBC tank and is sent to a secondary clarifier for solids removal.

60 https://en.wikipedia.org/wiki/Rotating_biological_contactor.

Breaking of the shaft has been a major cause of failure for RBC units and smaller units have provided higher removal efficiencies (USEPA 1984). Scaling up from pilot-plant units to full-scale plants has not always given the same BOD removal efficiency as needed.

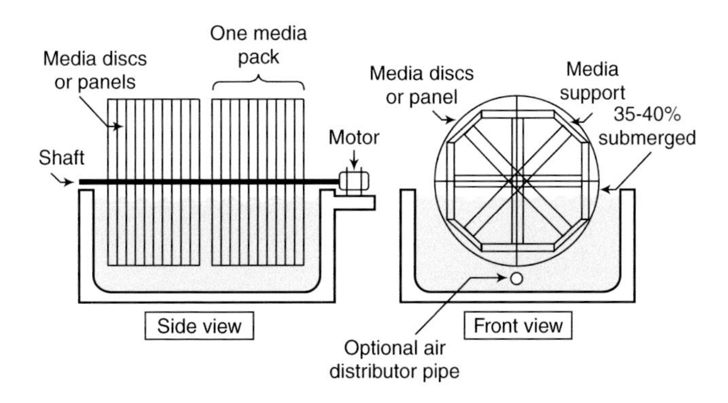

Figure 6.6 Rotating Biological Contactor (RBC) in front and side views (Source: Wikipedia).[61]

Staging of RBC units

RBC units can be arranged in series or parallel depending on the flow and treatment requirements. It was found that higher specific substrate removal rates were achieved in RBC biofilms at higher bulk liquid substrate concentrations. Therefore, disk area is reduced by using staged-RBC units. These function as CSTRs-in-series. Generally, two to four stages are used for BOD removal and six or more stages for nitrification. Stages are created using separate tanks in series or by using baffles in a single tank. Each stage receives effluent from the previous stage, i.e., the substrate concentration in the influent to the tank in series is less than its preceding stage. This configuration results in higher removal efficiencies than those achieved in a single tank. In small plants, the RBC drive shaft is parallel to the flow direction with baffle separation, while in large plants, shafts are oriented perpendicular to the flow direction.

Loading criteria

Performance is dependent on specific surface loading rate of total and soluble BOD for BOD removal and of NH_4-N for nitrification. Good results are achieved by keeping the loading rates less than the oxygen transfer capability of the system. When oxygen demand due to BOD exceeds the O_2 transfer capability, odor and sloughing problems are encountered resulting in poor efficiency. The first-stage soluble BOD (sBOD) loading can range from 12 to 15 g/m^2-d. If the soluble BOD concentration is not known, it can be calculated based on a ratio of sBOD/BOD of 0.5 to 0.75.

Process Design for BOD Removal

The soluble BOD (sBOD) concentration in each stage is given by

61 https://en.wikipedia.org/wiki/Rotating_biological_contactor.

$$S_n = \frac{-1 + \sqrt{1 + 4 * 0.00974 * \frac{A_s}{Q} * S_{n-1}}}{2 * 0.00974 * \frac{A_s}{Q}} \quad 6.3.13$$

where

S_n = sBOD concentration in stage n, mg/L

A_s = disk surface area in stage n, m^2 and

Q = flowrate, m^3/d

Process Design for Nitrification

For combined BOD removal and nitrification, nitrification is inhibited by sBOD concentration remaining which is dependent on the sBOD loading. The nitrification rate is affected as shown by the following relationship:

$$F_{r_n} = 1.00 - 0.1 \text{sBOD} \quad 6.3.14$$

where

Fr_n = fraction of nitrification rate possible without sBOD effect

sBOD = soluble sBOD loading, g/m^2–d (Metcalf and Eddy 2003)

PROBLEM 6.3.3

Design a rotating biological contactor for a flow of 10 MLD. Assume the following:

Total BOD = 200 mg/L; soluble BOD (sBOD) = 150 mg/L, desired effluent = 10 mg/L sBOD. Effluent concentration required is noted in the following table.

Solution

Flowrate = 10 MLD = 10000 m^3/d

Wastewater characteristics:

Parameter	Influent, mg/L	Required in Effluent, mg/L
BOD	200	20
sBOD	150	10
TSS	70	20

1. ***Determine number of RBC shafts for the first-stage***

Assume 1st-stage sBOD = 15.00 g/m^2-d

sBOD loading = sBOD*Q = 1500000.00 g/d

Disk Area required = sBOD loading/1st-stage sBOD = 100000.00 m^2

Suppose disk area per shaft is (As) = 9300.00 m^2/shaft

Number of shafts (n) = Disk area required/disk area per shaft = 10.75

Therefore, n = 11.00

Use 11 shafts for first-stage with disk area of 9300 m^2/shaft

2. ***If number of trains = number of shafts in 1st-stage***

Flowrate/train = 909.09 m^3/d

3. ***Calculate sBOD concentration in each stage based on equation 6.3.13***

Stage 1:

S^0 = 150 g/m^3

As /Q = 10.23 d/m (since (m^2/shaft)/(m^3/d))

S1 = 34.10 g/m^3

Repeat calculation until desired effluent quality is achieved

S2 = 14.15 g/m^3

S3 = 7.91 g/m^3, OK, since this is lower than the 10 mg/L sBOD required

Therefore, total 11 trains with 3 stages per train each are to be designed.

4. ***Determine the organic and hydraulic loadings***

a. First-stage organic loading = Lorg = Q*sBOD/n*As = 14.66 g sBOD/m^2-d

b. Overall organic loading, Torg = Q*BOD/(no. of stages * no. of shafts /stage *As) = 0.59 g BOD/m^2-d

c. Hydraulic loading to each unit

HLR = Q/(no. of shafts/ stage *A_s) = 0.098 m^3/m^2-d

Summary

No. of trains = 11.00

Flowrate/train = 909.09 m^3/d

Number of stages = 3.00

Total disk area/stage = 100000.00 m^2

First-stage sBOD loading = 14.66 g BOD/m^2-d

Total number of RBC units or shafts = 33.00

Overall organic loading = 0.59 g BOD/m^2-d

Hydraulic loading/shaft = 0.10 m^3/m^2-d

PROBLEM 6.3.4

Design a rotating biological contactor for a wastewater with the following characteristics:

a. Primary effluent flow rate = 20000 m^3/d, influent soluble BOD_5 = 150 mg/L

b. Effluent total BOD_5 = 30 mg/L and using a ratio of sBOD/BOD = 0.5, effluent soluble BOD = 15 mg/L
c. Use a shaft length of 7 m and disk diameter of 3.5 m for each module.

Determine the number of modules required, and the disk area based on graphs prepared by manufacturers. One such graph is shown in the textbook by Peavy, Rowe, and Tchobanoglous, (1985) – Figure 5-29. Similar graphs from other manufacturers are provided in a USEPA report (USEPA 1984).

Solution

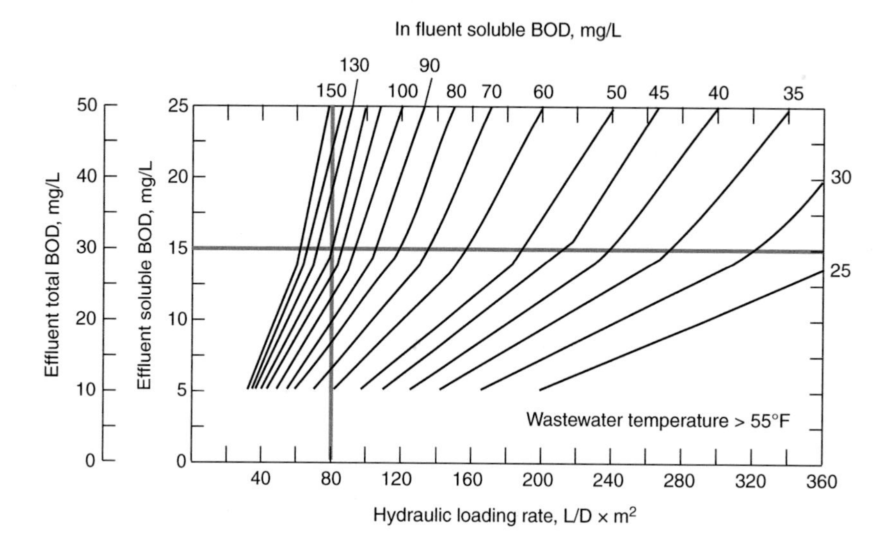

Efficiency and loading rate relationship for bio-surf medium treating municipal wastewater (courtesy of Autotrol Corp.).

Based on the preceding graph from Peavy, Rowe, and Tchobanoglous (1985):

For an influent soluble BOD = 150 mg/L and an effluent total BOD = 30 mg/L or an effluent soluble BOD = 15 mg/L, the hydraulic loading rate = 80 L/d-m^2 = 0.08 m^3/d-m^2.

Flow rate = 20,000 m^3/d

Therefore, surface area required = **2,50,000** m^2

Disk area per module

Length of shaft = 7 m

Spacing between disks = 30 mm

Thickness of disks = 10 mm

Surface area of each disk of diameter 3.5 m (both sides) = 19.24 m^2

Number of disks on each shaft = 175 disks

Total surface area/module = **3367.39 m^2**

Total number of modules required = 74.241 modules

Therefore, provide at least 75 modules.

6.3.3 Membrane Bioreactors (MBR)

Another new municipal wastewater treatment technology is the membrane bioreactor (MBR) which is gaining popularity and has been used successfully for full-scale municipal wastewater treatment. MBR is a suspended growth process where an ASP is followed by a membrane reactor. The reactor utilizes microporous membranes for solid/liquid separation instead of conventional secondary clarifiers. Effluent BOD_5 concentrations of 1 mg/L have been reported with this technology in at least one municipal wastewater treatment plant (CPCB 2013).

Advantages of MBR are that the amount of sludge produced in MBR is less than in an ASP with a conventional secondary clarifier. Excess sludge processing and disposal accounts for about 50–60 percent of the total cost of wastewater treatment and use of MBRs can reduce these costs. However, MBRs are high risk and costly compared to ASP or other more well-established technologies and are not in widespread use. Maintenance and operation of these bioreactors are major issues.

PROBLEM 6.3.5

Compare and contrast attached and suspended growth biological processes.

Solution

Parameter	Attached (fixed-film)	Suspended
Flow mode	Biomass is generally recycled in both types of processes; can be operated in plug flow or CSTR mode.	
Microbial growth	Biomass grows on surfaces like rock, stone or plastic	Biomass grows in suspension in aqueous medium
Land area required	Less	More
Sludge production	Less	More
Efficiency	Higher	Lower
Examples	TF and RBC	ASP, aerated lagoons, oxidation ponds

6.4 CLARIFICATION: TYPE 3 AND TYPE 4 SETTLING

Suspensions with higher solids concentrations such as sludge from biological processes can undergo Hindered settling (Type 3) or Compression settling (Type 4), respectively. As a thumb rule, solids concentrations <1000 mg/L TSS are considered dilute suspensions and concentrations above 1000 mg/L TSS are considered concentrated suspensions.

6.4.1 Hindered Settling (Type 3)

Hindered settling is observed in suspensions of intermediate concentration. In these suspensions, the magnitude of the inter-particle forces is appreciable and this affects the settling of particles. The entire mass tends to settle as a unit and the particles remain in the same position with respect to each other while settling.

6.4.2 Compression Settling (Type 4)

Compression settling is observed in suspensions of high concentration. The high concentration results in the formation of a visible structure. Therefore, further settling is possible only by compression of the previously formed structure. The cause of compression is the weight of suspended particles which are continuously deposited on the structure.

If a concentrated, uniform suspension such as sludge from an ASP is left standing in a column, the settling profile of solids over time is similar to that shown in Figure 6.7. Initially at t = 0, the suspension is uniformly distributed throughout the column. Particles move downwards by gravity and due to their high concentration, liquid moves up through their interstitial spaces. The particles settle as a 'zone', i.e., they remain in the same position with respect to each other while settling. As time passes, a clear water layer is formed above the hindered settling region and an interface or transition settling zone between the two regions can be identified.

With further settling, a visible structure is formed at the bottom of the column, indicating close physical contact between particles in this region. This is the compression settling region and further settling occurs by compression of this structure. The concentration of particles varies with height in the water column. Thus, the hindered settling region exhibits a gradation in solids concentration, low concentration at the interface with the clear liquid and increasing concentration near the compression region.

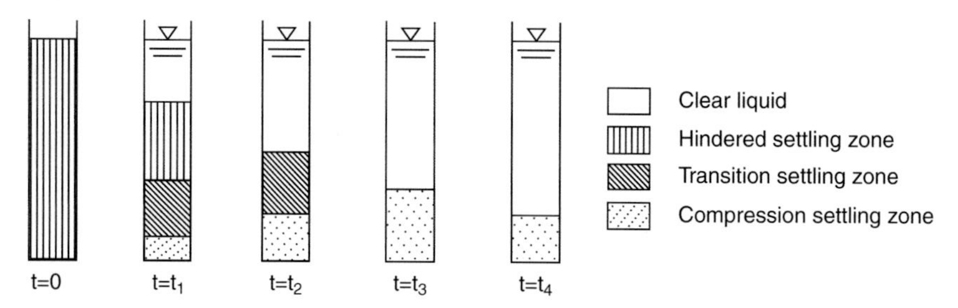

Figure 6.7 Different types of settling of ASP sludge solids in settling column analysis.

Column settling tests can be used to determine the area required for settling/thickening using two approaches:

1. Area requirement using Single-Batch Test results

In this approach, the results of one or more batch settling tests are used. Design of settling tanks is governed by the area needed for clarification and thickening and the rate of sludge withdrawal. The test results from settling columns give the area needed for settling, but it is not very useful as the area needed for thickening is usually greater. The area required for thickening can be found using a method developed by Talmage and Fitch (1955) (Metcalf and Eddy 2003).

In this method, a suspension of initial uniform concentration C_0 is filled in a column of height H_0. The position of the interface is plotted against time as shown by the line in red in Figure 6.8.

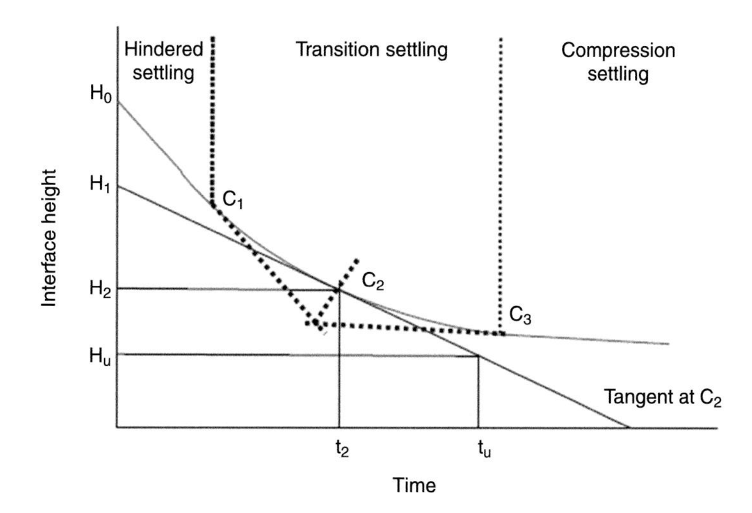

Figure 6.8 Settling column analysis for Type 3 and Type 4 settling (see color plate).

The area required for thickening is given by

$$A = \frac{Qt_u}{H_0} \tag{6.4.1}$$

where Q = flowrate into tank, H_0 = initial height of interface, t_u = time required to reach desired underflow concentration.

The point C_2 is the critical concentration which governs the solids handling capability of the tank. It corresponds to a height H_2. To locate point C_2 on the curve, tangents are drawn to the hindered settling and compression settling regions of the curve at points C_1 and C_3, respectively. At the point of intersection of the two tangents an angle is formed. Next, an angle bisector for this angle is drawn and the point at which it intersects the curve is the point C_2.

The depth H_u corresponding to the desired underflow concentration C_u is given by

$$H_u = \frac{C_0 H_0}{C_u} \tag{6.4.2}$$

To determine t_u, a tangent is drawn to the curve at C_2. The point where it intersects a horizontal line through H_u corresponds to t_u. Once t_u is known, the area needed for thickening can be calculated. It is compared with the area required for clarification. The greater area is the controlling factor.

2. Area requirement using solids flux method

In this approach, the results of several settling tests at various solids concentrations are used. The settling basin is assumed to operate at steady-state. Solids flux occurs due to gravity settling and due to the underflow being pumped out (Peavy et al. 1985). Existing facilities can be evaluated using this method and information for new designs can be obtained.

The total solids flux, $SF_t = SF_u + SF_g$

which is the sum of the flux due to underflow (SF_u) and flux due to gravity (SF_g).

Flux due to gravitational settling, $SF_g = v_i{*}C_i$

Where v_i = settling velocity of the solids, m/h, and

C_i = concentration of solids, mg/L or kg/m^3

Flux in the compression zone occurs as solids are removed with the underflow which has a flow rate Q_u and solids concentration, C_u.

The transport velocity,

$v_u = Q_u/A$

where Q_u = underflow rate, and

A = the surface area of the settling zone (perpendicular to Q_u).

Therefore, flux due to underflow, $SF_u = v_u{*}C_u$

Consolidation of sludge

The volume required for sludge in the compression region can also be obtained from settling tests. The rate of consolidation of sludge can be modeled on the same lines as a first-order decay function.

$$H_t - H_\infty = (H_2 - H_\infty)e^{-i(t-t_2)} \qquad 6.4.3$$

Where H_t = sludge height at time t, H_2 = sludge height at time t_2, H_∞ = sludge height after long time, i = constant for a given suspension.

PROBLEM 6.4.1

As calculated in Problem 5.4.1, Re is approximately 0.2 for a spherical particle of diameter 61 microns, i.e., Stokes' law is applicable to particles of size ≤61 microns as long as all other conditions are the same. For a settling column of depth 3 m, determine the time taken for discrete particles of size 1 micron, 10 microns, 15 microns, 20 microns and 30.5 microns to settle.

Solution

Use Stokes' law for determining the settling velocities of each particle size given as shown in the following table. Time required for 100 percent removal of particles is determined by dividing the depth of the settling column by the settling velocity.

Size of particle, micron	Settling velocity, m/h	Time to settle at depth of 3 m, hours
1	3.23×10^{-3}	928.8
10	0.323	9.29
15	0.728	4.12
20	1.3	2.3
30.5	3	1
61	12	0.25

PROBLEM 6.4.2

For the following conditions, determine underflow solids concentration, C_u.

Flow rate out of the aeration tank to the clarifier = 10000 m^3/d; 25 percent of the flow is recycled; assume there is no wasting of sludge

Solids content in the above flow = 5000 mg/L

Secondary clarifier has diameter of 20 m and depth of 4 m.

Settling column analysis data for wastewater from the aeration tank of an activated sludge process gave the following results data are from an unsolved problem in Peavy et al. (1985):

Solids concentration, C_i, mg/L	1,200	2,200	3,800	6,100	8,200	11,000
Settling velocity, v_i, m/h	5.8	3.2	1.6	0.6	0.4	0.09

Solution

Step 1: Set up a table as follows for solid flux due to gravity (SF_g), due to underflow (SF_u) and total solids flux (SF_t) where $SF_g = C_i v_i$, $SF_u = C_u Q_u/A = v_u C_i$ and $SF_t = SF_u + SF_g$

C_u = solids concentration in underflow, mg/L

$Q_u = Q_r$ = recycled flow which is 25 percent of influent flow; assume no flow is wasted, m^3/d

A = surface area of clarifier, m^2

Concentration, C_i, kg/m^3	v_i, m/h	SF_g, kg/m^2-h	SF_u, kg/m^2-h	SF_t	Tangent to total flux at minima
0					7.705
1.2	5.8	6.96	0.796	7.756	7.705
2.2	3.2	7.04	1.459	8.499	7.705
3.8	1.6	6.08	2.520	8.600	7.705

6.1	0.6	3.66	4.045	7.705	7.705
8.2	0.4	3.28	5.438	8.718	7.705
11	0.09	0.99	7.295	8.285	7.705
15			9.947		7.705

Step 2: Draw a graph for concentration versus each of the solids flux data as shown below.

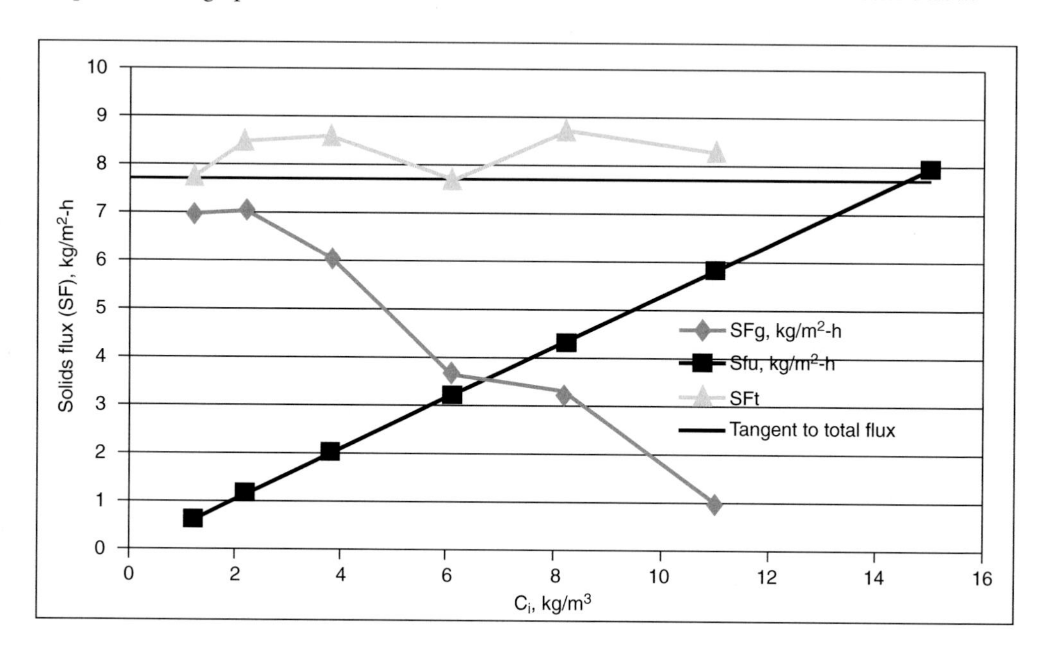

Step 3:

Solids loading to the clarifier = 5000 (g/m^3)*10000 (m^3/d) = 50000 kg/d

Surface area of the clarifier = $\pi(10)^2 = 314.16\ m^2$

Surface solids loading rate = (Qin + Qu)/A = 10000*1.25/314.16 = 39.8 m^3/m^2-d = 1.7 m^3/m^2-h

Underflow solids velocity, $v_u = SF_u/C_u = 0.66$ m/h

Solids concentration in the underflow, C_u is obtained from the point of intersection of SF_u and the tangent to the curve of total flux (SF_t).

Therefore, $C_u = 14.52\ kg/m^3$ for $SF_u = 7.705\ kg/m^2$-h

6.5 SLUDGE TREATMENT: THICKENING

Sludge treatment is a major challenge in all wastewater treatment. About 2 percent of the total volume of wastewater ends up as sludge and it generally contains about 1 to 5 percent solids (10,000 to 50,000 mg/L of solids). Sludge management takes up about 40-60 percent of the construction costs, 50 percent of the operating costs and most of the treatment plants operational problems.

Several sludge treatment processes are used in wastewater treatment plants and the treatment train is: concentration or thickening followed by digestion, dewatering, heating or drying and finally, disposal. Digestion, dewatering and disposal are always used in all conventional wastewater treatment plants regardless of the size of the plant. The other processes are likely to be used only in very large conventional wastewater treatment plants.

Thickening can be defined as increasing the solids concentration of sludge by reducing the liquid content. The reduction in sludge volume results in cost savings in terms of volume of reactors and equipment, heat energy in digesters and chemicals required for sludge conditioning. Sludge from an ASP typically has less than 1 percent solids and the sludge can be decreased in volume by 5-fold to achieve a solids content of 4 to 5 percent (Metcalf and Eddy 2003). Various methods can be used for sludge thickening and are described here.

6.5.1 Co-settling Thickening

Primary clarifiers are also used in sludge thickening. Thickening is achieved by creating a sludge blanket without allowing passage of clarified water. In this method, out of an assembly of clarifiers, only one is used for co-settling thickening (called the thickening clarifier). The solids underflow from the other clarifiers is discharged to this clarifier. The solids retention time is around 6 to 12 hours (Metcalf and Eddy 2003). Coagulants such as ferric chloride are added to give better results. It should be noted that very large solids retention time (SRT) lead to septic conditions and degasification. The underflow solids concentration of 3 to 5 percent can be achieved with this method.

6.5.2 Gravity Thickening

It is most effective for reducing the volume of primary sludge. Circular tanks, resembling sedimentation tanks are used for this purpose. A feed well in the center of a circular tank is used to feed dilute sludge in the clarifier in upflow mode. Settlement and compaction of the sludge takes place followed by withdrawal of the thickened sludge from the tank bottom. The thickened sludge is then sent to digesters for dewatering. Design parameters are solids loading, thickener overflow rate and sludge volume ratio (volume of sludge blanket divided by volume of sludge removed per day). However, high hydraulic loading causes too much solids carryover, whereas low hydraulic loadings result in septic conditions, foul odor and floating sludge.

6.5.3 Flotation Thickening

Flotation thickening is most effective for thickening of activated sludge. In this method, air is introduced into the suspension at high pressure. On lowering the pressure, the dissolved air escapes to the top as fine bubbles carrying sludge with them. The sludge is then removed from the surface. The floating solids concentration is governed by the air-to-solids ratio, the SVI (sludge volume index), the solids loading rate and polymer dose. The air-to-solids ratio is defined as weight of air available for flotation divided by weight of solids to be floated. Better performance is obtained at air to solids ratio of 2-4 percent, SVI < 200 and nominal polymer dosage [Metcalf and Eddy 2003].

Dissolved air flotation (DAF) is widely used in wastewater treatment for industries like oil refineries, petrochemical and chemical plants, natural gas processing plants, paper mills, and food and beverage industries. It can be used for removing total suspended solids (TSS), floating debris, fats, oil and grease, and other pollutants from municipal or industrial wastewaters making them suitable for reuse or discharge to a stream.

Air under pressure of several atmospheres is dissolved in the influent wastewater and held at high pressure for several minutes. The pressurized flow is then introduced into the flotation tank where the pressure is released back to atmospheric level. This results in the immediate formation of micro-bubbles as the dissolved air comes out of solution. These small bubbles attach to the suspended solids present in the wastewater causing them to move and float to the surface where they formed a froth layer and are subsequently removed by skimmers or skimming units. A portion of the DAF effluent (15 to 120 percent) can be recycled into the flotation tank. Solids that settle to the bottom of the flotation tank are removed using sludge scrapers into a sludge hopper.

Types of DAF units

Some DAF units use parallel plate packing material (like in lamella-clarifier) to provide more surface area for separation to enhance the removal or separation efficiency. DAF units can be circular or rectangular.

Open Tank DAFs

They are ideal for treating wastewater with high suspended solids concentration. This type of wastewater requires significant amounts of free surface area for flotation and separation. Open tank DAFs are normally rectangular in shape and are built wider and longer as the free separation area requirement increases.

Plate Pack DAFs

They are characterized by deeper tanks with inclined and corrugated plate packs. Water is introduced into the plate packs in a cross-flow configuration, reducing the distance solids have to float to be effectively separated. As they collide with an angled plate, light solids accelerate upward and heavy particles settle. Wastewater enters the plate pack heavily laden with flocculated contaminants and exits devoid of suspended and colloidal solids. A plate pack DAF unit is preferred for high hydraulic and low solids loading rates.[62, 63]

The major components of a DAF unit are a flotation tank, sludge dewatering grids, skimmers, flocculators and recycle pumps.[62, 63]

- **Flotation Tank** is where separation of suspended solids and water occur. Water flows slowly through the flotation tank to provide more time for heavy objects to settle down and lighter

62 http://frcsystems.com/pcl-dissolved-air-flotation-systems/.

63 https://www.jwce.com/knowledge-center/what-is-a-daf/.

objects to float to the top with the bubbles and flocculants. The flotation tank can vary in depth, width and length based on the application and the time needed for pollutants to be removed.

- **Sludge Dewatering Grid** is a rectangular framework of angular plates that trap sludge in place as it moves to the surface and provides for sludge thickening. When sludge has thickened enough to rise above the top edge of the grid it is skimmed by skimmers and pushed into the hoppers.
- **Skimmers** used most frequently in DAF are co-current skimmers and counter-current skimmers. Co-current skimmers push sludge across the entire length of the tank in the same direction as the wastewater flow. The counter-current skimmers consist of a skimmer assembly that rotates against the hydraulic flow of the water. The counter-current Skimmers design shortens the sludge skimming distance and eliminates solids carry-over.
- **Flocculators** are designed to provide mixing and retention time for coagulation and flocculation of solids.
- **DAF pumps** are a key component of all DAF Systems. In DAF, recycle pumps can be used in two ways. The first method is to provide a specialty white water pump. This pump not only pumps the water but also dissolves air into the water. These pumps are often expensive and difficult to find. Also, adding air in the pump creates the risk of cavitation, which causes internal damage and results in more-frequent-than-desired replacement of parts. Second way is to combine a standard ANSI pump with an angled air dissolving tube. In this case, the pump is used only for wastewater. It results in higher solids tolerance, operates at lower pressures, and reduces operational costs compared to the first method. White-water is generated in the air dissolving tube. The small size of recirculation pipes allows clarified effluent and small volume of compressed air to mix until the saturation is achieved. The angled alignment results in increased air-water interaction due to which saturation occurs almost instantly.

6.5.4 Centrifugal Thickening

It is generally used only for waste-activated sludge (Metcalf and Eddy 2003). The centrifuge used is called solid-bowl centrifuge. It consists of a long horizontal bowl, tapered at one end. The solids in the sludge tend to accumulate around the periphery. A helical scroll moves the accumulated sludge to the tapered end from where it is discharged as shown in Figure 6.9.

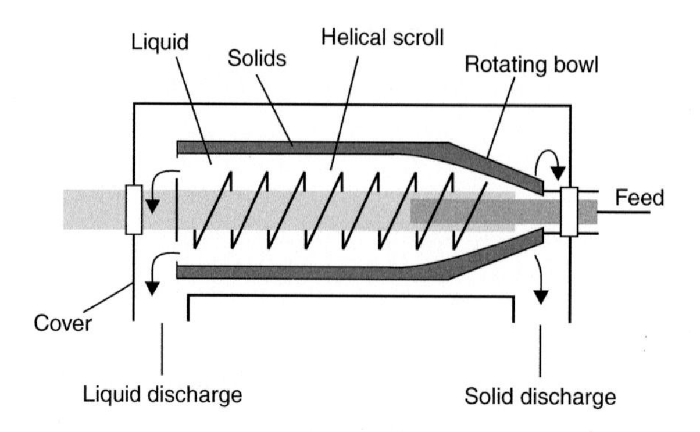

Figure 6.9 Solid bowl centrifuge (Source: Wikipedia)[64].

The performance is quantified by TSS recovery, defined as

$$R = \frac{TSS_P(TSS_F - TSS_C)}{TSS_F(TSS_P - TSS_C)} \times 100 \qquad 6.5.1$$

Where R = percent recovery,

TSS_P, TSS_F and TSS_C are the TSS concentrations in percent by weight in the thickened product, feed and centrate, respectively.

6.5.5 Gravity-Belt Thickening

It is effective for waste-activated sludge, but polymer addition is essential. In this method, a gravity belt is used which moves over rollers. A feed/distribution box takes in the sludge and distributes it evenly across the width of the belt. As the sludge moves towards the discharge end, water keeps draining through the belt. A series of plow blades facilitate the passage of water from the sludge through the belt. After the sludge is discharged, the belt undergoes a wash cycle. Solids in the feed water can be 0.5 to 1 percent and can be dewatered to a maximum concentration of 5 to 7 percent (Metcalf and Eddy 2003).

6.5.6 Rotary-Drum Thickening

In this method, sludge and polymer are mixed in a mixing and conditioning drum. The conditioned sludge is then fed to rotating-screen drums, where thickened sludge rolls out from the end while the water flows through the screens. Feed water solids with concentrations of 0.5 to 6 percent can be thickened to a concentration of 4 to 9 percent (Metcalf and Eddy 2003).

PROBLEM 6.5.1

In a conventional wastewater treatment plant of 1 MLD, the following units generate sludge that is combined in a gravity thickener. Flow rates and solids concentrations are provided. Determine

64 https://en.wikipedia.org/wiki/Solid_bowl_centrifuge.

the total sludge that is to be thickened and the combined solids concentration in the thickener. If the thickener increases the solids concentration to 6 percent solids, determine the volume of sludge generated, and volume of supernatant. Assume solids concentration in the supernatant is negligible compared to underflow concentration.

Treatment unit/ process	Flow rate, MLD	Solids concentration, percent
Primary clarifier underflow	0.1	4
Secondary clarifier underflow	0.3	2

Solution

Step 1: Add another column to the above table for solids concentration in mg/L as shown below.

Step 2: Calculate the solids concentration in the combined flow to the gravity thickener as follows: (0.1 MLD*4 percent + 0.3*2 percent)/(0.1 +0.3) = 2.5 percent for a total combined flow of 0.4 MLD.

Step 3: Since the sludge flow from the gravity thickener is thickened to a solids concentration of 6 percent, the volume of sludge and supernatant can be calculated by mass balances for water and solids as below:

$$Qin = Qout + Qsludge$$

$$Qin*Cin = Cout*Qout + Qsludge*Csludge$$

Since solids concentration in supernatant (Cout) << Csludge, it is ignored and

$$Qin*Cin = Qsludge*Csludge \text{ resulting in}$$

$$Qsludge = QinCin/Csludge = (0.4*2.5 \text{ percent})/6 \text{ percent} = 0.167 \text{ MLD}$$

Flow rate of sludge, Qsludge = 0.167 MLD with a solids concentration of 6 percent

Therefore, flow rate of supernatant = 0.4 – 0.167 = 0.233 MLD

Note: *Wastewater discharge standards in India require that TSS is < 100 mg/L which means supernatant concentration should be <100 mg/L.*

Treatment unit/ process	Flow rate, MLD	Solids conc., percent	Solids conc., mg/L
Primary clarifier underflow	0.1	4	40000
Secondary clarifier underflow	0.3	2	20000
Gravity thickener combined	0.4	2.5	25000
Gravity thickener underflow - sludge generated	0.2	6	60000

Treatment unit/ process	Flow rate, MLD	Solids conc., percent	Solids conc., mg/L
Supernatant	0.233	negligible	100

6.6 SLUDGE TREATMENT: DIGESTION

Sludge management (handling, treatment and disposal) is a very important part of water and wastewater treatment. The sludge generated from different unit processes is generally very high in water content (>95 percent) and low in solids content (1 to 5 percent). The objective of sludge digestion is to reduce the volume of sludge, and stabilize the biosolids in the sludge, i.e., reduce any potential risk due to the presence of pathogens in the bioactive solids.

Sludge digestion is of two types: aerobic or anaerobic. Anaerobic digestion is more popular since it allows energy recovery. Aerobic digestion is similar in design to the ASP. A comparison of major characteristics of aerobic and anaerobic processes is provided in Table 6.5.

Table 6.5 Differences between aerobic and anaerobic biological treatment processes

	Aerobic biological processes	Anaerobic Digestion
Biochemical pathways	Aerobic implies the terminal electron acceptor (TEA) is oxygen	Anaerobic digestion implies two possible biochemical pathways: anaerobic respiration where the TEA may be any compound but not oxygen and fermentation where the organic compound serves as both TEA and electron donor (ED). Fermentation is common in conventional wastewater treatment.
End-products	Carbon dioxide, water and biomass	Carbon dioxide, methane (in case of fermentation), and biomass. *Water is a reactant in fermentation.*
Organic loading	Low	High
Solids concentration	Low	High
Process objective	To remove soluble BOD	To decrease volume of sludge and generate biogas
Biomass or sludge generated	Higher amounts of biomass or sludge are generated	Lower amounts of biomass or sludge are generated
Biodiversity of process	High since bacteria, protozoa, rotifers, crustaceans exist together	Very few bacterial species are involved. No other organisms

	Aerobic biological processes	Anaerobic Digestion
Bacterial growth rates	Aerobic heterotrophic bacteria grow fast	Methanogenic bacteria have slow growth rates
Nutrient requirements	More	Less since sludge (biomass) production is less
Start-up time	Short	Long
Energy usage	Input of energy is high due to aeration requirement	Net output of energy in terms of biogas; heating may be required under some climatic conditions
Odor production	Less	More
Need for pH control	Less	More
Examples in municipal wastewater treatment	Activated sludge process, trickling filters and many more	Anaerobic digestion of sludge, biogas plants

6.6.1 Anaerobic Digestion

Anaerobic digestion is used for the stabilization of biosolids in the absence of oxygen. Major advantages of anaerobic digestion include the generation of biogas, energy savings and less sludge production in comparison to aerobic biological processes.

It is a complex 3-step process and is shown in Figure 6.10 (Rittmann and McCarty 2012).

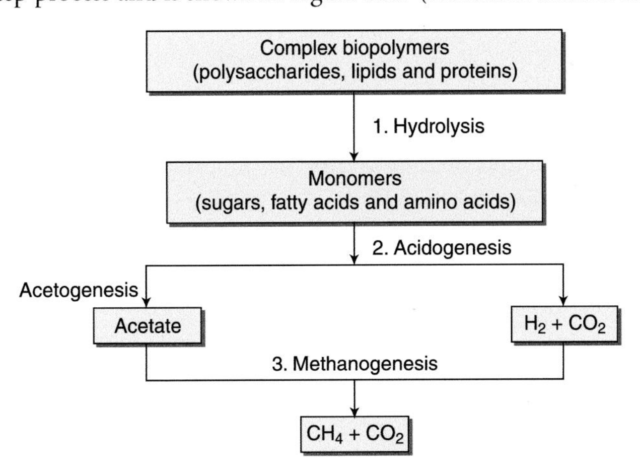

Figure 6.10 Anaerobic digestion and the generation of biogas (methane and carbon dioxide).

1. **Hydrolysis:** In this step, complex polymeric compounds are converted to monomeric forms which can then be absorbed by bacteria for growth and reproduction. Hydrolysis can occur either inside or outside the cell and this step is common to all aerobic and anaerobic processes. Bacteria secrete extracellular enzymes to breakdown complex organic matter.

2. **Acidogenesis:** The second step is the conversion of monomers to simple organic acids (C1 to C6 compounds) like acetate, propionate, and butyrate and hydrogen gas. Acidogenic bacteria have faster growth rates than methane formers (or methanogens). This step also includes acetogenesis (often shown as a fourth step in many literature sources). Hydrogen generated in this step is used by methanogens in the next step. If acidogenic activity is high, acids can accumulate resulting in a decrease in pH which inhibits methanogenic activity and can destroy methanogenic bacteria. The process becomes unbalanced and methane production can cease.
3. **Methanogenesis:** The last step is the fermentation of organic acids to H_2, CO_2, and CH_4. These reactions are carried out by two distinct groups of methanogenic bacteria: acetate fermenters and hydrogen oxidizers as shown below:
   i. *Acetate fermenters* that breakdown acetate; Y = 0.04 g VSS/g acetate:

$$CH_3COOH + H_2O \rightarrow CO_2 + CH_4 + H_2CO_3$$

   ii. *Hydrogen oxidizers* that are responsible for the utilization of hydrogen; Y = 0.45 g VSS/ g H_2:

$$4H_2 + CO_2 \rightarrow CH_4 + 2H_2O$$

The bulk of methane is formed by acetate fermenters (70 percent) while the remaining is formed by hydrogen oxidizers (30 percent) (Bouwer 1991).

Maintaining low H_2 concentration in the reactor is of critical importance for running an efficient anaerobic digester. Based on thermodynamic considerations, conversion of propionate to methane is favorable only when the partial pressure of hydrogen lies between 10^{-6} to 10^{-4} atm, while the conversion of ethanol to methane is favorable when the partial pressure of hydrogen lies between 10^{-6} to $10^{-0.5}$ atm.

Biogas generated in this process is a combination of carbon dioxide and methane. For obtaining useful fuel, i.e., methane, another step for separating the two gases is necessary since methane is flammable while CO_2 is a fire retardant. If the gases are not separated, the biogas flame sputters continuously and is not very useful. Various methods are available for separating methane from CO_2 and include cryogenic separation of gases, adsorption of carbon dioxide on activated carbon, and bubbling of the gas mixture through lime slurry.

Some of the important factors in anaerobic digestion are:

Hydraulic retention time (HRT): It is the average time for which the influent remains in the digestion process. HRT = Volume/Flow rate.

Solids retention time (SRT): It is the average time for which solids remain in a digestion process. Change in SRT results in change in the extent to which the above reactions are completed. Each of these reactions has a minimum SRT. If the SRT provided is lower than the minimum SRT, then bacteria cannot grow rapidly enough and the digestion process fails. In general, a fraction of the biosolids generated is recycled back into the process. Therefore, SRT is usually much greater than the hydraulic retention time (HRT). However, if there is no recycling of biosolids, then SRT = HRT.

pH: The optimum pH range for anaerobic digestion is 6.8 to 7.5 while there is no inhibition in the range of 6 to 8. Any variation in pH outside this range can result in a build-up of hydrogen in the reactor and if the pH falls below 6, it is called a 'pickled digester'.

Temperature: Temperature affects the reaction rates (bacterial growth, and digestion) as well as the gas transfer rates and the settling of biosolids. Anaerobic systems can be operated in the mesophilic temperature range (optimum: 37-41 °C; natural or ambient conditions: 20 – 45 °C) or in the thermophilic temperature range (optimum: 50 to 52 °C; can tolerate up to 70 °C). The required digestion period decreases with increase in temperature.

Mesophilic systems are more stable than thermophilic digestion systems and are easy to start since there are more mesophilic species than thermophilic species in the environment. Mesophilic bacteria are more tolerant to changes in environmental conditions than thermophiles.

Thermophilic: Thermophilic systems require heat inputs, but they have higher reaction rates and faster gas yields. Also, their digestate is easier to sterilize prior to discharge.

Alkalinity: Since maintaining near neutral pH is essential for running an efficient anaerobic digester, alkalinity is an important operating parameter. The optimum operating range for alkalinity is 1000 to 5000 mg/L as $CaCO_3$. Carbon dioxide production is a function of pH and alkalinity and ranges from 25 to 45 percent of the total biogas produced while the remaining is mainly methane with trace amounts of hydrogen.

Nutrients: Methanogenic bacteria have yields that are 1/3 to 1/10 of the yields of aerobic bacteria which means the nutrient requirements for methanogens are much less than for aerobic bacteria. The stoichiometric formula for methanogenic bacteria is estimated to be $C_5H_7O_2NP_{0.06}S_{0.1}$ while that for aerobic bacteria is generally estimated as $C_5H_7O_2N$. This difference in P and S requirements is attributed to the nature of enzymes used by methanogenic bacteria. Typically, S requirements for methanogens are 1 to 2 orders of magnitude higher than aerobes (Bouwer 1991). Trace metals like Fe, Ni and Co are important micronutrients for methanogens since they act as electron carriers in the fermentation process.

Anaerobic bacteria are more sensitive to toxic compounds and heavy metals. However, if the anaerobic cultures are gradually exposed to increasing concentrations of toxins, they can tolerate 20 to 50 times higher concentrations of the toxins compared to unacclimated anaerobes.

Different types of anaerobic digestion processes are in use and can be categorized as follows:

1. Batch or continuous-flow process which may be either CSTR or PFR
2. Temperature: Mesophilic or thermophilic reactors
3. Solids content: high (>22 percent solids) or low
4. Complexity: single-stage or multiple-stages.

6.6.1.1 Plug flow anaerobic digesters

Plug flow digesters are capable of generating biogas and treating sludge with a solids content of 11-14 percent. They are long and narrow, typically with a length to width ratio >5:1, i.e., they are more than 5 times as long as they are wide. These reactors are insulated and heated tanks made of reinforced concrete, steel or fiberglass or plastic materials with a gas tight cover to capture the biogas. They can be operated at mesophilic or thermophilic temperatures and SRT = HRT with a range of 15 to 20 days.

6.6.1.2 Low solids anaerobic digesters

Low solids digesters are popular for generating methane from human, animal and agricultural waste (as shown in Figure 6.11) with a low solids content of 4 to 8 percent solids in the waste stream. Depending on the nature of the waste, nutrients and pH adjustment may be necessary. The waste may be mixed and heated to a temperature of 55–60 °C. Gas collection and storage systems have to be designed and separation of the two main components of biogas: carbon dioxide and methane is necessary prior to distribution.

A photo of a plug flow low solids biogas digester for treating agricultural waste (mainly animal excreta along with municipal wastewater) is shown in Figure 6.11. The reactors are made of flexible plastic materials that are inflatable. The reactor in the foreground is fully inflated with biogas (it is in the mature stage) while all other reactors behind it are at different stages of maturity as indicated by the degree of inflation. Black rubber layers are provided on the reactors to insulate the reactors for retaining heat generated and absorbing radiant sunlight.

Figure 6.11 Plug flow digesters in a wastewater treatment plant in India (see color plate).

Another popular example of a low solids digester is the upflow anaerobic sludge blanket (UASB) digester.

6.6.1.3 High solids anaerobic digesters

Fermentation of sludge with a solids content >22 percent can be done in high solids anaerobic digesters. These digesters have lower water requirements, higher gas production/unit volume of reactor, and can operate in either mesophilic or thermophilic ranges of temperatures.

6.6.1.4 Thermophilic anaerobic digesters

The basic advantage of thermophilic digesters is that solids destruction capability is higher, dewatering is better and pathogen destruction is greater than mesophilic digesters. Some of its disadvantages include higher energy requirement for heating, poor quality of supernatant liquid containing large quantity of dissolved solids, and odor problems.

These reactors can be operated as single-stage or multi-stage reactors. Two phase anaerobic digestion can be done in two ways: thermophilic followed by mesophilic or mesophilic followed by thermophilic mode. In this approach, advantages of thermophilic digestion are incorporated with greater stabilization and greater thermophilic digestion rate (generally 4 times faster than mesophilic digestion). It also absorbs shock-loading better. The system tends to have greater VSS with less foaming. The main advantages of mesophilic phase are destruction of odorous compounds and improved stability of the digestion process.

6.6.1.5 Single-stage digesters

There are two basic types of single-stage anaerobic digesters: standard rate or high rate.

Single-stage standard rate digester: Generally, solids are not recycled and SRT ranges from 30 to 60 days. Since the concentration of solids is high, stratification and separation of solids and liquids happens naturally. This results in an active reactor volume that is less than the total volume of the reactor. Even though no mixing is provided, some mixing can occur due to bubbling of methane gas. Due to the combined effects of stratification and mixing, the actual SRT ranges from 15 to 25 days. These digesters give better effluent quality in comparison to high-rate digesters since settling of solids is better.

Single-stage high-rate digester: This is characterized by heating, auxiliary mixing, uniform feeding and thickening. Initially the sludge is mixed with the help of gas recirculation, pumping or draft tube mixer. Then it is heated to achieve optimum digestion rates. In this process, uniform feeding is important and the digested sludge should be withdrawn before adding feed sludge. Sequential batch processing is more effective for killing pathogens (in feed sludge) in large numbers in comparison to a continuous-flow mode where the feed sludge is used to displace digested sludge. Since mixing is provided without recycle, SRT = HRT and ranges from 15 to 30 days. These digesters provide higher methane production in comparison to standard rate digesters but the effluent quality is not as good due to poorer settling of solids.

6.6.1.6 Multi-stage digesters

Two-stage digestion: In this process, a high rate digester is coupled with a second tank. The first tank is used for digestion which involves heating and mixing of sludge, and the second tank is used for storage. Tanks can be fixed roofs or floating covers.

Separate sludge digestion: In general, primary and secondary sludge is treated in a common anaerobic digester. Separation of solids and liquids in primary sludge decreases when even small amounts of biosolids are added, and this also decreases the reaction rate under anaerobic conditions. Therefore, separate sludge digestion is generally recommended if it is affordable. In that case, digestion of primary and biological sludge is accomplished in separate tanks resulting in excellent dewatering characteristics of digested sludge and optimum process control conditions.

Multi-staged mesophilic digestion produces more stable, less odorous biosolids that are easy to dewater.

***Acid or gas-phase digestion*:** In this process, anaerobic digestion proceeds in a two-stage process through each of the 3 distinct phases of hydrolysis, fermentation (acetogenesis) and methanogenesis. In the first-stage, solubilisation of particulate matter occurs and volatile acids are formed. It has pH less than 6 with short SRT to produce high concentration of volatile acids. The second phase is conducted at neutral pH and longer SRT, which is favorable for methane-generating bacteria and maximizes gas production. This method gives greater volatile solid reduction, digester foam can be controlled, and the two stages can be operated in the mesophilic or thermophilic temperature range.

6.6.2 Design Considerations

Process design for mesophilic anaerobic digestion is based on the following factors:

Solids Retention Time (SRT): For a complete mix, high rate digester without recycling, mass of biosolids synthesized, P_x, can be estimated by

$$P_x = \frac{YQ(S_o - S) * (10^3 \text{ g/kg})^{-1}}{1 + k_d(SRT)}$$

Where Y = yield coefficient, g VSS/ g biodegradable COD; k_d = endogenous coefficient, 1/d (values range from 0.02 to 0.04). All other terms are the same as in ASP design.

Loading factors: The most important factors are the (i) Mass of volatile solids added per day per unit volume of digester capacity, and (ii) Mass of volatile solids added to the digester each day per mass of volatile solids in the digester.

The first factor is the preferred one. The upper limit of volatile solids loading rate is typically determined by the rate of accumulation of toxic materials like ammonia or washout of methanogens. Excessively low volatile solids loading rate can result in a design that is costly to build and difficult to operate.

Volatile solids destruction estimation: In a high rate, complete mix digester, the amount of volatile solids destroyed is given by an empirical equation (Liptak 1974):

$$V_d = 13.7\ln(SRT) + 18.9$$

where V_d is volatile solids destruction (percent), SRT is time of digestion (d)

Population basis: Digestion tanks are also designed on a volumetric basis by allowing a certain number of cubic meters per capita.

PROBLEM 6.6.1

A wastewater sludge comprising of biosolids has the following empirical formula $C_{10}H_{19}O_3N$. The sludge is to be digested anaerobically (fermented). Based on Section 6.12 of Sawyer et al. (2003), determine the following:

1. Maximum amount of methane that can be generated per mole of wastewater
2. Maximum amount of methane that can be generated per g of wastewater
3. Molar and weight ratios of methane to carbon dioxide produced.

Solution

Any biological process has two aspects to it: energy generation for the organisms (in this case bacteria) and biomass production. The coupling of two half reactions: one for an electron donor and another for an electron acceptor results in energy generation and biomass production.

The overall equation for any biological process is given by

$$R = f_s R_c + f_e R_a - R_d$$

where f_s = fraction of electron acceptor that is utilized for cell synthesis (biomass generation) and f_e is the fraction of electron acceptor that is utilized for energy generation. The sum of f_e and f_s is always 1. An $f_{s,max}$ value of 0.14 was computed based on equal proportions of carbohydrates, proteins, and fatty acids as given in Table 6.5 of Sawyer, McCarty, and Parkin (2003). An overall reaction can be written for wastewater sludge, where wastewater sludge is represented by $C_{10}H_{19}O_3N$. Since wastewater and sludge have ammonia not nitrate in it, hence, ammonia is assumed to be the nitrogen source.

$$R = f_s R_c + f_e R_a - R_d$$

$$f_s{*}R_c = 0.14 * \left[\frac{1}{5}CO_2 + \frac{1}{20}HCO_3^- + \frac{1}{20}NH_4^+ + H^+ + e^- \rightarrow \frac{1}{20}C_5H_7O_2N + \frac{9}{20}H_2O\right]$$

$$f_e{*}R_a = 0.86 * \left[\frac{1}{8}CO_2 + H^+ + e^- \rightarrow \frac{1}{8}CH_4 + \frac{1}{4}H_2O\right]$$

$$-R_d = \frac{9}{50}CO_2 + \frac{1}{50}HCO_3^- + \frac{1}{50}NH_4^+ + H^+ + e^- \rightarrow \frac{1}{50}C_{10}H_{19}O_3N + \frac{9}{25}H_2O$$

Final balanced reaction:

$$R = \frac{1}{50}C_{10}H_{19}O_3N + 0.082\,H_2O \rightarrow 0.007C_5H_7O_2N + 0.0445\,CO_2 + 0.1075\,CH_4 + 0.013\,HCO_3^- + 0.013\,NH_4^+$$

Normalize the reaction to 1 mole of wastewater gives:

$$R = C_{10}H_{19}O_3N + 4.1\,H_2O \rightarrow 0.35C_5H_7O_2N + 2.225\,CO_2 + 5.375\,CH_4 + 0.65\,HCO_3^- + 0.65\,NH_4^+$$

Therefore,

1. Methane production per mole of wastewater is 5.375 mole of methane/ mole of wastewater sludge
2. Methane production per g of wastewater is 5.375*16/201 = 0.428 g methane/ g wastewater sludge
3. Molar ratio of methane to carbon dioxide is 2.416 moles of methane to 1 mole of carbon dioxide

   Weight ratio of methane to carbon dioxide is 0.88 g methane to 1 g carbon dioxide.

PROBLEM 6.6.2

A wastewater comprising mainly of starch which is a polymer of glucose can be fermented. Assume the nitrogen source is ammonia. Based on Section 6.12 of Sawyer et al. (2003), determine the following:

1. Maximum amount of methane that can be generated per mole of wastewater
2. Maximum amount of methane that can be generated per g of wastewater
3. Molar and weight ratios of methane to carbon dioxide.

Solution

Based on the overall equation for any biological process noted below and $f_{s,max}$ value of 0.28 (based on carbohydrates (CH_2O) as electron donor and CO_2 as electron acceptor, Table 6.5 of Sawyer et al. (2003)), an overall reaction can be written. Ammonia serves as the nitrogen source.

$$R = f_sR_c + f_eR_a - R_d$$

$$f_s{*}R_c = 0.28 * \left[\frac{1}{5}CO_2 + \frac{1}{20}HCO_3^- + \frac{1}{20}NH_4^+ + H^+ + e^- \rightarrow \frac{1}{20}C_5H_7O_2N + \frac{9}{20}H_2O\right]$$

$$f_e{*}R_a = 0.72 * \left[\frac{1}{8}CO_2 + H^+ + e^- \rightarrow \frac{1}{8}CH_4 + \frac{1}{4}H_2O\right]$$

$$-R_d = \frac{1}{4}CO_2 + H^+ + e^- \rightarrow \frac{1}{4}CH_2O + \frac{1}{4}H_2O$$

Final balanced reaction:

$$R = 0.25CH_2O + 0.014HCO_3^- + 0.014NH_4^+ \rightarrow 0.014\,C_5H_7O_2N + 0.104\,CO_2 + 0.09\,CH_4 + 0.056\,H_2O$$

Normalize the reaction to 1 mole of wastewater gives:

$$R = CH_2O + 0.056\,HCO_3^- + 0.056\,NH_4^+ \rightarrow 0.056\,C_5H_7O_2N + 0.416\,CO_2 + 0.36\,CH_4 + 0.224\,H_2O$$

Starch can be represented as $(Glucose)_n$. Glucose can be written as $6CH_2O$ and the calculations for every mole of wastewater in this case are expressed in equivalents of glucose.

Therefore,

1. Methane production per mole of wastewater is 0.36*6 = 2.16 mole of methane/ mole of glucose
2. Methane production per g of wastewater = 0.192*6 = 1.152 g methane/g glucose
3. Molar ratio of methane to carbon dioxide is 0.865 moles of methane to 1 mole of carbon dioxide

   Weight ratio of methane to carbon dioxide is 0.315 g methane to 1 g carbon dioxide.

PROBLEM 6.6.3

Biogas is a mixture of methane and carbon dioxide. Answer the following questions based on your knowledge of the stoichiometry of the fermentation process:

1. What is the theoretical fraction of methane to carbon dioxide in a biogas mixture?
2. Most literature reports state that methane is found from 55 to 75 percent while carbon dioxide ranges from 25 to 45 percent by volume in biogas. Explain the difference in fractionation predicted by theory versus that observed.

Solution

1. Biogas is produced by fermentation of organic material. Fermentation occurs under strictly anaerobic conditions where part of the organic carbon is completely oxidized to CO_2 and the remaining is completely reduced to CH_4.

   The solution to Problem 6.6.1 shows that for municipal wastewater where biomass production was also accounted, the molar and weight fractions of methane: CO_2 are 2.42:1 and 0.88:1, respectively. It is important to note that biogas production and composition are primarily related to the feed composition, i.e., municipal wastewater, or a wastewater with one or more well-defined compounds.
2. Carbon dioxide is highly soluble in water and its aqueous concentration is a function of pH, alkalinity and temperature. Methane is relatively insoluble in water and remains mostly in gas phase. The difference between the observed fractions of gases versus the predicted fractions is due to the solubility of CO_2 in the aqueous media.

6.6.3 Aerobic Digestion

Aerobic digestion is one of the processes used in sludge treatment for reducing the mass of sludge solids. It is similar to the activated sludge process except that the biosolids are maintained in endogenous phase. As substrate gets depleted, microorganisms start consuming their own protoplasm to obtain energy which is referred to as the decay or endogenous phase. Cell tissue is oxidized aerobically to form carbon dioxide, water and ammonia as shown in equation 6.6.1. About 75-80 percent cell tissue can be oxidized while inert or non-biodegradable compounds, which are measured as nbVSS, are the remaining fraction. Thus, nbVSS are the end products of aerobic sludge digestion. Ammonia released during the aerobic endogenous phase is subsequently oxidized to nitrate as digestion proceeds. When organic nitrogen is converted to nitrate, it results in an increase in the hydrogen ion concentration and decrease in pH if the sludge does not have adequate buffering capacity. Biochemical changes in the aerobic digester can be represented by the following equations:

Biomass destruction (1st term in reaction represents bacterial cells or biomass):

$$C_5H_7O_2N + 5O_2 \rightarrow 5CO_2 + 2H_2O + NH_3 \qquad 6.6.1$$

Nitrification of released ammonia nitrogen:

$$NH_3 + 2O_2 \rightarrow HNO_3 + H_2O \qquad 6.6.2$$

Overall equation with complete nitrification:

$$C_5H_7NO_2 + 7O_2 \rightarrow 5CO_2 + 3H_2O + HNO_3 \qquad 6.6.3$$

If the above process is combined with the strictly anaerobic denitrification process, then the following reactions will occur.

Using nitrate-nitrogen as electron acceptor (denitrification):

$$C_5H_7NO_2 + 4NO_3^- + H_2O \rightarrow NH_4^+ + 5HCO_3^- + 2NO_2 \qquad 6.6.4$$

With complete nitrification/ denitrification:

$$2C_5H_7NO_2 + 11.5O_2 \rightarrow 10CO_2 + 7H_2O + 2N_2 \qquad 6.6.5$$

The three most commonly used aerobic digestion processes are:

1. Conventional aerobic digestion

The following factors are to be kept in mind:

Temperature: Aerobic digesters are generally open tanks. As a result, digester temperatures are dependent on atmospheric temperatures which may fluctuate. Reaction rate increases as temperature increases. The digester tanks should be designed such that sludge is stabilized at lowest liquid operating temperature, and maximum oxygen is provided at maximum liquid operating temperature.

Volatile Solids Reduction: Solids treatment and volume reduction is the major objective of aerobic digestion. Volatile solids are reduced up to 35–50 percent in aerobic digestion. The change in biodegradable volatile solids is represented by:

$$\frac{dM}{dt} = -k_d M$$ where M is the biodegradable volatile solids mass at time t.

Solids reduction is a function of both liquid temperature and SRT.

Tank volume and detention time - Digester tank volume (V) can be given by

$$V = \frac{Q(X + YS)}{X\left(k_d P_v + \frac{1}{SRT}\right)}$$ where Q is average flow rate of the influent, X is influent suspended solid (mg/L), Y is fraction of influent BOD consisting of raw materials, S is influent BOD_5, P_v is volatile fraction of digester suspended solids (in decimal).

Feed solid concentration: If thickening precedes aerobic digestion, higher feed solids concentration will result in higher oxygen input levels per digester volume, longer SRTs, smaller digester volume requirement, easier process control (less decanting in batch operating system), and subsequently increased levels of volatile solids destruction.

2. High-purity oxygen, aerobic digestion:

In this process, high purity oxygen is used instead of air, otherwise the process is similar to conventional aerobic digestion.

3. Auto-thermal aerobic digestion (ATAD):

In this process, the feed sludge is pre-thickened and reactors are insulated to conserve energy produced from oxidation of volatile solids. Within the reactor, aerobic microorganisms are allowed to degrade organic matter to carbon dioxide, water and nitrogen by-products. Main advantages of this process are:

- Retention time is reduced significantly
- Operation is simple with greater reduction of bacteria and viruses

Aerobic digestion is mainly used in the treatment of sludge from activated sludge processes, trickling filters and primary sludge.

Advantages of aerobic over anaerobic digestion are:

- BOD concentration is lower in supernatant liquor.
- Operation is easy and at less cost; no heating requirements unlike anaerobic digestion.
- It is suitable for digesting nutrient-rich biosolids.

Disadvantages of aerobic over anaerobic digestion are:

- Higher cost of aerobic process, if oxygen is supplied instead of air.
- It is significantly affected by temperature, type of mixing/ aeration device, and type of tank material.
- No useful by-product unlike anaerobic processes where a fuel like methane is generated.

PROBLEM 6.6.4

A wastewater sludge comprising of biosolids has the following empirical formula $C_{10}H_{19}O_3N$. The sludge is to be digested aerobically. Based on Section 6.12 of Sawyer et al. (2003), determine the following:

1. Maximum amount of biomass that can be generated per mole of wastewater
2. Maximum amount of biomass that can be generated per g of wastewater

Solution

Based on the overall equation for any biological process noted below and fs,max value of 0.65 (based on computing an average value for equal proportions of carbohydrates, proteins, and fatty acids given in Table 6.5 (SMP 2003), an overall reaction can be written. Ammonia serves as the nitrogen source.

$$R = f_sR_c + f_eR_a - R_d$$

$$f_s{*}R_c = 0.65 * \left[\frac{1}{5}CO_2 + \frac{1}{20}HCO_3^- + \frac{1}{20}NH_4^+ + H^+ + e^- \rightarrow \frac{1}{20}C_5H_7O_2N + \frac{9}{20}H_2O\right]$$

$$f_e{*}R_a = 0.35 * \left[\frac{1}{4}O_2 + H^+ + e^- \rightarrow \frac{1}{2}H_2O\right]$$

$$-R_d = \frac{9}{50}CO_2 + \frac{1}{50}HCO_3^- + \frac{1}{50}NH_4^+ + H^+ + e^- \rightarrow \frac{1}{50}C_{10}H_{19}O_3N + \frac{9}{25}H_2O$$

Final balanced reaction:

$$R = \frac{1}{50}C_{10}H_{19}O_3N + 0.0875O_2 + 0.0125\,HCO_3^-$$

$$+ 0.0125\,NH_4^+ \rightarrow 0.0325C_5H_7O_2N + 0.05\,CO_2 + 0.1075\,H_2O$$

Normalize the reaction to 1 mole of wastewater gives:

$$R = C_{10}H_{19}O_3N + 4.375O_2 + 0.625\,HCO_3^- + 0.625\,NH_4^+ \rightarrow 1.625\,C_5H_7O_2N$$

$$+ 2.5\,CO_2 + 5.375\,H_2O$$

Therefore,

1. Biomass production per mole of wastewater is 1.625 mole of biomass/ mole of wastewater.
2. Biomass production per g of wastewater is 1.625*113/201 = 0.913 g biomass/ g wastewater solids.

6.7 SLUDGE DEWATERING AND DISPOSAL

Sludge dewatering and final disposal are the last two steps in sludge treatment and management.

6.7.1 Sludge Dewatering

The objective of sludge dewatering is to reduce the volume of water in sludge, i.e., its moisture content. The solids content of dewatered sludge ranges from 20 to 45 percent (Metcalf and Eddy 2003).

Some of the reasons for dewatering sludge are:

i. Handling of dewatered sludge is easier than thickened sludge.

ii. Volume of dewatered sludge is less so it can be transported easily, thus, reducing costs.

iii. It helps to increase the calorific value of sludge and is done before incineration of sludge.

iv. It is done before composting to minimize bulking agent requirements.

v. It precedes landfilling of sludge to reduce leachate production at the landfill site.

Processes used for dewatering are:

1. Centrifugation

Two types of centrifuges used are Solid-Bowl Centrifuge and High-Solids Centrifuge (Metcalf and Eddy 2003).

The solid-bowl centrifuge (described earlier in sludge thickening) separates the sludge into two parts - a sludge cake (10-30 percent solids) and a centrate. The centrate is composed of fine solids and is sent back to the treatment system. Polymer addition enhances performance. The high-solids centrifuge is a modification of solid-bowl centrifuge. It has a longer bowl, a modified scroll with pressing action, a lower bowl speed and therefore, gives a dryer sludge cake.

Performance is affected by feed flowrate, speed of scroll, and particle properties. Increase in residence time or chemical conditioning results in better capture of solids. Addition of lime helps in odor control. This process requires less area and has low initial cost. However, considerable power costs offset the low initial cost.

2. Belt-Filter Press

In this method, conditioned sludge is thickened in a gravity drainage section. This is followed by the squeezing action of porous cloth belts in a low pressure section. Further dewatering can be achieved by subjecting the sludge to shearing forces in a high pressure region. Drier cakes are obtained with feed sludge of higher solids concentration.

3. Filter Presses

In this method, high pressure is used for dewatering, giving high concentration cake solids and good solids capture. Its limitations are complexity of equipment, high chemical and labor costs. It is of two types:

A. **Fixed-Volume, Recessed-Plate Filter Press:** In this filter press, an array of rectangular plates, having recesses on both sides, are mounted vertically on a frame with a fixed and movable head. There is a filter cloth over each plate. Hydraulic rams hold the plates together while pressure is applied. The water comes out through the filter cloth and plate outlets.

B. **Variable-Volume, Recessed-Plate Filter Press/ Diaphragm Press:** In this filter press, the same arrangement as the fixed-volume filter press is used except that a rubber diaphragm is placed behind the filter media. The expansion of the rubber diaphragm during the compression step reduces the cake volume.

4. Sludge Drying Beds

It is one of the most widely used methods for dewatering. The main advantages of using this method are high solids content in the dried sludge, low cost, and less skill requirement. Its limitations are large land and labor requirement, climate dependence and odor problem.

The following types of drying beds are used:

A. **Conventional Sand Drying Beds:** They are generally used for small communities. Sludge is placed on a 230-300mm deep sand bed in a layer of thickness 200-300mm. Individual beds are usually 6m wide and 6-20m long (Metcalf and Eddy 2003). Dewatering occurs by evaporation into the air and by drainage through the sand which calls for the presence of an under-drainage system. The drainage lines should have a covering of coarse gravel.

   Sludge removal is done after it has become sufficiently dry. It is done using shovels (manual), scrapers, or mechanical sludge removers. Open beds are used in isolated areas to avoid odor complaints and where land is available. Covered beds with enclosures are used when dewatering is to be done throughout the year.

B. **Paved Drying Beds:** They are of two types: drainage type and decanting type. The drainage type relies on an under-drainage system and sludge removal is done using front-end loaders. The decanting type depends on greater evaporation by decanting of a supernatant and mixing of sludge.

C. **Artificial-Media Drying Beds:** They are of two types: steel wedge-wire type and polyurethane type. In the former, a horizontal open drainage medium of stainless steel wedge-shaped bars is used. The flat part of the wedge lies on the top. This method has the benefits of higher throughput, rapid drainage, easier maintenance and no clogging. In the latter, 300 mm interlocking panels of polyurethane are installed in steel trays. Each panel has an inbuilt underdrain system and 8 percent open area for dewatering (Metcalf and Eddy 2003). Even dilute sludge can be dewatered by this method and front end loaders can easily clean the units.

D. **Solar Drying Beds:** These beds consist of a rectangular base, a translucent covering, sensors to measure drying conditions, ventilation fans, a mobile device called mole which agitates the sludge solids and a microprocessor which regulates the drying environment. Solar radiation is the driving energy. The microprocessor, depending upon the climate, activates operations to optimize moisture absorption.

5. Reed Beds

Reed beds are made up of layers of sand and gravel. Reeds are planted in the gravel region below the sand. Sludge is applied after the reeds have grown sufficiently high. The reeds absorb water from the sludge. Their growth ensures availability of drainage paths. Besides, biological stabilization of the sludge takes place during oxygen transfer from the roots.

6. Lagoons

They can be used for dewatering of digested sludge. They are not recommended for dewatering untreated or limed sludge. Dewatering depends mainly on evaporation and therefore, climate plays a significant role in the performance of the lagoons. Subsurface drainage is restricted by environmental regulations. Lagoons may have to be lined if an aquifer used for water supply is located below them.

PROBLEM 6.7.1

Determine the density of sludge containing 5 percent solids. Assume density of solids is 1400 kg/m^3 and density of water is 1000 kg/m^3.

Solution

Density of sludge = 1400 kg/m^3*0.05 + 1000 kg/m^3*0.95 = 1020 kg/m^3

PROBLEM 6.7.2

The sludge underflow from a gravity thickener has a flow rate of 0.2 MLD and a solids concentration of 40,000 mg/L. The sludge was centrifuged and the resultant total solids concentration of the solid waste generated was 0.5 g/g. Centrifugation removes 90 percent of the solids. Determine the moisture content of the biosolids coming out of the centrifuge, concentration of biosolids in clarified water and volume of water clarified by centrifugation. Assume density of sludge is 1.016 kg/L.

Solution

Step 1: SS in sludge from gravity thickener underflow = 40,000 mg/L

Flow to centrifuge, Q^0 = 0.2 MLD = 2,00,000 L/d

Mass of sludge to centrifuge = 2,00,000 L/d *1.016 kg/L = 2,03,200 kg/d

Mass of solids to centrifuge (dry weight) = Q^0C^0 = 2,00,000 L/d*40,000 mg/L * 1 kg/10^6 mg = 8,000 kg/d

Step 2: Moisture content of solids coming out of centrifuge = 50% since 0.5 g solids/g waste is given.

Step 3: Mass of solids removed by centrifugation, 90% efficiency = 7,200 kg/d

Mass of centrifugate (wet weight) which contains water and solids = 7,200 kg/d / (0.5 g/g) = 14,400 kg/d

Therefore, volume of centrifugate = 14,400 kg/d/(1.016 kg/L) = 14,173.3 L/d

Therefore, volume of water clarified by centrifugation = 2,00,000 – 14,173.3 = 1,85,527 L/d

Concentration of biosolids in clarified water = $800 * 10^6$ mg/d/(1,85,527 L/d) = 4,312 mg/L

6.7.2 Sludge Disposal

There are several options for disposal of dried sludge solids. These solids can be incinerated with or without energy recovery, they can be used as compost if there is no contamination by toxic organic or inorganic compounds (pesticides, heavy metals or others), or they can be used as intermediate landfill cover.

There are several benefits associated with land application of sludge solids since these solids are rich in nutrients and can serve as fertilizer supplement and soil conditioner. However, there are health concerns about ingesting crops grown in waste sludge, and toxicity due to heavy metals, nitrates and pesticides. For these reasons, sludge solids are used for the production of forage crops only.

6.8 DISINFECTION

The primary objective of disinfection is the destruction or removal of vegetative pathogens. Disinfection is different from sterilization which implies destruction of all life forms, viz. microbes, pores, cysts, viruses. Disinfection is the final water treatment process and is used in all *drinking water treatment* plants where water has to be stored and distributed to consumers. Many *wastewater treatment plants* disinfect treated wastewater prior to discharge into a receiving water body.

6.8.1 Disinfection of Drinking Water

Disinfection can be done using physical or chemical methods. Common physical methods include membrane filtration, UV or X-ray radiation, and chemical methods include the use of chemicals like chlorine, chloramine, chlorine dioxide, ozone, and potassium permanganate.

Physical disinfection

Membrane filtration and UV radiation are common physical disinfection methods used in drinking water treatment. Membrane filtration was covered in the previous chapter and UV radiation is described in brief later. X-ray radiation has no known application in drinking water treatment and is used mostly in the food processing industry, mainly for grain storage.

Chemical disinfection

In chemical disinfection, chemicals are used to oxidize microbes present in water. Therefore, chemicals with high oxidation potentials are expected to be the most effective disinfectants. A summary of the oxidation potentials of different chemicals is provided in Table 6.6.

Table 6.6 Oxidation potential of various chemicals

Disinfectant	Oxidation potential (Volts)
Fluorine	–3.06
Hydroxyl free radical (OH^*)	–2.80
Oxygen (atomic)	–2.42
Ozone (O_3)	–2.07
Hypobromous acid (HOBr)	–1.59
Hypochlorous acid (HOCl)	–1.49
Chlorine (Cl_2)	–1.36
Oxygen (molecular)	–1.23
Bromine (Br_2)	–1.07
Chlorine dioxide (ClO_2)	–0.95
Monochloramine (NH_2Cl)	–0.75
Dichloramine ($NHCl_2$)	–0.74

Fluorine which is the strongest oxidizing agent in the table is hazardous to health in high concentrations and cannot be used as a disinfectant in drinking water treatment. Hydrogen and oxygen free radicals are formed during ozonation and are highly unstable. Ozone has an extremely short half-life ranging from seconds to hours in drinking water depending on the nature and concentration of disinfectant demand compounds (van Gunten 2003). Due to this highly unstable nature, it cannot be used to maintain a residual during storage and distribution of drinking water. Therefore, it is often used as a pre-disinfectant but never as a final disinfectant. The remaining halogen compounds listed in Table 6.6 are used as disinfectants but they are less effective as compared to chlorine. Only chlorine, chloramine and ozone are discussed in this section.

6.8.1.1 *Chlorine*

The most popular and commonly used final disinfectant in drinking water treatment is chlorine; it has been in use in public drinking water treatment systems since 1908 in the US. Chlorine has a relatively high oxidation potential and allows maintenance of a residual in the distribution system. Chlorine oxidizes natural organic matter (NOM) and removes compounds causing taste and odor problems. It also prevents the growth of algae and bacteria in storage reservoirs and water supply systems. Chlorine also helps in the removal of iron and manganese by oxidizing the reduced forms of these elements. Synthetic organic compounds can also be removed by chlorine.

There are some disadvantages associated with chlorine. These disadvantages include: high hazard potential due to its corrosive and explosive potential which makes it difficult to handle, store and transport. Chlorine has a pungent smell and a disagreeable taste. It also causes dermal and eye irritation. Moreover, it is more effective against bacteria rather than spores, cysts and viral particles. Use of chlorine also results in the formation of disinfection by-products (DBPs) which are potentially mutagenic, carcinogenic, and teratogenic.

Chlorine is a strong oxidizing agent and when added to water, hypochlorous acid (HOCl) is formed. Chlorine can be applied as a gas (reaction 1), as a liquid in the form of bleach (reaction 2) or in solid (powder) form (reaction 3) as shown in the following reactions. In all cases, hypochlorous acid (HOCl) or hypochlorite ion (OCl^-) are formed.

1. $Cl_2 + H_2O \rightarrow H^+ + HOCl$ $\qquad pK_a = 3.39$

2. $NaOCl \rightarrow Na + OCl^-$

3. $Ca(OCl)_2 \rightarrow Ca^{2+} + 2OCl^-$

Hypochlorous acid (HOCl) and hypochlorite ion (OCl^-) are in equilibrium with each other and their respective fractions will change depending on the solution pH. The two species are approximately equal in concentration at a pH of 7.57.

$$HOCl \leftrightarrow H^+ + OCl^- \qquad pK_a = 7.57$$

The hypochlorite ion then degrades to a mixture of chloride and chlorate ions:

$$3\,OCl^- \rightarrow 2\,Cl^- + ClO_3^-$$

Sodium hypochlorite can decompose to chlorite (ClO_2^-) and chlorate (ClO_3^-) ions as well as chloride during storage. Disinfection efficiencies for all microbial species: bacteria, *Giardia* cysts and viruses, are known to increase with decrease in pH. Since HOCl is the predominant species below pH 7.57, it is considered to be a better disinfectant than hypochlorite ion.

Chlorine demand

When chlorine is added in small increments to any water sample, the chlorine residuals measured are likely to follow the same trend as shown in Figure 6.12. No residual chlorine is formed on addition of chlorine initially since the chlorine added is consumed in reactions with NOM and other reduced inorganic compounds present in water. After Point A, a small increase in combined chlorine residual is observed which increases in direct proportion to the applied chlorine dose up to point B. This is due to the formation of chlorinated organic compounds which result in a measurable combined chlorine residual while the free chlorine residual remains zero. From point B onwards, a decrease in combined chlorine residual is observed and is attributed to the destruction and loss of these chlorinated organics mainly by volatilization. Further increase in chlorine dose will result in point C after which there is an increase in free chlorine residual and the combined chlorine residual remains constant. Point C is the break-point which is so termed since free chlorine residual is formed only after this point. Therefore, for maintenance of a free chlorine residual for long-term protection of drinking water during storage and supply, it is essential to apply chlorine beyond the breakpoint C.

This curve is called the breakpoint chlorination curve and it illustrates the loss of residual chlorine due to chlorine demand compounds present in any drinking water. Other disinfectants are also subject to disinfectant demand for similar reasons. Measurements of free chlorine residual include hypochlorous acid and hypochlorite ions. Total chlorine residuals can be measured and these include all chlorinated compounds formed after chlorination. The difference between total and free chlorine is the combined chlorine residual.

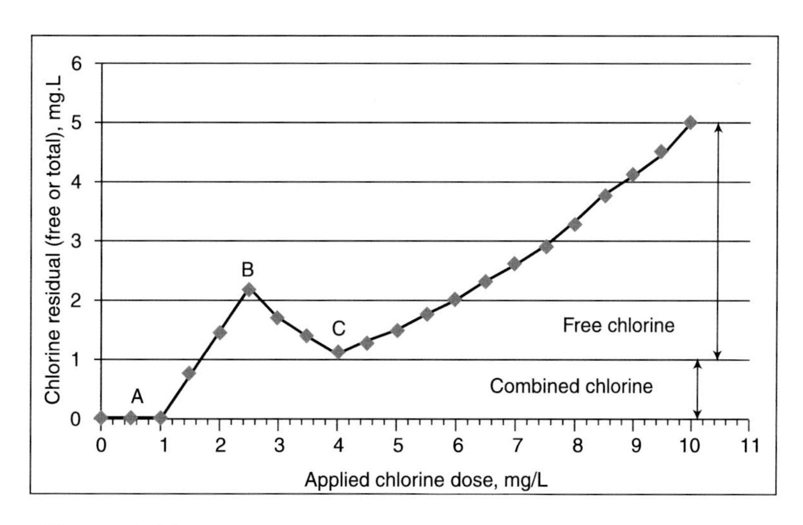

Figure 6.12 Chlorine residuals as a function of applied chlorine dose showing chlorine demand.

6.8.1.2 Chloramine

Chlorine can be mixed with ammonia at the site of application to generate chloramines. Reactions of chlorine with ammonia are noted below:

$$NH_3 + HOCl \rightarrow NH_2Cl + H_2O \qquad \text{formation of monochloramine}$$

$$NH_2Cl + HOCl \rightarrow NHCl_2 + H_2O \qquad \text{formation of dichloramine}$$

$$NHCl + HOCl \rightarrow NCl_3 + H_2O \qquad \text{formation of nitrogen trichloride}$$

The amount of chloramine formed is a function of pH and chlorine to ammonia ratio. Chloramine is a much weaker disinfectant and has much slower reaction rates compared to chlorine but also leaves a durable residual in water systems. It has been used in drinking water treatment for this reason. It is not capable of disinfecting virus and *Giardia* effectively (Krasner et al. 2006). Due to its slow reaction rates, it is capable of penetrating deeper into biofilms, allowing their removal. Chlorine is a stronger oxidizing agent and may not be able to penetrate deep into the biofilm layers as it reacts with the surface layer more easily and is consumed or depleted there.

Problems with chloramine include the formation of DBPs. Even though fewer chlorinated DBPs are formed, higher brominated and iodinated DBPs can be formed and these compounds are often more toxic than chlorinated DBPs (Krasner et al. 2006). Other problems associated with chloramine are difficulty in handling ammonia, odor problems with ammonia, toxicity of ammonia to fish, and during kidney dialysis. Chloramines can also act as final disinfectants but their oxidation potential is less than chlorine and therefore, they are less effective.

6.8.1.3 Ozone

Ozone, like chlorine, is a very strong oxidizing agent and works by direct oxidation of cellular material. It destroys the cell wall resulting in the leakage of cellular constituents out of the cell.

Oxygen free radicals are generated by breaking the diatomic molecule of oxygen gas which in turn combines with molecular oxygen to generate ozone:

$$O_2 \rightarrow 2O^{\bullet}$$

$$O_2 + O^{\bullet} \rightarrow O_3$$

Ozone is often used as a pre-disinfectant as it is effective against all kinds of pathogens, unlike chlorine which is not effective against viruses, cysts, and spores. It has a very short half-life in water ranging from seconds to hours and low solubility in water, and cannot be used as a final disinfectant since it does not leave a stable residual like chlorine. Since it is a powerful oxidizing agent (stronger than chlorine), it is capable of inactivating resistant strains of bacteria, cysts of protozoa and viruses.

Problems associated with the use of ozone include greater bacterial regrowth problems since ozone reacts with NOM and reduces the complex organic compounds to smaller, simpler molecules that are easily biodegraded by bacteria and other microbes. Ozonation can be combined with biofiltration as a treatment process prior to final disinfection to solve this problem. Ozone mineralizes a significant fraction of NOM in water and helps to reduce total DBP formation after chlorination.

6.8.1.4 *Factors influencing disinfection*

Several factors are known to influence disinfection and include contact time, concentration of disinfectant, temperature, number of organisms, types of organisms, and composition of the solution. Disinfection efficiency is defined in terms of log removal of microorganisms.

***Number of organisms or concentration of microbes*:** Disinfection efficiency is inversely proportional to the initial microbial concentration while the reaction rate constant remains the same, regardless of the initial microbial concentration.

***Contact time*:** As contact time increases, the degree of inactivation of microbes increases, i.e., disinfection efficiency increases.

***Concentration of disinfectant*:** As the concentration of the disinfectant increases, disinfection efficiency increases.

***Temperature*:** Increase in temperature results in an increase in the reaction rates leading to higher disinfection efficiency.

***Microbial species*:** Different microbial species have different reaction rate constants under the same disinfection conditions of concentration of disinfectant and microbes, temperature and solution composition (Peavy et al. 1985).

***Composition of solution*:** Solution composition affects disinfection efficiency in two ways:

(a) Organic matter and other reduced inorganic compounds exert a disinfectant demand since the disinfectant is consumed in oxidizing these reduced compounds as illustrated by the breakpoint chlorination curve. Thus, the concentration of disinfectant that a microbial organism is effectively exposed to is reduced and so is the disinfection efficiency.

(b) The availability of nutrients including organic matter affects the growth of microbes in solution and can impact their physiology. This in turn can affect their response when exposed to a disinfectant (Goel and Bouwer 2004; LeChevallier et al. 1988). However, nutrient availability did not impact bacterial response to chloramine to the same extent as it did to chlorine. For example, bacteria grown under low nutrient conditions are more resistant to chlorine than those grown under high nutrient conditions.

6.8.1.5 Disinfection model

The earliest model used to describe disinfection was Harriette Chick's law in (1908) (Chick 1908) where the decrease in microbial concentration was directly proportionate to the concentration of microbes in solution. If N is the number of microbes (in this case, bacterial cells) and t = time, then,

$$\frac{dN}{dt} = -kN \tag{6.8.1}$$

Integration of this equation yields: $N = N_0 \exp(-kt)$

Where k is the inactivation rate constant (1/unit time) and t is time of contact. This equation does not account for the concomitant decrease in disinfectant and microbial concentration during disinfection as observed in all experimental studies. This was soon modified to the Chick–Watson's law (1908) after publication of Watson (1908) where the change in disinfectant concentration C was given by the equation:

$$\text{Log } t + n \log C = \text{constant} \tag{6.8.2}$$

Values of n were found to be 5.5 for phenol, 3.8 for mercuric chloride and 0.86 for silver nitrate and the bacteria disinfected in all cases was *Bacillus paratyphosus*.

Equation 6.8.1 can then be modified to

$$\ln (N/N_0) = -k'C^n t \tag{6.8.3}$$

where k′ = inactivation constant for the Chick–Watson model that combines C and N, 1/min

C = disinfectant concentration, mg/L and n = the value determined by fitting data to equation 6.8.2.

Often, *n* in the equation is assumed to be 1 in drinking water treatment practice. Experimental studies show that the change in bacterial concentration is a function of contact time and disinfectant concentration (C*t). C*t values for 2-, 3- or higher log removal are derived from lab-scale studies and are used for designing disinfection processes in water and wastewater treatment.

6.8.1.6 Disinfection by-products (DBPs)

Disinfection by-products (DBPs) are toxic compounds formed due to reaction of disinfectants with natural organic matter (NOM) as shown below:

Natural organic matter + Chlorine (or other disinfectant) + bromine + iodine = DBPs

Many of these DBPs may be cytotoxic and/or genotoxic (including mutagenic and carcinogenic). Some DBPs also cause developmental and reproductive defects such as spontaneous abortions, infertility or low birth weight.

Milestones in the history of DBP monitoring and regulations

First DBP (chloroform – $CHCl_3$) discovered by Rook	1974
USEPA national survey found THMs in all chlorinated drinking waters	1976
National Cancer Institute published results linking chloroform with cancers in laboratory animals	1976
Regulation of THMs to a maximum contaminant level of 100 micro-g/L	1979
Promulgation of disinfectants/DBP rule in the US	1998
Four THMs regulated in India	2012

Major DBPs formed due to the reaction of chlorine with NOM are trihalomethanes (THMs), haloacetic acids (HAAs), haloacetonitriles (HANs), chlorite, cyanogen chloride, and Mutagen X (MX) which is considered to be the most toxic (Richardson 2003). Several DBPs have been monitored and regulated in most developed countries and their maximum allowable concentrations are summarized in Table 6.7.

Table 6.7 DBP regulations in different countries and WHO guidelines; Concentrations are in μg/L

DBP	WHO	USA	EU	India
Total THMs		80	50	
Chloroform	300	70	–	200
Bromoform	100			100
Bromodichloromethane (BDCM)	60			60
Dibromochloromethane (DBCM)	100			100
Bromate	10	10	100	–
Chlorite	700	1000	–	–
Cyanogen chloride				
Haloacetic acids			–	–
Monochloroacetic acid			800	
Dichloroacetic acid			1,500	
Trichloroacetic acid			8,000	

The two major groups of DBPs generally found in most drinking water systems are THMs and HAAs with most of the total organic halogen (TOX) remaining unaccounted for (Richardson 2003). The concentrations of THMs and HAAs are specific to the source water and while it may be possible to find a correlation between the concentrations of THMs and HAAs for one source water, the same correlations may not be applicable to other waters (Malliarou, Collins, Graham, and Nieuwenhuijsen 2005). Examples of DBPs that are currently regulated include trihalomethanes (THMs): chloroform, bromoform, bromodichloromethane and chlorodibromomethane; haloacetic acids (HAAs): monochloroacetic acid, dichloroacetic acid and trichloroacetic acid; and haloacetonitriles (HANs): dichloroacetonitriles and dibromoacetonitriles. Some inorganic compounds that are also regulated as DBPs include chlorite, chlorate, bromate, and cyanogen chloride.

DBP formation is proportionate to its precursor concentrations. Precursors for currently known DBPs include NOM (measured as TOC), chlorine, bromine, and iodine. Surface waters tend to have higher concentrations of NOM than groundwater sources resulting in much higher concentrations of DBPs in chlorinated surface waters rather than in groundwaters (Krasner et al. 2006; Cantor 1982). Three THMs that are regulated contain bromine, and brominated DBPs tend to occur in high concentrations when waters with high levels of bromide are disinfected with chlorine or ozone (Chang, Lin, and Chiang 2001). Brominated DBPs form at a much faster rate than non-brominated DBPs (Westerhoff, Chao and Mash 2004) and are generally more toxic than chlorinated DBPs (Richardson 2003; Richardson, Plewa, Wagner, Schoeny, and Demarini 2007). Also, groundwaters tend to have higher bromine concentrations and this can lead to higher concentrations of brominated DBPs in groundwater supply systems. Coastal areas impacted by saltwater intrusion tend to have high bromine and iodine concentrations in their surface and groundwaters and therefore, are likely to have higher brominated and iodinated DBP concentrations.

Chloramines produce halogenated DBPs as well but not to the same extent as chlorine. They also produce iodinated and brominated compounds which are even more toxic than the chlorinated compounds. Chlorine dioxide also produces DBPs like chlorite, THMs, HAAs, while DBPs formed with ozone include bromates, aldo-ketoacids and aldehydes (Richardson 2003).

Ozone helps to reduce the formation of THMs and HAAs during final chlorination by mineralizing a fraction of the NOM. However, it may result in higher iodinated THM levels (Krasner et al. 2006). Another DBP that is formed by ozone is bromate which is a potent carcinogen and is regulated in the US and EU and guidelines are provided by WHO.

One of the objectives of current drinking water treatment is to minimize DBP formation. The formation of DBPs can be reduced by reducing chlorine requirements. Chlorine requirement can be reduced by pre-treatment for reducing precursor concentrations before chlorination. This can be achieved by:

- Coagulation–flocculation and settling which can be optimized for turbidity and TOC (total organic carbon) removal and is termed enhanced coagulation.
- Treatment of water to remove bromine and iodine.
- Ozonation and biofiltration – Enhanced TOC removal.
- Membrane filtration

Other disinfectants include irradiation with ultraviolet light. Like ozone, it does not produce any residual and can efficiently inactivate bacteria, cysts and viruses. UV radiation produces very few DBPs unlike halogenated disinfectants and is the most popular end-of-pipe treatment method.

Several other disinfectants like iodine, bromine, potassium permanganate, gamma ray irradiation, sonification and some metals like copper and silver are being evaluated as disinfectants and are restricted to small-scale use.

A comparison of microbial versus chemical risks as a function of chlorine added (chlorination level) is shown in Figure 6.13. It is important to note that risk of adverse health effects due to microbial diseases is much greater and acute in comparison to risk associated with exposure to DBPs over a long-term due to chronic exposure to these compounds. Small amounts of chlorine can reduce risks due to microbial diseases in a big way while higher concentrations of chlorine can lead to a small increase in chemical risks (Morris 1978).

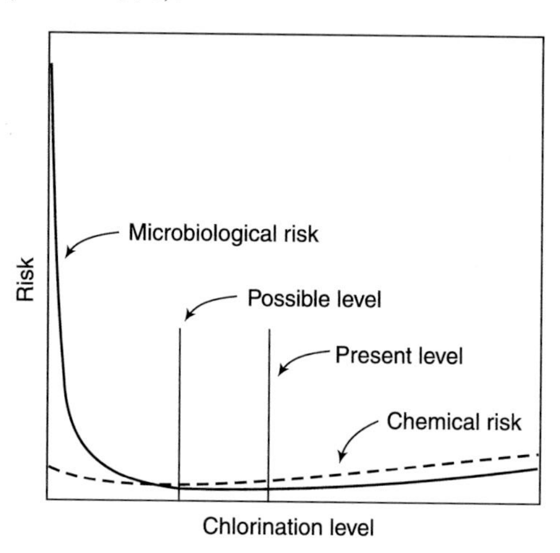

Figure 6.13 Microbial and chemical risks versus level of chlorination (Morris 1978).

6.8.2 Disinfection of Wastewater

Treated wastewater effluent has very high concentrations of bacteria and other microbes including pathogenic species. Therefore, it is necessary to disinfect wastewater prior to discharging it on land or in water. Ambient water quality criteria are based on coliforms and disinfection of wastewater is important for achieving these water quality standards.

Wastewater Pathogens

Untreated wastewater or even secondary treated effluents contain bacteria (mainly enteric bacteria), viruses, protozoa, and helminths, some of which may be pathogenic. These pathogenic organisms pose a potential risk to the health of humans, livestock and other organisms exposed to them. Some of these organisms are discussed here and a list of common pathogens in each category is provided in Table 6.8.

Table 6.8 Diseases caused by some common pathogenic species found in wastewater (Hammer and Hammer Jr. 2008)

Organism group	Pathogenic species found in sewage	Disease caused
Virus	Adenoviruses	Respiratory and eye infections
	Hepatitis A virus	Infectious hepatitis
	Poliovirus	Aseptic meningitis, poliomyelitis
	Calicivirus	Diarrhea
Bacteria	*Salmonella typhi*	Typhoid
	Vibrio cholera	Cholera
Protozoa	*Entamoeba histolytica*	Amoebic dysentery
	Cryptosporidium species	Cryptosporidiosis, diarrhea
	Giardia lamblia	Giardiasis, diarrhea
Helminths	*Ascaris lumbricoides* (roundworm)	Ascariasis
	Taenia solium	Taeniasis
	Trichuris trichiura	Trichuriasis
	Necator americanus, Ancylostoma duodenale (hookworm)	Hookworm disease

Enteric viruses are commonly found in sewage and one of the most virulent is Hepatitis A virus which causes infectious hepatitis. Several other viruses found in human excreta and therefore, sewage, are listed in Table 6.8. Viruses range in size from a few nm to hundreds of nm.

The most common microbial pathogens found in wastewater are *Bacteria* and two examples are provided in Table 6.8. Treated wastewater and ambient water quality standards are based on total coliforms which serve as indirect indicators of fecal contamination and as a means for assessing the efficiency of treatment and disinfection methods. Chlorine is highly effective as a disinfectant against *Bacteria* in comparison to viruses, and cysts of protozoa.

Protozoan cysts are quite similar to helminths in terms of resistance to chlorination and can be infectious even at small doses. Cryptosporidiosis can occur by ingestion of 10 to 30 oocysts of Cryptosporidium species (Tortora, Funke, and Case 2004).

Helminths and their eggs are generally found in small numbers in wastewater and sludge as compared to other organisms. Since the worms do not multiply in the human host, any infection is related to ingestion of helminthic eggs. It takes only one egg to enter the gastro-intestinal tract and cause an infection (Hammer and Hammer Jr. 2008; WHO 1989). In a recent review of soil-transmitted helminths, concentrations of helminthic eggs/L varied widely for different countries and numbers were much higher in wastewater and sludge from developing countries as compared to developed countries (Amoah, Adegoke, and Stenström 2018). Maximum concentration of 16,000 eggs/L in wastewater was reported from Vietnam and 2,300 eggs/L in sludge from China. Helminthic eggs are known to be resistant to low doses of chlorine (USEPA 1999a). Conventional treatment methods such as chlorination are not particularly effective in reducing pathogenic helminths to an acceptable risk level.

Disinfection Processes

Wastewater has to be disinfected prior to discharge into surface water bodies or even on land. Different disinfectants can be used and their effectiveness against different pathogenic species is summarized in Table 6.9.

Table 6.9 Effectiveness of different disinfectants to various pathogenic species

Pathogenic organisms	Chlorine (USEPA 1999a)	Ozone (USEPA 1999b)	UV Radiation (USEPA 1999c)
Helminths	Not effective	No information	No information
Bacteria	Effective	Effective	Effective
Virus	Not effective	Effective	Effective
Protozoan cysts	Not effective	Not effective at low doses	Effective

6.8.2.1 *Chlorination*

Chlorine is the most widely used disinfectant for municipal wastewater. Chlorine is used to disinfect wastewater in the form of chlorine gas, sodium hypochlorite solutions (liquid) or calcium hypochlorite (solid form). All forms of chlorine react with water resulting in the generation of hypochlorous acid (HOCl) and hydrochloric acid. HOCl rapidly dissociates to form the hypochlorite ion which in turn decomposes to chlorite, chlorate and chloride (WHO 2005).

The presence of high concentrations of chlorine-demanding compounds, i.e., organic or inorganic compounds in reduced form requires a high chlorine dose and the chlorine residual is often unstable (USEPA 1999a). Since aquatic ecosystems may be sensitive to chlorine residuals, dechlorination of wastewater is the final step prior to discharge of wastewater. Cysts of parasites like *Cryptosporidium parvum*, *Entamoeba histolytica* and *Giardia lamblia and* helminthic eggs are known to be resistant to chlorine at low doses. Viruses are also not inactivated to the same extent as bacteria at the same chlorine doses (Qasim, Motley, and Zhu 2000).

6.8.2.2 *Ozonation*

Most wastewater treatment plants generate ozone *in situ* by passing a high voltage current (6 to 20 kilovolts) across a dielectric discharge gap that contains an oxygen-bearing gas (USEPA 1999b). The gas stream generated using air contains 0.5 to 3 percent ozone. The gas is transferred to water immediately in ozone contactors which use bubble diffusers, mechanical agitation, pressure injection or packed tower methods. Contact time in the contactor ranges from 10 to 30 minutes. Off-gases containing ozone from a contactor have to be quenched prior to release to the atmosphere. If pure oxygen is used as feed instead of air, the off-gas can be recycled back to the contactor.

Ozone is a very strong oxidant. When ozone decomposes in water, perhydroxyl radical (HO_2) and hydroxyl (OH) free radicals are formed. These free radicals have high oxidizing capacity and are excellent disinfectants. Microbes are destroyed by disintegration of the cell wall (cell lysis). The effectiveness of ozone disinfection depends on the targeted organism, contact time and concentration of ozone. Ozone disinfection is generally used in medium to large sized plants after secondary treatment. The half-life of ozone in water is a function of pH, temperature and the presence of ozone-demanding compounds. At a pH of 7 and temperature of 20 °C, the half-life of ozone in different aqueous solutions ranged from approximately 500 to 5000 seconds (Gardoni, Vailati, and Canziani 2012).

Disinfection of anaerobic sanitary wastewater effluent with ozone was done at doses of 5.0, 8.0, and 10.0 mg O_3/L for contact times of 5, 10, and 15 min. Total coliforms were removed by 2.00 to 4.06 logs, and the inactivation range for *Escherichia coli* was 2.41 – 4.65 logs (USEPA 1999b).

Ozone is more effective than chlorine in destroying viruses and bacteria. There are no harmful residues that need to be removed after ozonation because ozone decomposes rapidly. Also, there are fewer safety problems associated with shipping and handling since ozone is generated on-site. Ozonation can also increase the dissolved oxygen (DO) levels in wastewater thereby eliminating the need for recreation of wastewater prior to discharge.

Ozonation has some disadvantages as well. It may not be economical due to the need for complex equipment and efficient contacting systems (USEPA 1999b). Its reactive and corrosive nature requires the use of corrosion-resistant material such as stainless steel. Also, it is not economical for wastewater with high levels of suspended solids (SS), biochemical oxygen demand (BOD), chemical oxygen demand (COD), or total organic carbon (TOC). Ozone is extremely irritating and toxic, so off-gases from the contactor must be destroyed to prevent worker exposure. The cost of treatment can be relatively high in terms of capital costs and energy usage.

6.8.2.3 *Ultraviolet radiation*

Electromagnetic energy from a mercury arc lamp is transferred to an organism's genetic material (DNA and RNA) thus destroying the cell's ability to reproduce. The effectiveness of UV disinfection is dependent upon the characteristics of the wastewater, the intensity of UV radiation, amount of exposure time and reactor configuration (USEPA 1999c).

The advantages of UV disinfection are that it is effective in inactivating most virus, spores, and cysts. Also, UV disinfection is a physical process which makes it user-friendly for the operator by eliminating problems with handling and transportation of chemicals. Also, it requires less space and less contact time (20 to 30 sec for low-pressure lamps) and leaves no harmful residuals.

UV disinfection has some drawbacks. At low doses, it may not be effective in inactivating some viruses, spores and cysts. Further, some microbes can repair and reverse the destructive effects of UV through a repair mechanism known as photo-reactivation or in the absence of light it is known as 'dark repair'. Sometimes, turbidity and total suspended solids in the wastewater can render UV disinfection ineffective. UV disinfection with low-pressure lamps is not as effective for secondary effluent with TSS levels above 30 mg/L.

6.8.2.4 *Membrane filtration*

Membrane technologies are highly effective in removing various microbial species from treated wastewater by physically filtering out microorganisms (EPA Victoria 2002). Microfiltration is the most commercially viable technology for the disinfection of treated wastewater. The wastewater passes through membrane fibres or hollow cylinders with millions of microscopic pores. This disinfection process does not require addition of reactive chemicals and no toxic disinfection by-products are formed. Microfiltration efficiently reduces the concentrations of particulate matter, bacteria, viruses, algae, and protozoa. Protozoa are generally larger than 0.2 microns and are removed effectively by microfiltration, giving this method an advantage over other technologies. Viruses larger than 0.2 micron (which includes most enteric viruses) are also reduced effectively. The main disadvantages associated with microfiltration include the potentially high capital costs, the resultant concentrate with significant microbial contamination, and the handling and management of contaminated chemicals produced by periodic cleaning of the membranes.

Reverse osmosis can also be used for disinfecting wastewater but is too expensive in terms of capital and operating costs. It may be a viable option where the wastewater is to be reused rather than discharged to a water body.

6.8.2.5 *Lagoons*

The storage of secondary treated wastewater in pond systems with a detention time of 30 days allows natural disinfection to take place before discharging or reusing the treated wastewater (EPA Victoria 2002). Natural disinfection can occur via sunlight or natural microbial die-off. Natural disinfection processes are affected by a number of factors: turbidity of the wastewater as it affects sunlight penetration; and the ineffectiveness of sunlight in seawater as compared to freshwater. Temperature, pH, adsorption, and sedimentation further influence natural disinfection of wastewater stored in lagoons. Re-infection of ponds by bird populations can also pose a problem for operators. Algal blooms in the ponds during summer will also reduce the efficiency of the natural disinfection process.

PROBLEM 6.8.1

Bacteria killed during disinfection are generally modeled as a first-order decay reaction where $\ln(N/N_0) = -kt$ and N and N_0 are bacterial concentrations at time t and t_0, respectively. If bacterial concentration is halved in 15 min, how long will it take to achieve 3-log removal and 2-log removal.

Solution

Microbial disinfection is described by Chick's law: $N = N_0 * \exp(-kt)$

Where N = microbial concentration at time t; k = inactivation or disinfection rate constant

N_0 = Microbial concentration at t = 0

For $t = 15$ min, $N = 0.5N_0$

Therefore solving for k, using $\log(N/N_0) = -kt$ results in k = 0.02 1/min

Using this k value in the above equation, we can determine t for 2- and 3-log removal.

For 2-log removal, t = 100 min

For 3-log removal, t = 150 min.

PROBLEM 6.8.2

The following data for microbial concentrations were obtained during disinfection experiments. Determine k, assuming the reaction order is 1st-order. This problem is best solved graphically.

time, min	N, cfu/mL
0	7.00×10^5
0.5	5.00×10^4
1	4.16×10^3
2	1.08×10^3
3	3.60×10^2

Solution

Step 1: Plot the above data in a spreadsheet and determine the k value assuming exponential decrease in microbial cell concentration as shown in the following graph.

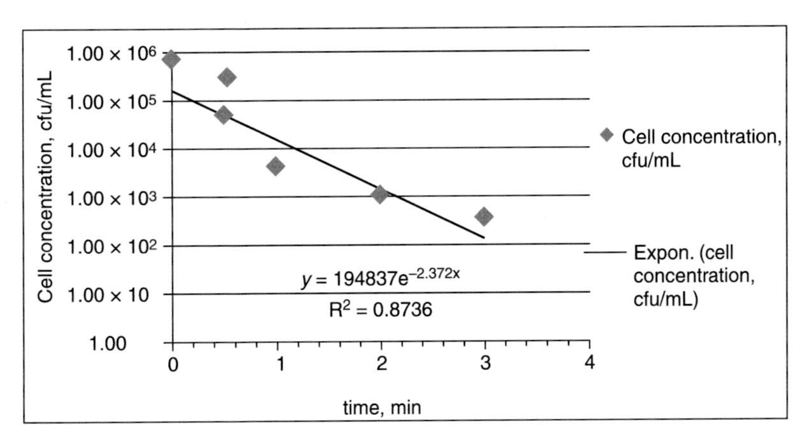

Step 2: From the exponential equation determined by curve fitting, k = –2.372 1/min.

PROBLEM 6.8.3

A water treatment plant is designed for a flow of 1 MLD. The contact time in a chlorine contactor is 20 min, chlorine demand of water is 8 mg/L and chlorine residual to be maintained is 2 mg/L. Calculate the volume of the contactor and the amount of chlorine required/day.

Solution

Flow = 1.00×10^6 L/d = 6.94×10^2 L/min

Detention time = 20 min

Volume of contactor = 1.39×10^4 L = 13.89 m^3

Applied chlorine dose = 10 mg/L

Amount of chlorine required = 1.00×10^7 mg/d = 10.00 kg/d

PROBLEM 6.8.4

A water treatment plant is designed for a flow of 1 MLD. Pre-ozonation is to be provided followed by coagulation–flocculation–settling, slow sand filtration, and final disinfection with chlorine. Detention time in the ozone contactor should be 5 min, ozone demand of water is 2 mg/L. Calculate the volume of the contactor and the amount of ozone required/day.

Solution

Flow = 1.00×10^6 L/d = 6.94×10^2 L/min

Contact time = 5 min

Volume of contactor = 3.47×10^3 L = 3.47 m^3

Height of contactor, if it is assumed to be cylindrical with diameter of 1 m, H = 4.421 m

Applied ozone dose = 2 mg/L

Amount of ozone required/d = 2.00×10^6 mg/d = 2.00 kg/d

PROBLEM 6.8.5

Residual chlorine and chlorination data are provided below. Plot the chlorination curve and determine breakpoint chlorine dose. A free chlorine residual of 2 mg/L is to be maintained in the effluent of the water treatment plant. What is the initial chlorine demand and the total chlorine demand (kg/d) for a flow rate of 1 MLD?

Chlorine applied, mg/L	Chlorine residual (free or total), mg/L
1	0
2	1
3	2
4	3
5	2
6	1.7
7	2.3
8	3
9	3.75
10	4.5

Solution

Step 1: Plot a graph of chlorine residual versus chlorine dose as shown in the following figure.

Step 2: Breakpoint chlorine dose = 6 mg/L

Free chlorine residual to be maintained = 2 mg/L

Applied chlorine dose for maintaining a residual of 2 mg/L can be read from the graph. In this case, it is 9 mg/L.

Initial chlorine demand = 1 mg/L

Total chlorine demand = chlorine applied - free chlorine residual = 7 mg/L

Q, flow = 1.00×10^6 L/d

Initial chlorine demand = 10,00,000 mg/d = 1 kg/d

Total chlorine required = 90,00,000 = 9 kg/d

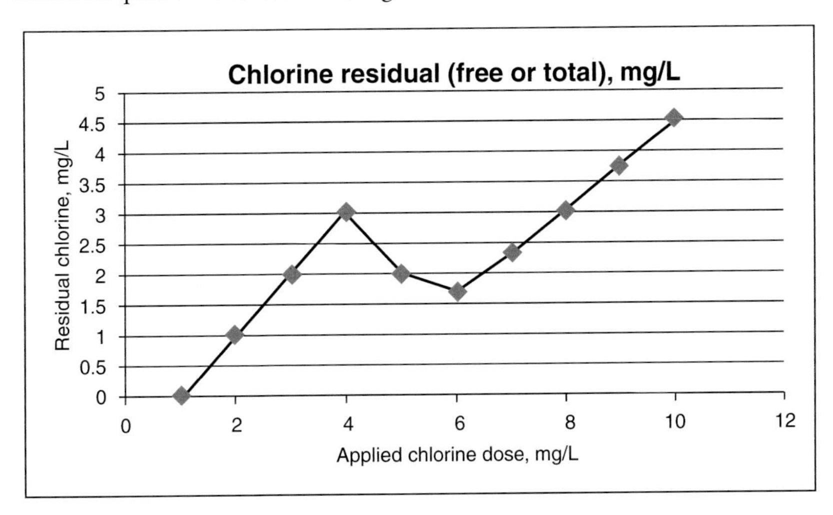

Study Outline

Wastewater characteristics

Water supplied that is returned as wastewater is termed sewage. In general, 'full flushing' sewerage systems are provided only in urban centers. Characteristics of wastewaters in these systems are the basis for the design of wastewater treatment plants.

Quantity

Due to increase in population as well as increase in per capita water consumption rates, there has been a tremendous increase in the quantity of water used and therefore, sewage generated. Wastewater generation can fluctuate on a daily as well as annual or seasonal basis.

Wastewater collection and treatment systems can be of two types: combined and separate wastewater systems.

Quality

Basic water quality parameters that are generally used to characterize municipal wastewater are BOD, COD, TS, TSS, TDS, pH, temperature, turbidity, TKN, and phosphate.

Secondary treatment: suspended growth processes

The objective of biological treatment of wastewater is BOD removal. If the biomass is in suspension, the process is a suspended growth process. Examples of aerobic suspended growth processes are activated sludge process (ASP), sequencing batch reactors (SBR), aerated lagoons (AL), oxidation ponds (OP), and waste stabilization ponds (WSP).

Activated sludge process

Activated sludge process (ASP) is so named because it involves the production of an active mass of microbes which stabilize waste under aerobic conditions.

Conventional ASP is a two-stage process with an aeration tank followed by a secondary clarifier. In the aeration tank, contact time is provided for mixing of the influent with suspended biomass solids (microbes) and aeration of the influent leading to BOD removal. The biomass generated in the aeration tank is removed by settling in the secondary clarifier. The clarified effluent at the top of the clarifier is discharged after disinfection to a receiving water body. A fraction of the solids removed from the clarifier is taken for sludge treatment (digestion and dewatering) prior to disposal in a landfill, incinerator, or for use as compost. The remaining solids are recycled back to the head of the aeration tank to maintain constant MLVSS concentration in the tank.

Sequential batch reactor

Unlike an ASP which is a continuous flow system, SBRs are run in batch mode with complete mixing of the contents of the reactor. Higher BOD removal efficiencies can be achieved with SBRs compared to CSTR processes like ASP.

Aerated lagoons

Aerated lagoons are similar to extended aeration ASPs with an important difference. These are continuous flow-through systems which can be operated without sludge recycle. Oxygen or air is provided by surface or diffused aerators.

Secondary treatment: fixed film processes

In fixed-film or attached growth processes, microorganisms are attached to an inert medium or substratum such as plastic or rock forming a biofilm (layer of microbial cells). These processes require less energy, are easier to control and have better sludge thickening properties. Examples of attached growth processes include trickling filters, biofilters, rotating biological contactors, fluidized bed reactors, and roughing filters. In all cases, a secondary clarifier follows the main unit for removal of biomass generated in the wastewater.

Trickling filter

Trickling filters are attached growth aerobic biological processes where the media is not submerged in water. The reactor is packed with plastic or rock materials and wastewater is distributed over the packing continuously using a rotary distributor. Large void spaces in the reactor allow air to circulate freely and naturally (natural draft) or air is provided mechanically (forced draught). Sludge from the process is taken for digestion and treated wastewater is available for further treatment, discharge, or reuse purposes.

Rotating biological contactor

An RBC is followed by a secondary clarifier and serves a similar purpose as the aeration tank in ASP. It has a tank with a central shaft on which several closely spaced circular disks made of high density polypropylene, polystyrene, or polyvinyl chloride are mounted. Corrugations on the disks increase the available surface area for biofilm formation. Wastewater flows into the tank at one end resulting in submergence of 35 to 40 percent of the disk area. The disks rotate at speeds of 1–2 rpm and are attached to mechanical drive units. Due to continuous rotation, the biofilm formed on the surface of the disk is alternately submerged and then exposed to air resulting in aerobic growth of biomass and degradation of wastewater BOD. Breaking of the shaft has been a major cause of failure for RBC units while smaller units have provided higher removal efficiencies than larger units.

Membrane bioreactor

A relatively new municipal wastewater treatment technology is the membrane bioreactor (MBR). MBR is a suspended growth process where an aeration tank is followed by a membrane reactor. The reactor utilizes microporous membranes for solid/liquid separation instead of conventional secondary clarifiers.

Clarification – Type 3 and Type 4 settling

Suspensions with high solids concentrations such as sludge from biological processes can undergo Hindered settling (Type 3) or Compression settling (Type 4), respectively.

Hindered settling - Type 3

When the entire mass of particles tends to settle as a unit and the particles remain in the same position with respect to each other while settling it is termed Hindered settling.

Compression settling – Type 4

In suspensions with very high concentrations of particles a visible structure is formed. Therefore, settling is possible only by compression of the previously formed visible structure and is called compression settling. The cause of compression is the weight of suspended particles which are continuously deposited on the previous structure.

Sludge treatment – thickening

Sludge treatment is a major challenge in wastewater treatment. About 2 percent of the total volume of wastewater ends up as sludge and it generally contains about 1 to 5 percent solids (10,000 to 50,000 mg/L of solids).

Several sludge treatment processes are used in wastewater treatment plants and the treatment train is concentration or thickening followed by digestion, dewatering, heating or drying and finally, disposal.

Thickening can be defined as increasing the solids concentration of sludge by reducing the liquid content. Various methods used for thickening include co-settling thickening, gravity thickening, flotation thickening, centrifugation, gravity belt thickening, and rotary drum thickening.

Sludge treatment – digestion

The objective of sludge digestion is to reduce the volume of sludge, and stabilize the biosolids in the sludge, i.e., reduce any potential risk due to pathogen presence in the bioactive solids.

Sludge digestion is of two types: aerobic or anaerobic. Anaerobic digestion is more popular since it allows energy recovery. Aerobic digestion is similar in design to the ASP.

Anaerobic digestion

Anaerobic digestion is used for the stabilization of biosolids in the absence of oxygen. Major advantages of anaerobic digestion include the generation of biogas, energy savings, and less sludge production in comparison to aerobic biological processes.

It is a complex 3-step process which includes hydrolysis, acidogenesis and methanogenesis. Maintaining low H_2 concentration is critical for running an anaerobic digester efficiently. Biogas generated in this process is a combination of carbon dioxide and methane. Separation of the two gases is essential for energy recovery.

Factors important in the design of anaerobic digesters are hydraulic detention time, solids retention time, pH, temperature, alkalinity, and nutrients.

Anaerobic digesters can be run in batch or continuous-flow as CSTRs or PFRs. They can be run as mesophilic or thermophilic reactors, or as high solids digesters (>22 percent solids), or low solids digesters. These digesters can be single-stage or multi-stage digesters.

Aerobic digestion

Aerobic digestion is used for reducing the mass of sludge solids. It is similar to the activated sludge process except that the biosolids are maintained in endogenous phase. As substrate gets depleted, microorganisms start consuming their own protoplasm to obtain energy which is referred to as the decay or endogenous phase.

Sludge treatment – dewatering and disposal

Sludge dewatering

The objective of sludge dewatering is to reduce the volume of water in sludge, i.e., its moisture content. The solids content of dewatered sludge ranges from 20 to 45 percent.

Processes used for dewatering are centrifugation, belt-filter press, filter presses, sludge drying beds, reed beds, and lagoons.

Sludge disposal

There are several options for disposal of dried sludge solids. These solids can be incinerated with or without energy recovery, they can be used as compost if there is no contamination by toxic organic or inorganic compounds (pesticides, heavy metals or others), or they can be used as intermediate landfill cover.

Disinfection

The primary objective of disinfection is the destruction of vegetative pathogens.

Disinfection of water

Disinfection is the final water treatment process and is used in all *drinking water treatment* plants where water has to be stored and distributed to consumers. Disinfection can be done using physical or chemical methods. Common physical methods include membrane filtration, UV or X-ray radiation, and chemical methods include the use of chemicals like chlorine, chloramine, chlorine dioxide, ozone, and potassium permanganate.

The most popular model for describing disinfection is the Chick–Watson model.

A breakpoint disinfection curve has to be developed for each individual water sample to determine its disinfectant demand.

Three disinfectants are commonly used in drinking water treatment plants: chlorine, chloramine and ozone. Chlorine remains the most popular disinfectant since free chlorine residuals can be maintained in the storage and distribution system unlike with ozone.

Chloramine is used as an alternative disinfectant to chlorine but is weaker than chlorine. The amount of chloramine formed is a function of pH and chlorine to ammonia ratio.

Ozone is the strongest of the three disinfectants and is generated *in situ*. It is used only as a pre-disinfectant since it is highly unstable and no residual is available.

Disinfection of wastewater

Untreated wastewater or even secondary treated effluents contain bacteria (mainly enteric bacteria), viruses, protozoa, and helminths, some of which may be pathogenic. These pathogenic organisms pose a potential risk to the health of humans, livestock and other organisms exposed to them. Therefore, it is necessary to disinfect wastewater prior to discharging it on land or in water.

Ambient water quality criteria are based on coliforms and disinfection of wastewater is important for achieving these water quality standards.

Disinfection methods that can be used for wastewater include chlorination, ozonation, membrane filtration, UV radiation, and lagoons.

Study Questions

1. Why should municipal wastewater be treated prior to discharge? Where and under what conditions can wastewater be discharged?
2. What is the importance of coliforms, BOD, COD, and TSS for wastewater discharge standards? Why are these parameters associated with ambient water quality?

3. What is the average COD: BOD_5 ratio for sewage in India? Why does this ratio increase after biological treatment?
5. What are total solids and how can they be categorized into different fractions? What is the importance of each fraction of total solids?
6. How does wastewater flow vary on a daily and an annual basis?
7. Define attached and suspended growth biological processes and provide examples of each.
8. What are the most important design parameters for activated sludge process design?
9. Why are tanks in series similar to PFRs in terms of process efficiency?
10. All biological processes require clarification of treated wastewater – why?
11. Compare and contrast a trickling filter with a rotating biological contactor.
12. Define Type 3 and Type 4 settling.
13. Why is it necessary to reduce sludge volume and what methods are available for doing so?
14. Compare and contrast aerobic and anaerobic digestion for sludge treatment.
15. What methods are available for sludge disposal?
16. Why is disinfection of treated water and wastewater necessary?
17. What are disinfection by-products and how are they formed?
18. Compare microbial and chemical risks due to disinfection of water or wastewater.
19. Compare and contrast the different disinfectants used in water or wastewater treatment.
20. What pathogens are likely to be present in raw water sources and sewage? List names of species and diseases that they are associated with.

CHAPTER 7

Strategies for Water and Wastewater Treatment

Learning Objectives

- *Estimate the quantity and quality of wastewater generated in a community*
- *Analyze water quality data and determine treatment needs for different water uses*
- *Identify contaminants in untreated water or wastewater and the extent of removal required based on prevalent standards*
- *Describe the objective of each treatment unit to be used in water or wastewater treatment*

The first step in the design of water and wastewater treatment plants is identification of appropriate treatment processes for meeting the required standards or treatment objectives. This requires identification of contaminants in water or wastewater based on regulatory standards that are applicable and an understanding of the extent to which treatment objectives can be achieved by a given treatment process. In many cases, treatment objectives cannot be fulfilled by one treatment process and a combination of two or more processes becomes necessary. All sections in this chapter address the issue of treatment objectives and the choice of treatment processes that can be used to achieve these objectives. Conventional drinking water treatment processes for surface and groundwater are addressed first, followed by non-conventional water treatment. Specific treatment methods (combinations of unit processes) for removal of specific pollutants like arsenic, fluoride, and nitrate are discussed in this section. The last section deals with conventional and higher levels of municipal wastewater treatment.

7.1 CONVENTIONAL DRINKING WATER TREATMENT SCHEMES

The main objective of any water supply scheme is to provide safe and adequate amount of potable water to all residents in the service area. This requires 'sourcing' of water so that it is of good quality to begin with. Protection of raw water quality for maintaining its quality in the future and ensuring that it remains a sustainable source is the next important aspect of a water supply scheme. The source water then has to be treated to the level of the prevailing drinking water standards. Water treatment has two major outputs: finished water which is provided to the consumer through the distribution system and wastewater including sludge which requires further management. Wastewater generated at the drinking water treatment plant, includes backwash water, wastewater generated from housekeeping in the treatment plant, and supernatant from the sludge drying beds. These wastewater streams are used for irrigation or horticulture or in some cases, taken back to the head of the water treatment plant and mixed with fresh water so as to minimize water losses. Sludge from the drying beds is generally transported to a landfill or if it is of adequate quality, it can be used as compost.

7.1.1 Conventional Treatment for Surface Water

Surface water sources include rivers or streams (flowing water), or lakes or reservoirs (standing water). Often, river water quality tends to be lower than lake or reservoir water quality. Further, lakes and reservoirs often have uniform quantity and quality of water and therefore, require less treatment compared to river waters which have large fluctuations in both, quality and quantity. However, lakes and reservoirs which are standing water bodies can experience large-scale growth of algae. These algal blooms cause an increase in the turbidity of water, produce pungent smell and odor, and release toxic compounds which can be difficult to remove. The extent of treatment required for any source water depends on the raw water quality and use for which the water is intended. For example, drinking water or water supply for residential or domestic uses requires high level of treatment while irrigation uses of water may require little or no treatment depending on the source water quality. River water supply normally requires more extensive treatment facilities with operational flexibility to handle day-to-day fluctuations in raw water quality as compared to groundwater.

Conventional water treatment processes for municipal supplies (mainly residential uses) are coagulation–flocculation, sedimentation or settling, filtration and finally, disinfection and are shown in Figure 7.1.

Depending on the nature of the intake from the river or stream, extensive or little treatment may be required. For example in Kharagpur city, river water is withdrawn through radial collector wells which draw water through the river bed. The natural filtration inherent in this process results in high quality river water. Water is currently supplied to the city without any treatment except disinfection. Therefore, it is important to remember that each of the unit processes is generally designed with a single treatment objective, which is the removal of a single group of contaminants.

Screening or pre-sedimentation is necessary for most surface waters and the objective is removal of large objects and particulate matter (in the visible range). Screens are generally provided at the intake point to prevent large objects like twigs and branches, stones and rocks, etc., from entering the system (pipes and pumps) and damaging them. Pre-sedimentation reduces silt and settleable organic matter prior to chemical treatment. Sometimes, pre-disinfection is also provided at this point to prevent biological growth (algal and bacterial) in the water treatment and supply system. Chlorine or a combination of chlorine and ammonia (monochloramines) can be added in this process to oxidize organics and arrest biological activity in downstream processes. However, formation of halogenated disinfection by-products (DBPs) can be a problem and pre-ozonation can be used instead of chlorine or chloramine to reduce the formation of DBPs. Use of ozone can lead to increase in the biodegradability of organic matter in water necessitating TOC removal during or after filtration. Sludge from this process is removed periodically and disposed in landfills.

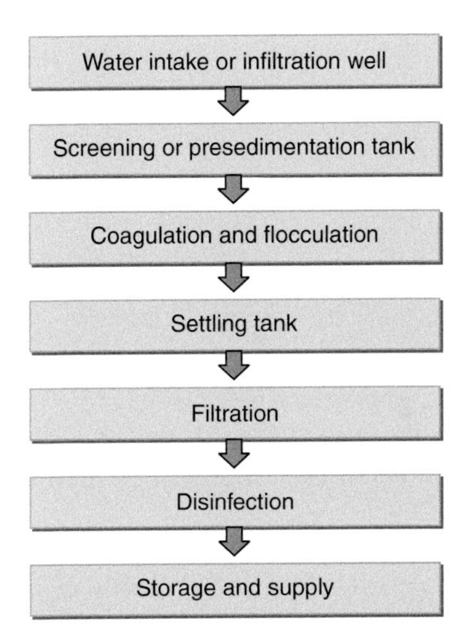

Figure 7.1 Flowchart for a conventional drinking water treatment plant for treating surface water.

The objective of **coagulation–flocculation–settling** is the removal of turbidity which is a measure of the concentration of colloidal particles in the water. Coagulants are added to water to chemically condition 'stable' particles and reduce the net repulsive interaction energy between particles. The most popular coagulant used in drinking water treatment is alum while other coagulants include ferric chloride and ferric sulphate. Flocculation relies on mixing and turbulence to promote collisions between particles. Increase in the number of collisions results in greater probability of particle aggregation and subsequent settling by gravity. Various cationic synthetic organic polymers (SOPs) like polyethylene imine, and polydiallydimethyl ammonium chloride (Cat-Floc) can also be used as coagulants (Amirtharajah and O'Melia 1990).

Filtration is required to remove suspended particles or floc that could not be removed during coagulation–flocculation–settling. Generally, single medium filters, i.e., slow or rapid sand filters are used for removing floc and SS. However, multi-media filters which combine different materials like sand, gravel, anthracite and garnet for higher removal efficiencies can also be used. Rapid sand filtration requires regular backwashing to maintain removal efficiencies through the filters and to prevent excessive head loss.

Disinfection is used to destroy pathogens. Generally, chlorine is used as the final disinfectant. A good disinfectant must be toxic to microorganisms at a concentration well below the toxic threshold for humans and animals. It should have a fast rate of killing and should be persistent enough, i.e., it should be possible to maintain a residue to prevent regrowth of microorganisms including pathogens in the distribution system.

PROBLEM 7.1.1

River Kasai serves as a water source for the city of Kharagpur. A complete analysis of the river water is provided below. Based on the data, draw a treatment flowchart and justify the need for each treatment unit.

Turbidity = 13 NTU

TDS = 133 mg/L

pH = 8.12

Alkalinity = 80 mg/L of $CaCO_3$

Hardness = 85 mg/l of $CaCO_3$

Total coliforms = 1000 cfu/mL

Fecal coliforms = 100 cfu/mL

Solution

Each water quality parameter and its measured value is compared to the current drinking water standards (BIS 2012) and the treatment processes required for achieving standards are summarized in the following table. A flowchart for treating this raw water is provided after that.

Water quality parameter	Measured value	Drinking water standards (BIS 2012)	Treatment requirement
Turbidity	13 NTU	1 NTU	Coagulation–flocculation and settling, filtration
TDS	133 mg/L	500 mg/L	No removal required
pH	8.12	6.5 to 8.5	No treatment required
Alkalinity	80 mg/L as $CaCO_3$	200 mg/L as $CaCO_3$	No treatment required
Hardness	85 mg/l as $CaCO_3$	200 mg/L as $CaCO_3$	No treatment required
Total coliforms	1000 cfu/mL	0 cfu/mL	Final disinfection with chlorine
Fecal coliforms	100 cfu/mL	0 cfu/mL	Final disinfection with chlorine

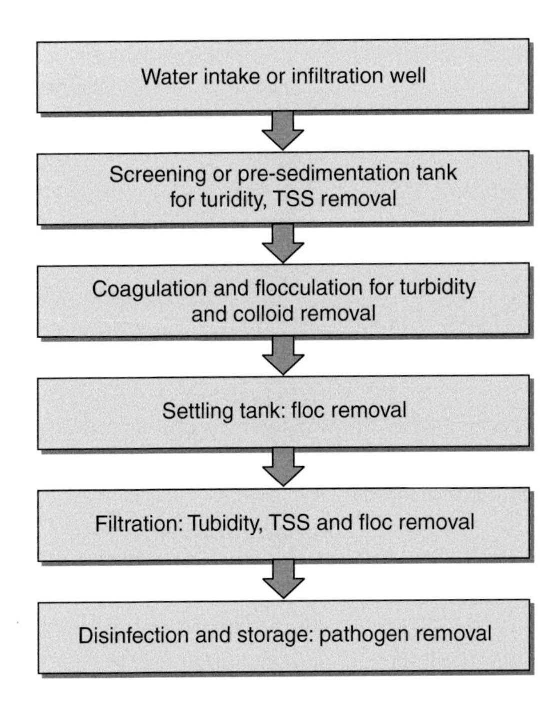

PROBLEM 7.1.2

A pristine lake in the Himalayas is the only source of water for a small city in the hills. What contaminants need to be monitored on a regular basis and what measures or treatment processes are needed prior to serving this water to the city. Explain the need for each process.

Solution

Since the area is assumed to be unpolluted and there are no point discharges of wastewater to the lake, very few contaminants are likely to be present in this water. Nevertheless, regular monitoring for pH, DO, turbidity, TOC, and coliforms due to non-point discharges like surface runoff is a minimum requirement based on regulations.

If all contaminant levels are acceptable based on standards, then the minimum treatment required is disinfection. NOM or TOC levels, and chlorine demand need to be determined prior to deciding on an appropriate chlorine dose required to maintain an adequate chlorine residual of >0.2 mg/L. If possible, trihalomethanes(THM) formation potential should be determined as well.

7.1.2 Conventional Treatment for Groundwater

Shallow wells and tubewells are the main sources of groundwater and generally yield water of uniform quality throughout the year for multiple uses. Magnesium, calcium, manganese and iron are the main non-toxic minerals in water, and Mg and Ca contribute to its hardness. In agricultural regions, groundwater may also have nitrates due to leaching of fertilizers and leakage from septic tanks and cess pits into the water table. Groundwater in industrial areas may have arsenic, chromium or radionuclides (which are highly toxic) because of industrial wastes. Keeping the following points in mind, treatment processes in different regions have to be designed accordingly. Some of the basic processes include:

Aeration: Aeration is the first step in the treatment of groundwater to strip out dissolved gases like carbon dioxide and hydrogen sulphide, and to add oxygen. Increasing oxygen in water helps to oxidize reduced inorganic pollutants like Fe, Mn, and As, as well as to reduce taste and odor problems by stripping out volatile odorous organic compounds.

Softening: Softening can be done using chemicals like lime and soda ash (chemical softening), or by ion exchange. Sludge in the form of calcium carbonate and magnesium hydroxide is removed at the end of lime softening. Carbon dioxide is applied to stabilize the softened water prior to final filtration. If ion exchange is used, some pre-treatment will be required depending on the feed (or raw) water quality.

Filtration: Filtration can be done using either porous media filters or membrane processes. If the raw water has low turbidity levels and coagulation–flocculation–settling is not necessary then direct filtration using porous media filters can be done (Amirtharajah and O'Melia 1990). Increasingly, membrane filtration processes are being used for desalination of water with high TDS content.

Disinfection: Chlorine is added for disinfection of water as it removes pathogens and other disease-causing organisms while providing a final residue for long-term protection.

Storage: Storage of treated water is essential and the capacity of the storage reservoirs should be sufficient to meet demand during peak periods.

In arid and semi-arid regions, groundwater sources are frequently not of potable quality due to their high TDS content. No other major pollutant may be present in such waters. In such cases, the only treatment requirement is desalination followed by disinfection. Various treatment methods are available for desalination and include different types of membrane processes (pressure and electric current driven), distillation and freeze-thawing.

Examples of membrane processes used in desalination include electrodialysis and electrodialysis reversal, reverse osmosis, ultrafiltration, microfiltration and nanofiltration. Distillation processes include multi-stage flash distillation, multiple-effect evaporation, vapor compression and solar distillation.

An example flow chart for treating groundwater is provided in Figure 7.2.

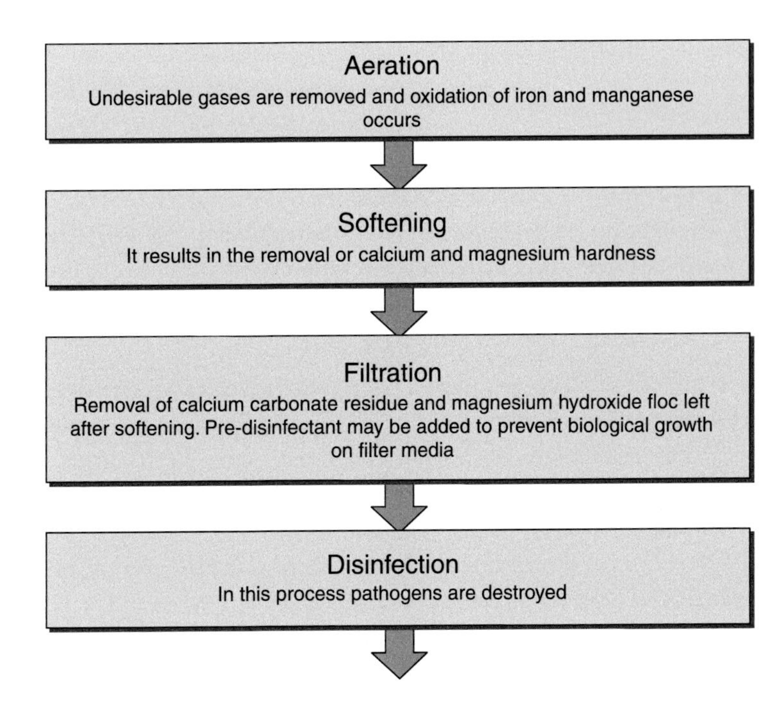

Figure 7.2 Example treatment scheme for groundwater.

7.2 NON-CONVENTIONAL WATER TREATMENT SCHEMES

Natural pollutants in groundwater include arsenic and fluoride which are present due to their dissolution from aquifer solids or sediments. These pollutants tend not to 'travel' in groundwater in the same manner as anthropogenic contaminants like pesticides or petroleum compounds. Nitrates and organic matter measured as TOC may be present in groundwater at very high concentrations and can be natural or anthropogenic in origin.

7.2.1 Arsenic in Groundwater and Its Removal

Arsenic is a major geogenic pollutant that is found in all parts of the environment and in various parts of the world including India, Bangladesh, Taiwan, Japan, Canada and the USA. Twenty-three cases of large-scale presence of arsenic in groundwater have been reported by Mandal and Suzuki (2002). It is estimated that over 137 million people in more than 70 countries are affected by arsenic poisoning of drinking water alone. In decreasing order of magnitude, these countries/regions are Bangladesh, West Bengal – India, Inner Mongolia – P. R. China, and Taiwan. In all these countries, groundwater withdrawals for irrigation are increasing steadily leading to increased health risks not only from drinking water but also from food grown with

As-containing water. Chatterjee et al. (1995) analyzed groundwater from six districts of West Bengal (India) and found that mean total arsenic levels ranged from 193 to 737 micro-g/L with a maximum value of 3700 micro-g/L. Mean arsenite [As(III)] levels in groundwater were around 50 percent of the total arsenic. Nine districts in West Bengal, India and 47 districts in Bangladesh have arsenic levels in groundwater above the WHO guideline value of 10 micro-g/L (Das et al. 2002). The World Health Organization's current guideline for arsenic in drinking water is 10 μg/L, but many developing countries affected by polluted groundwater are still struggling to keep up with the previous WHO guideline value of 50 μg/L. The drinking water standard for arsenic in India was lowered to 10 ppb in 2012 (BIS 2012).

Arsenic is found in soil and groundwater mainly due to weathering of rocks and minerals, volcanic emissions and anthropogenic sources followed by subsequent leaching and runoff (Smedley and Kinniburgh 2002). The main source of arsenic pollution is geogenic in origin. Anthropogenic activities like oil and coal burning in power plants, waste incineration, cement works, disinfectants, surplus arsenic from animal feeding, household waste disposal, glassware production, electronics industries, ore production and processing, pesticides and insecticides, and pharmaceutical works are also responsible for the presence of arsenic in the environment (Matschullat 2000).

7.2.1.1 *Arsenic in the environment*

The presence of arsenic in different parts of the environment is well-documented. It is present in air, soil, sediments, surface and groundwater. Further, it also tends to accumulate in plant and animal tissue and is ingested as food. Drinking water was identified as the first major route of exposure for humans leading to As-related diseases as long ago as the later part of the nineteenth century. Since then arsenic monitoring is a part of several water quality monitoring programs around the world. Subsequent monitoring of other environmental media has shown the presence of arsenic in all parts of the environment. A summary of arsenic concentrations in different environmental samples is provided in Table 7.1 (Escobar, Hue, and Cutler 2006).

Table 7.1 Arsenic concentrations in environmental media (Escobar, Hue, and Cutler 2006)

Environmental media	Arsenic concentration range
Air, ng/m^3	1.5-53
Rain from unpolluted ocean air, μg/L (ppb)	0.019
Rain from terrestrial air, μg/L	0.46
Rivers, μg/L	0.20-264
Lakes, μg/L	0.38-1,000
Ground (well) water, μg/L	1.0-1,000
Seawater, μg/L	0.15-6.0
Soil, mg/kg	0.1-1,000
Stream/river sediment, mg/kg	5.0-4,000

(*Contd.*)

Environmental media	Arsenic concentration range
Lake sediment, mg/kg	2.0-300
Igneous rock, mg/kg	0.3-113
Metamorphic rock, mg/kg	0.0-143
Sedimentary rock, mg/kg	0.1-490
Biota – green algae, mg/kg	0.5-5.0
Biota – brown algae, mg/kg	30

Arsenic has been monitored in different environmental media like air, water, soil, food and biota in India as well and a summary is provided here.

Drinking water

As of 2008, arsenic polluted groundwater has been detected in the states of West Bengal, Jharkhand, Bihar, Uttar Pradesh in the Ganga river basin; Assam, Manipur, Tripura, Nagaland and Arunachal Pradesh in the Brahmaputra and Imphal river basins and Rajnandgaon village in Chhattisgarh state (SOES Jadavpur University 2015). Based on data compiled from different sources, the total arsenic affected area in these states is 2,25,830 km^2 with a population of around 103 million people. The most severely affected areas are Murshidabad, Malda, North and South 24 Paraganas districts of West Bengal, Bhojpur district of Bihar, Rajnandgaon district of Chattisgarh, and Ballia, Varanasi and Ghazipur districts of Uttar Pradesh. Maximum concentration of arsenic in groundwater in West Bengal was 3700 ppb while in Haldi village of Ballia district in UP it was found to be 621 ppb.

Soil and sediments

Soil concentrations have been monitored in areas known to be affected by arsenic and range from 1.1 mg/kg in Varanasi to 31.6 mg/kg in the Domkal block of Murshidabad district of West Bengal (Roychowdhury et al. 2002; Raju 2012). Only one study reported arsenic concentrations in sediments from the Ganga river bed in Karkatpur village of Ghazipur district in UP (Saxena, Kumar, and Goel 2014). Maximum concentration observed in this study was 470 mg/kg and points to the accumulation of arsenic in river sediments. The implications for arsenic levels in riverine biota are significant due to the mobility of arsenic under reduced conditions found in anaerobic sediments and its uptake and bioaccumulation by plants and fish in reduced form.

Food and biota

Several studies and reviews have been published regarding arsenic levels in different foods, plants and other tissues. Arsenic can be transferred from water and/or soil to plants. Since rice constitutes a major source of calories for most people living in arsenic affected areas, many studies have concentrated on evaluating arsenic concentrations in paddy fields and different parts of the rice plant (Meharg 2003). Paddy fields and rice plants are a major source of arsenic due to high levels of soluble arsenic and reducing conditions found in the standing water in these fields. Concentrations observed in white polished rice ranged from 0.05 to 0.18 mg/kg with a mean concentration of

0.07 mg/kg (Meharg et al. 2009). In another study, roots of rice plants were found to have arsenic concentrations ranging from 23 to 155 mg/kg while concentrations in rice grain ranged from 0.4 to 1.68 mg/kg (Dwivedi et al. 2010).

In studies with wheat, tea and medicinal plants, arsenic concentrations were found to be highest in the root zone of the plant studied and decreased in the order: root > shoot > leaves > grain (Tripathi et al. 2012; Karak et al. 2011; Kundu et al. 2012). Arsenic levels were highest in skins of various tubers like potatoes, arum, and turmeric (Roychowdhury et al. 2002; Al Rmalli et al. 2005). In another study of arsenic concentrations in vegetables, highest concentrations were found in vegetables like radish which grow below the ground surface and are in contact with soil (Mishra, Dubey, and Usham 2014). Juicy vegetables like tomatoes also had high As levels. The concentrations of arsenic in decreasing order were: root vegetables > juicy vegetables > least juicy vegetables.

7.2.1.2 *Chemistry of arsenic*

Arsenic can be found in solid, liquid and gaseous forms in the environment and in four different oxidation states: -3, 0, +3 and +5. Arsine [AsH_3] gas has an oxidation state of -3 and is the most reduced form of As and can exist in equilibrium with water. Elemental As is insoluble while the most soluble forms of As are arsenite [As(III)] and arsenate [As(V)]. As(III) and As(V) are the two predominant forms found in groundwater and soil, while it is mainly found as As(V) (60 percent) in minerals (Mandal and Suzuki 2002). As(III) is the dominant form under reducing or anoxic conditions (groundwaters are often anoxic) and As(V) dominates under oxidizing conditions (in the presence of oxygen as in most surface waters or in the presence of other oxidizing agents like chlorine, ozone, potassium permanganate) (Edwards 1994). The solubilities and speciation of arsenic salts are dependent on pH, redox conditions and the ionic environment. Methylated forms of As can exist but are at far lower concentrations compared to the inorganic forms (Chatterjee et al. 1995).

The speciation of As(III) and As(V) in water is governed by the following reactions and their equilibrium constants are provided here. Graphs showing the fraction of different arsenic species as a function of pH are plotted in Figure 7.3.

For arsenite – As(III), the deprotonation reactions and their constants are as follows:

$$H_3AsO_3 \leftrightarrow H_2AsO_3^- + H^+ \qquad pKa_1 = 9.1$$

$$H_2AsO_3^- \leftrightarrow HAsO_3^{2-} + H^+ \qquad pKa_2 = 12.1$$

$$HAsO_3^{2-} \leftrightarrow AsO_3^{3-} + H^+ \qquad pKa_3 = 13.4$$

For arsenate – As(V), the deprotonation reactions and their constants are as follows:

$$H_3AsO_4 \leftrightarrow H_2AsO_4^- + H^+ \qquad pKa_1 = 2.1$$

$$H_2AsO_4^- \leftrightarrow HAsO_4^{2-} + H^+ \qquad pKa_2 = 6.7$$

$$HAsO_4^{2-} \leftrightarrow AsO_4^{3-} + H^+ \qquad pKa_3 = 11.2$$

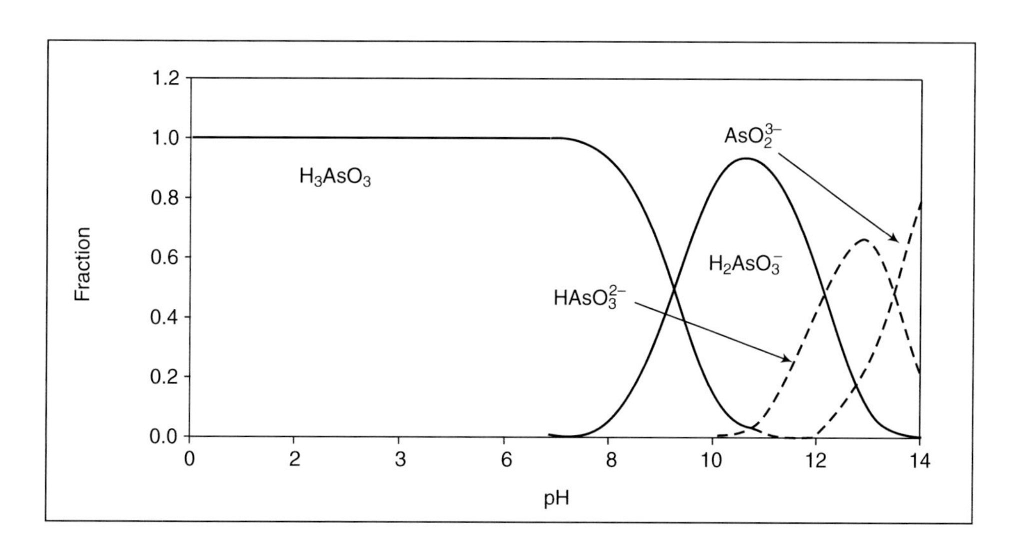

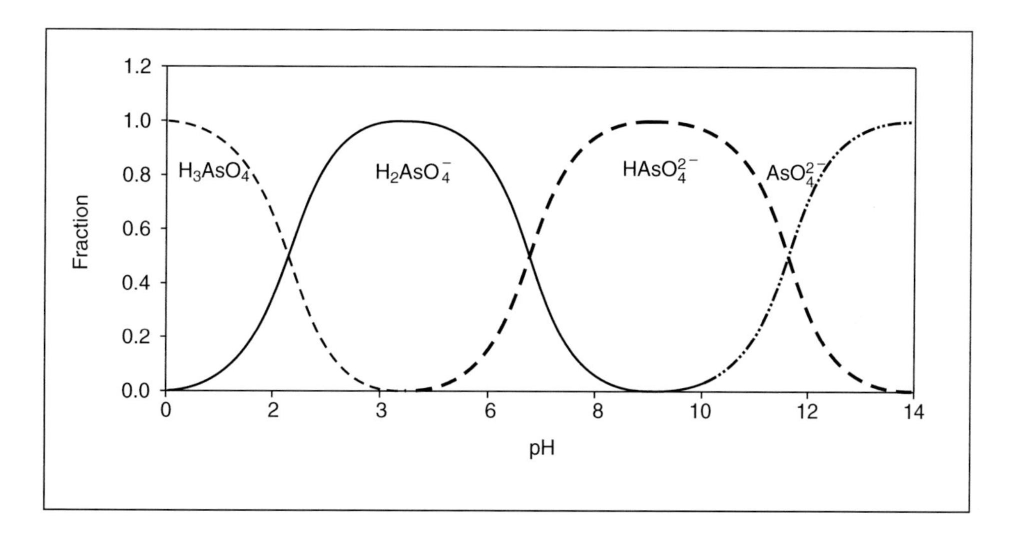

Figure 7.3 Speciation of As(III) and As(V) as a function of pH (Mondal 2008).

The solubility of arsenic in water is a function of pH, ionic strength, redox conditions and the presence of other ions. The presence of other ions like aluminum, iron, barium and chromium can increase or decrease the solubility of arsenic. Calculated values of aluminum arsenate and ferric arsenate are shown in Table 7.2 (Wagemann 1978). These values are for solutions that are supersaturated with aluminum hydroxide and ferric hydroxide, respectively.

Table 7.2 Dissolved arsenic concentration (mg/L) under different pH conditions (Wagemann 1978)

	Total Dissolved Arsenic in equilibrium with typical freshwater at 25 °C			
Salt	pH = 8.5	pH = 8.0	pH = 7.5	pH = 7.0
$AlAsO_4$	220	710	2.6×10^3	$>7.5 \times 10^3$
$FeAsO_4$	0.17	0.055	0.2	0.86

Oxidation of As(III) to As(V): Conversion of As(III) to As(V) is thermodynamically favorable under oxic conditions, but the rate of oxidation may vary from seconds to weeks and months depending on various factors like pH levels , light, oxygen, presence of oxyanions, metals, other oxidants, and unknown catalysts (Edwards 1994; Bissen et al. 2001; Dutta et al. 2005). Oxidation of As(III) is generally of the order of seconds to minutes in the presence of light, catalysts, or oxidants (Bissen et al. 2001; Dutta et al. 2005). In the Bissen et al. (2001) study, no oxidation of As(III) was observed in the dark even in one week, while under solar simulated conditions, only 40 percent of the initial As(III) was oxidized after 25 min. Addition of a catalyst like titanium dioxide increased the oxidation rate to get complete oxidation in 200 seconds. Light wavelength and intensity are factors that were found to influence the oxidation rate of As(III). Atmospheric oxygen can serve as an oxidant, but the reaction rate was found to be of the order of weeks (Pierce and Moore 1982).

7.2.1.3 *Health effects of arsenic*

The toxic properties of arsenic have been known for centuries since arsenic has been used as a rat poison, ant poison, pesticide, and preservative for wood (ATSDR 2007). The earliest medical report from India regarding the impact of ingesting arsenic was published in 1976 suggesting that there may be a relationship between high As levels in drinking water or medicines and non-cirrhotic portal fibrosis (Datta et al. 1979; SOES Jadavpur University 2015). This study was carried out in Chandigarh and a few villages of Patiala district in Punjab.

Occupational exposure to arsenic has been documented in workers in mines, smelters and those working with wood preserved with arsenic compounds. In the general population, chronic exposure to arsenic can occur due to its presence in food, water, soil and air (ATSDR 2000). Food containing arsenic accounts for more than 80 percent of the dietary intake of arsenic mainly from fish, meat and poultry. Of the three major routes of exposure (inhalation, ingestion and dermal contact), adverse health effects due to inhalation and ingestion are well-documented while there is little evidence of adverse health effects due to dermal contact. Symptoms of chronic exposure to groundwater polluted with arsenic at concentrations significantly >50 μg/L include skin, cardiovascular, renal, hepatological and respiratory disorders.

Toxicity Assessment for arsenic

Due to the gravity of arsenic poisoning, several epidemiological studies have been conducted all over the world and a large amount of data for 'weight-of-evidence' analyzes are available for arsenic Silvera, Navarro, and Rohan 2007; ATSDR 2007). No animal toxicity studies need to be referred to in cases where there is sufficient evidence of carcinogenicity from epidemiological studies. Arsenic is classified as a Group A carcinogen and oral and inhalation potency factors are available. Similarly, non-carcinogenic effects of arsenic are also well-documented and reference doses for the same are also available from USEPA.

Carcinogenic effects

Based on ATSDR (2007), the following health effects of arsenic are summarized here.

***Inhalation route*:** ATSDR reviews note that inorganic arsenic is associated with respiratory system cancers, especially lung cancers by the inhalation route. These effects have been observed in people exposed to arsenic due to occupational exposure such as miners and those working in copper smelters. Residents in areas around chemical manufacturing industries where arsenic is used may also be at risk.

***Ingestion route*:** Ingestion of inorganic arsenic is associated with skin cancer. Long-term studies where arsenic exposure has occurred are often associated with bladder cancer (despite the fact that most arsenic in well-nourished humans is excreted via urine or removed from the body via nails and hair).

***Dermal contact*:** Dermal contact with inorganic arsenic does not seem to result in any obvious effects. However, similar exposure to organic arsenic is associated with contact dermatitis. Only one study reported that the ability of arsenic to penetrate skin is dependent on its speciation. Arsenite and dimethyl arsenic(V) acid were able to penetrate human skin 29 to 59 times faster than arsenate or arseno-sugars (Ouypornkochagorn and Feldmann 2010).

Non-carcinogenic effects

Occupational exposure to inorganic arsenic is generally associated with peripheral neuropathy through both inhalation and ingestion routes of exposure. Short-term and long-term exposure to inorganic arsenic is associated with irritation of the gastrointestinal tract, the effects of which often diminish or end when there is no further exposure, i.e., the effects are reversible.

Respiratory system effects have been observed in those exposed occupationally as well as in acute high-dose arsenic exposures. In general, chronic low-dose exposures do not result in respiratory system effects. Several cardiac effects due to acute and chronic arsenic exposure through ingestion are documented and include altered myocardial depolarization, cardiac arrhythmias, and ischemic heart disease. Blackfoot disease, observed in Taiwan only, is a dramatic manifestation of chronic exposure to arsenic and results in loss of circulation in hands and feet, followed by necrosis and gangrene.

Arsenicosis is generally characterized by skin lesions associated with chronic oral exposure to arsenic levels. While these dermal effects are dramatic in cases of arsenic exposure, they are not observed in animals but only in humans. Other indicators of arsenicosis are elevated levels of arsenic in hair, nails and urine. One study reported 2.4 mg/kg in hair, 5.2 mg/kg in nails and 0.214 mg/L in urine (Rahman et al. 2014). The authors further suggest that exposure to drinking water for 2 years with ≥ 300 ug/L of arsenic can result in skin lesions. Dermal effects of high arsenic uptake can be apparent in children as young as 2 years. However, in Bangladesh and West Bengal, dermal manifestations in children were not apparent in those ≤ 11 years even though other bioindicators showed high arsenic levels.

It is important to note that malnourishment is a major factor contributing to adverse health effects due to arsenic exposure (Roychowdhury et al. 2002). Where people are well-nourished, they are resistant to arsenic and do not show any of the adverse health effects despite elevated As levels in hair, nails and urine.

Arsenic in water, soil and food

As mentioned before, water was long suspected to be the most important route of exposure to arsenic. While both surface water and groundwater can contain As, concentrations of arsenic in groundwater tend to be far greater than those in surface water (Patel et al. 2005). Exposure to As-containing water can occur when water is used for drinking (ingestion), for food preparation (ingestion), and for irrigation especially in paddy fields (ingestion of rice, and dermal contact for farmers working in the fields).

Transfer factors relating arsenic concentrations in soil to those in various plants have been calculated in several studies. Calculations of daily dietary intake of arsenic in the Bengal region showed that the highest body burden from food ingestion is due to rice, followed by vegetables and spices (Roychowdhury et al. 2002; Signes-Pastor et al. 2008). Further, seasonal and gender variations in intake of water can lead to significant differences in total arsenic ingested (Biswas, Deb, and Ghose 2014).

7.2.1.4 Treatment processes for arsenic removal

There are two ways to tackle arsenic in of drinking water: the first option is to find an alternative, safe water source and the second option is to remove arsenic from the water. Substitution of the drinking water source may not be possible in all regions or throughout the year or it may be too expensive. Arsenic removal from water is often a better way and sometimes the only available option rather than substitution of the water source. The current drinking water standard in India for arsenic is a maximum concentration of 10 ppb.

Many treatment strategies have been developed and tested for the removal of arsenic. Many of these strategies can be applied at the large-scale municipal treatment plant level or at the community or household level and are all based on physico-chemical processes. The most common treatment strategies are conventional filtration, aeration, oxidation-reduction, precipitation, coagulation–flocculation, adsorption, ion exchange, and membrane filtration.

7.2.1.4.1 Oxidation by aeration or chemical addition

Aeration is the first process used in many water treatment plants where the raw water contains reduced forms of inorganic compounds such as arsenic along with iron and manganese. However, only aeration is generally inadequate for achieving drinking water standards for arsenic.

For raw water containing high levels of arsenic, advanced oxidation processes can be used where strong oxidizing agents like gaseous chlorine, ozone, potassium permanganate, hydrogen peroxide, and Fenton's reagent are added to the water to convert arsenite to arsenate. Also, solids such as manganese oxide can be used to oxidize arsenic. Any oxidation process has to be followed by a removal process such as coagulation, adsorption, membrane filtration or ion exchange to remove arsenate from solution.

A relatively new treatment strategy is to use electrocoagulation for arsenic removal. Oxidation of arsenite to arsenate can be achieved by electrocoagulation (EC) where the two processes of electro-oxidation and coagulation occur simultaneously (Sharma, Adapureddy, and Goel 2014). In laboratory experiments with air, mixing, light and electrocoagulation, arsenite was oxidized to the greatest extent with EC using distilled water or groundwater solutions of As(III) as shown in Table 7.3.

Table 7.3 Results for oxidation of As(III) to As(V) under different experimental conditions (Sharma, Adapureddy, and Goel 2014)

Experimental conditions	Conversion of As(III) to As(V without EC, percent		Conversion of As(III) to As(V) with EC, percent	
	Double distilled water	Ground water	Double distilled water	Ground water
Air + Light with mixing	25.97	29.4	97	70.91
Air + Light without mixing	20.95	25.8	95	56.37
Air + Dark with mixing	17.92	23.54	92	59.74
Air + Dark without mixing	5.90	12	90	53.61

7.2.1.4.2 Coagulation, flocculation and filtration

The primary objective of coagulation, flocculation and settling followed by filtration (generally, rapid sand filtration) is removal of turbidity from raw water sources. However, simultaneous removal of various other pollutants including arsenic, iron, manganese, phosphate, nitrate, natural organic matter and fluoride can be achieved. Significant reduction is also possible in odor, color and trihalomethane formation which are associated with the presence of natural organic matter. Thus, coagulation–flocculation–settling and filtration can also improve other water quality parameters.

Alum, ferric chloride and ferric sulphate can be used as coagulants for arsenic removal and are the most popular method in developing countries due to their low cost and low technical skill requirement (Mohanty 2017). Floc generated during coagulation and flocculation can be removed by direct filtration or in the conventional manner, after settling and filtration. Arsenic does not leach out of the solids generated from the coagulation process (Odell 2016).

Arsenite is found in neutral state at pH ≤7 while arsenate is found in anionic form in the pH range of 6 to 8. Therefore, in the desired pH range of 6.5 to 8.5 in treated drinking waters, arsenite will be in neutral form while arsenate will be present in anionic form. The anionic form of arsenic – As(V) was easier to remove by coagulation as compared to the neutral arsenite – As(III) using ferric chloride (Hering et al. 1996). Removal of total arsenic declines at pH >8.5 and silica can act as an interfering ion at pH >7.5. In general, ferric salts are better at removing arsenic in comparison to alum and this may be attributed to co-precipitation of arsenic with Fe(III) and the lower solubility of ferric arsenate as compared to aluminum arsenate as shown in Table 7.2.

In two case studies (Plant A and Plant B; both plants use surface water) where conventional drinking water treatment was provided, drinking water standards for arsenic were achieved and pertinent results are summarized in Table 7.4 (Fields, Chen, and Wang 2000). The treatment train in Plant A was screening, ozonation, coagulation with 1 to 2 mg/L ferric chloride and 1 to 5 mg/L cationic polymer, followed by filtration through anthracite and a thin layer of pea gravel and finally, chlorination. The treatment train in Plant B was screening, chlorination, coagulation with 25 to 30 mg/L alum and 0.75 mg/L cationic polymer only when water temperature was <10 °C, followed by filtration through anthracite and sand and post-chloramination.

The treatment train in Plant C which uses groundwater was aeration, chlorination, precipitation using lime, chlorination, filtration through graded gravel and sand and final chlorination. Results for this plant are also provided in Table 7.4 and are discussed later in Section 7.2.1.4.6.

Table 7.4 Average Arsenic concentrations (in ppb) in two drinking water treatment plants (Fields, Chen, and Wang 2000)

	Inlet			After filtration		
	Total As	As(III)	As(V)	Total As	As(III)	As(V)
Plant A	7.5	0.7	6.9	3.5	0.6	3.0
Plant B	19.1	0.6	19.7	4.0	0.4	4.2
Plant C	32.0	30.0	3.9	16.6	0.4	16.7

7.2.1.4.3 Ion exchange resins

Synthetic anion exchange resins can be used for removing arsenic from water (Duarte, Cardoso, and Alçada 2009). For arsenic removal, the resin is loaded with chloride ions which are exchanged with arsenic from solution. It was found that the adsorption capacity of these ion exchange resins was orders of magnitude higher for arsenate compared to arsenite. Chlorine was effective in completely converting arsenite to arsenate which is then exchanged with chloride ions on these synthetic ion exchange resins (Kartinen and Martin 1995). Pre-oxidation with chlorine which is necessary for converting As(III) to As(V) has a detrimental effect on the resin. In low sulphate waters, ion exchange resin can easily remove over 95 percent of arsenate. According to the USEPA, strong-base anion exchange resins are most effective in waters with low sulphate concentrations (<25 mg/L) and should not be used in water with sulphate >120 mg/L and TDS >500 mg/L (USEPA 2002).

7.2.1.4.4 Activated alumina

Activated alumina is a granulated form of aluminum oxide with very high internal surface area (200-300 m^2/g) (Johnston and Heijnen 2001). The high surface area gives the material a very large number of sites where adsorption can occur. The process of arsenic removal by alumina is slower than the ion exchange method and some arsenic leakage often occurs in activated alumina systems. However, arsenic removal efficiency is excellent (typically > 95 percent) but arsenic capacity varies significantly with pH and influent arsenic concentration. Removal efficiency is best in a narrow pH range of 5.5 to 6 where the alumina surface is protonated. In general, activated alumina has zero surface charge at a pH of 8.2. Below pH of 8.2 the surface is positively charged and arsenic anions are easily adsorbed. Above this pH, very little arsenic is adsorbed. Other ions in solution can also be removed by activated alumina in the order:

$$OH^- > H_2AsO_4^- > H_3SiO_4^- > F^- > HSeO_3^- > TOC > SO_4^{2-} >> H_3AsO_3$$

Activated alumina can treat thousands of bed volumes before breakthrough, and filters can be operated for a year or more before the media needs to be changed or regenerated (USEPA 2002). Regeneration of exhausted activated alumina is necessary and is done in a four-step process of backwashing, regeneration, neutralization and rinsing. Regeneration is done using a concentrated solution of sodium hydroxide and neutralization is done using sulphuric acid.

The advantage of using activated alumina is that it is simple and can be used even at the household level without any chemical addition. Disadvantages include the possibility that the media will be fouled or clogged by precipitated iron, the narrow pH range for optimal operation, and the relative difficulty in regeneration due to the use of hazardous chemicals (Odell 2016).

7.2.1.4.5 Membrane filtration methods

Synthetic membranes are available which are selectively permeable, i.e., the structure of the membrane is such that some molecules can pass through, while others are excluded, or rejected. Membrane filtration has the advantage of removing many pollutants from water, including bacteria, salts, and various heavy metals. Two classes of membrane filtration can be considered: low-pressure membranes, such as microfiltration and ultrafiltration; and high-pressure membranes such as nanofiltration and reverse osmosis. In recent years, a new generation of RO and NF membranes has been developed that is less expensive and operates at lower pressures, yet allows improved flux and is capable of efficient rejection of both arsenate and arsenite (Johnston and Heijnen 2001). Solution pH had a direct impact on removal of arsenite and arsenate by membrane filtration (Kang et al. 2000).

Membrane filtration has the advantage of lowering the concentrations of many other constituents of water in addition to arsenic. Even ultrafiltration (UF) membranes are able to remove over 99.9 percent of bacteria, *Giardia* and viruses. Further, there is no accumulation of arsenic on the membrane; the rejected arsenic is removed along with the rejectate from the membrane process. Operation and maintenance requirements are minimal and no sludge is generated. Membranes can be cleaned by backflushing or with dilute acid solutions to remove buildup of scales on the membranes. On the other hand, water recovery can be very poor, permeate or product water may be only 10-20 percent of the feed water. Other disadvantages of this process are high operating pressures, relatively high capital and operating costs, and the risk of membrane fouling. Also, particularly with RO, the treated water has very low concentrations of dissolved solids and can be corrosive, and deficient in minerals which are important micronutrients for humans.

7.2.1.4.6 Precipitation processes

Alum and ferric salts like ferric chloride and ferric sulphate have been used for removing arsenic from water in developing countries for several decades. Oxidation of As(III) is essential for the success of any other treatment method including precipitation. Arsenic removal efficiencies in different precipitation processes, with and without chlorine are summarized in Table 7.5.

1. Alum precipitation

Alum precipitation can be used to remove particles (colloids) and dissolved metals. Alum is most effective when the pH is reduced to less than 7 and chlorine is added ahead of the flocculator and clarifier (Choong et al. 2007).

2. Iron precipitation

In this process, an iron compound is added to untreated water (Choong et al. 2007). Arsenic coprecipitates with iron oxyhydroxide due to the combination of arsenic and iron and is removed in the clarifier by settling. Best results are obtained at a pH less than 8.5, with or without chlorine. This treatment strategy is the simplest and it is versatile and inexpensive.

3. Lime softening

It is well known that lime softening can remove hardness in the form of Ca and Mg from water. Arsenic can also be removed using lime softening by raising the pH to higher than 11. However, drinking water standards of 10 ppb total As could not be achieved with this method (Plant C described in Section 7.2.1.4.2) (Fields, Chen, and Wang 2000). The average inlet pH was 7.2 and the pH after filtration was 8.6. In another study, the final pH was 10.5 and 90 percent As removal with lime and pre-chlorination was obtained (Kartinen and Martin 1995). It is evident that pH >11 is important for achieving drinking water standards and neutralization of pH will be necessary in such cases.

4. Combined with iron (and manganese) removal

Chlorine is injected into raw water containing iron and/or manganese and allowed to react with iron and/or manganese in a reaction vessel for a short time — a minute or two (Kartinen and

Martin 1995). Following the chlorine reaction vessel, sulphur dioxide may also be injected into the water and allowed to react for a short period of time. The water is then discharged into one or more filter vessels which contain proprietary media for arsenic removal.

Table 7.5 Summary of Technologies for Arsenic Removal by Precipitation processes (Kartinen and Martin 1995)

Parameters		pH	With Cl_2	Without Cl_2
Precipitation processes	Alum precipitation	<6.5	90 percent	20 percent
	Iron precipitation	6-8	90 percent	60 percent
	Lime softening	>10.5	90 percent	80 percent
	Combined with iron (and manganese)	>7	40 percent–90 percent	–

In summary, several treatment methods and schemes are available for removing arsenic from drinking water and removal efficiencies are shown in Table 7.6.

Table 7.6 Summary of Technologies for Arsenic Removal

Treatment method	Maximum removal efficiency (percent) with real water			Reference
	Total As	As(III)	As(V)	
Electrocoagulation	–	–	71	(Sharma, Adapureddy, and Goel 2014)
Coagulation with $FeCl_3$	53.3	14.3	56.5	(Fields, Chen, and Wang 2000)
Coagulation with alum	79	33.3	78.7	(Fields, Chen, and Wang 2000)
Ion exchange		<30	80-95	(Duarte, Cardoso, and Alçada 2009)
Activated alumina		>95	>95	(Johnston and Heijnen 2001)

PROBLEM 7.2.1

Water quality analyzes were done for a groundwater source with the following results: turbidity 25 NTU, Arsenic 400 micro-g/L, TDS 500 mg/L, pH 7.8, total coliforms 10 cfu/mL, fecal coliforms 0 cfu/mL. Draw a flowchart for treating this water prior to supply and justify your choice of each treatment process in 1-2 sentences.

Solution

The following treatment processes can be used for achieving IS 10500 standards for drinking water. Other variations can be used with appropriate justification.

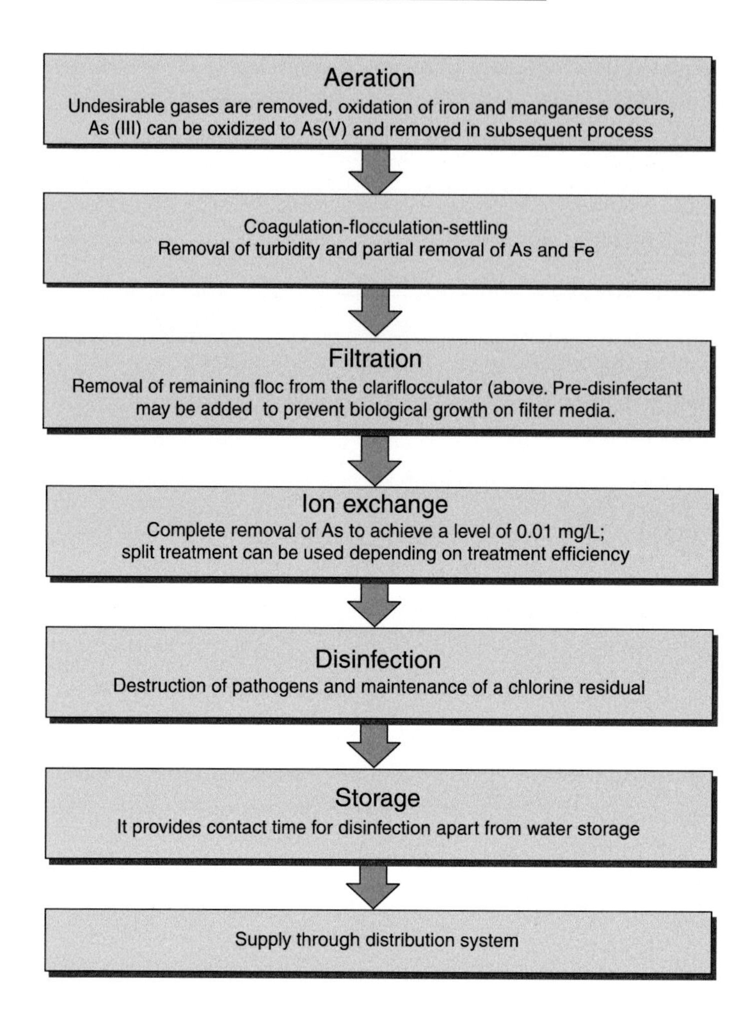

Water quality parameters that are in exceedence of IS10500 standards are: turbidity, As, coliforms are OK as such but not more than 5 percent samples of the samples can have 10 cfu/mL. Therefore, the treatment scheme for this source will have to be: aeration or oxidation, coagulation–flocculation–settling, filtration, ion exchange and disinfection.

Aeration or oxidation using ozone or chlorine is required for oxidation of As(III) and subsequent removal of As(V).

Coagulation, flocculation and settling should be done with $FeCl_3$. This will result in removal of turbidity and As by co-precipitation with Fe.

Filtration is necessary for the removal of floc remaining from the previous process along with other SS, and is necessary for achieving drinking water standards for turbidity.

Ion exchange is necessary to meet drinking water standards for As, and activated alumina or synthetic resins can be used.

Disinfection with chlorine is necessary to achieve the required coliform levels in the final treated water and maintain an adequate residual.

7.2.2 Fluoride in Groundwater and Its Removal

Fluoride is a common geogenic pollutant found in groundwater in many parts of the world including India. In a CGWB study of 15,640 wells in shallow aquifers all over India, 927 wells had water in exceedance of the maximum permissible fluoride concentration of 1.5 mg/L (CGWB 2010). The average fluoride concentration in these wells was 2.9 mg/L with a standard deviation of 2.77 mg/L. The highest fluoride concentration detected was 38 mg/L in Bikaner, Rajasthan.

7.2.2.1 *Chemistry of fluoride*

Fluoride is a very strong oxidizing agent but a weak acid since it has a dissociation constant (pKa of 3.17) that is lower than that of other hydrohalic acids. Hydrofluoric acid does not fully ionize in dilute aqueous solutions, as it forms fluoride ions and hydrofluoric acid in aqueous solution (Apshankar 2018). Fluorite (CaF_2) is a common fluoride containing mineral with very low solubility ($Ksp = 3 \times 10^{-11}$) which results in a solubility limit of 7.6 mg/L in the presence of excess calcium. This also implies that high fluoride concentrations are likely to occur only in groundwaters with low Ca concentrations.

Fluoride forms strong complexes with aluminum, beryllium and ferric ions and a series of mixed fluoride-hydroxide complexes are possible with boron. Cations present in high quantities, especially aluminum, bind to fluoride very strongly and allow removal of fluoride from solution.

7.2.2.2 *Health effects of fluoride*

Low concentrations of fluoride (<1 mg/L) are considered to protect against dental cavities while high concentrations of fluoride (>1.5 mg/L) can lead to tooth mottling, dental and skeletal fluorosis, kidney and thyroid injury, and death (in order of increasingly severe effects) (Fawell et al. 2006). Fluorosis is irreversible and no treatment is available for it. Fluoride concentrations >1 mg/L have been reported from 29 countries and many people are affected by dental and skeletal fluorosis (Fawell et al. 2006). China and India are the most severely affected countries.

Fluorosis is endemic in India and was first documented in 1937 Shortt, Pandit, and Raghavachari (1937). Long-term exposures and use of groundwater having high fluoride concentrations in excess of 1.5 mg/L result in severe dental, skeletal and non-skeletal fluorosis. Correlations between fluoride concentrations in three districts of Rajasthan (Banswara, Dungarpur and Udaipur) and the prevalence of dental fluorosis (DF) in people <16 years of age and skeletal fluorosis (SF) in people > 21 years of age are shown in Figure 7.4 Choubisa 2001). As is evident from the graphs, average fluoride concentrations are correlated exponentially with the prevalence of dental fluorosis in children (those who are <16 years of age) and linearly with skeletal fluorosis in adults (>21 years of age).

An estimated 26 million people are affected by fluorosis through drinking water and another 16.5 million people are affected by fluorosis due to coal smoke pollution in China (Fawell et al. 2006). It is estimated that 66 million people are affected by high fluoride concentrations in India and another 5 million in Mexico. Nutrient deficiency or malnourishment is known to exacerbate fluorosis (S. L. Choubisa, Choubisa, and Choubisa 2009).

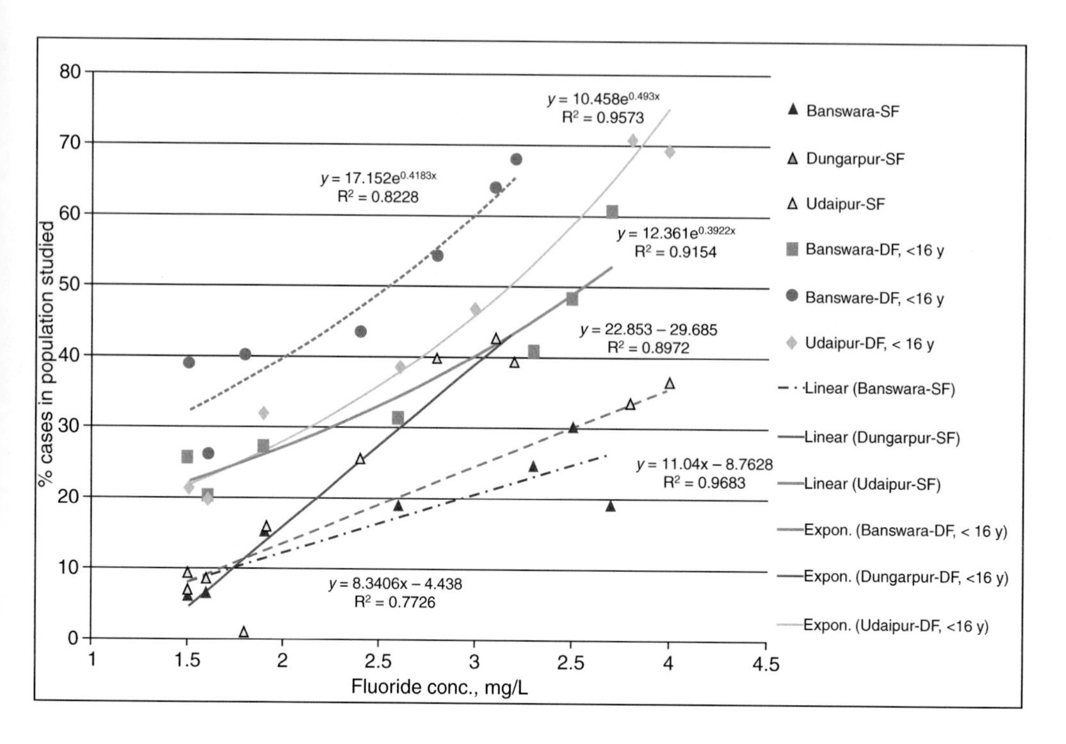

Figure 7.4 Correlations between average fluoride concentrations versus dental fluorosis in children (<16 years of age) and skeletal fluorosis in adults (>21 years of age) (based on Choubisa 2001).

7.2.2.3 *Fluoride in the environment*

Fluoride is present in soil, plants, phosphate fertilizers, rocks and subsurface materials. It can enter groundwater due to weathering of aquifer materials and rocks, or along with volcanic ash, flyash and fertilizers (Brindha and Elango 2011). Mineral dissolution of compounds such as sodium fluoride and fluorosilicates into groundwater are a major source of fluoride. Discharge from phosphate fertilizers or aluminum factories and addition of fluoride to drinking water by communities to promote dental health are other reasons for high fluoride concentrations in drinking water. In groundwater, the natural concentration of fluoride depends on the geological, chemical and physical characteristics of the aquifer, the porosity and acidity of the soil and rocks, the temperature, the action of other chemical elements, and the depth of the aquifer.

7.2.2.4 *Treatment processes for fluoride removal*

Defluoridation of groundwater is essential for meeting drinking water standards. In India, the desired fluoride level is 1 mg/L, and 1.5 mg/L is permissible in the absence of an alternative source. WHO guidelines for maximum fluoride concentration in drinking water is 1.5 mg/L. Many different defluoridation methods have been employed all over the world with varying efficiencies.

The cheapest, oldest and most common method is coagulation with alum and since the 1980s, coagulation-precipitation with alum and lime (Nalgonda method) have been popular in India and Africa. Adsorption with activated alumina and ion exchange using zeolites and synthetic resins are also popular now in developed countries due to their high efficiency. Membrane filtration processes like ED or EDR, nanofiltration and reverse osmosis can also be used for defluoridation (Singh et al. 2013).

7.2.2.4.1 Coagulation-precipitation

Coagulation-precipitation or the Nalgonda method is the most popular defluoridation method in India. Alum and lime are added to fluoride containing water and the following reactions are known to occur (Fawell et al. 2006):

Dissolution of alum:

$$\text{Alum} - Al_2(SO_4)_3.18H_2O \rightarrow 2Al^{3+} + 3SO_4^{2-} + 18H_2O$$

Precipitation of aluminum hydroxide leading to decrease in pH and formation of strong complexes with F:

$$2Al^{3+} + 6H_2O \rightarrow 2Al(OH)_3 + 6H^+$$

Lime addition for increasing pH:

$$3Ca(OH)_2 + 6H^+ \rightarrow 6H_2O + 3Ca^{2+}$$

Aluminum is quite efficient in removing fluoride from solution due to its ability to form strong complexes with fluoride. Since fluoride precipitates easily with calcium forming CaF_2, addition of lime is an effective method for lowering dissolved fluoride concentration to its solubility limit of 7.6 mg/L (Masters and Ela 2008). Further removal of fluoride in this case is by aluminium and settling of floc. The drinking water standard was achieved for initial fluoride concentrations of 10 mg/L but not higher fluoride concentrations (Singh et al. 2013). For greater removal efficiency, this method can serve as a pre-treatment method to be followed by adsorption or ion exchange or membrane filtration.

7.2.2.4.2 Adsorption or ion exchange

The removal of dissolved compounds from solution by accumulating them on the surface of solid materials is termed adsorption and was described in Chapter 5. In this interfacial process, raw water is passed through a bed containing adsorbent which is used for defluoridation. The adsorbent will retain fluoride either by physical, chemical or ion exchange mechanisms.

Major advantages of using adsorption are the high removal efficiency (>90 percent), higher water recovery than membrane processes, and its cost-effectiveness. Limitations of this process are that the efficiency of this process is pH-dependent and optimum pH range is 5 to 6. Regeneration of the adsorbent is required after every 4–5 months. Adsorbents that can be used for defluoridation include bone charcoal, activated alumina and clay minerals like kaolinite, bentonite and mica and their modifications. Fluoride can also be removed from water supplies using natural and synthetic zeolites and synthetic organic resins. Several studies show that activated alumina can give >90 percent removal at pH ≤ 7 (Loganathan et al. 2013).

PROBLEM 7.2.2

A groundwater source has the following characteristics: pH = 6.8, turbidity = 25 NTU, F = 15 mg/L, TDS = 150 mg/L, total coliforms/100 mL = 2, fecal coliforms - 0/ 100 mL. Draw a flowchart of the treatment units required and justify the need for each process.

Solution

The following treatment processes can be used for achieving IS10500 standards for drinking water. Other variations can be used with appropriate justification.

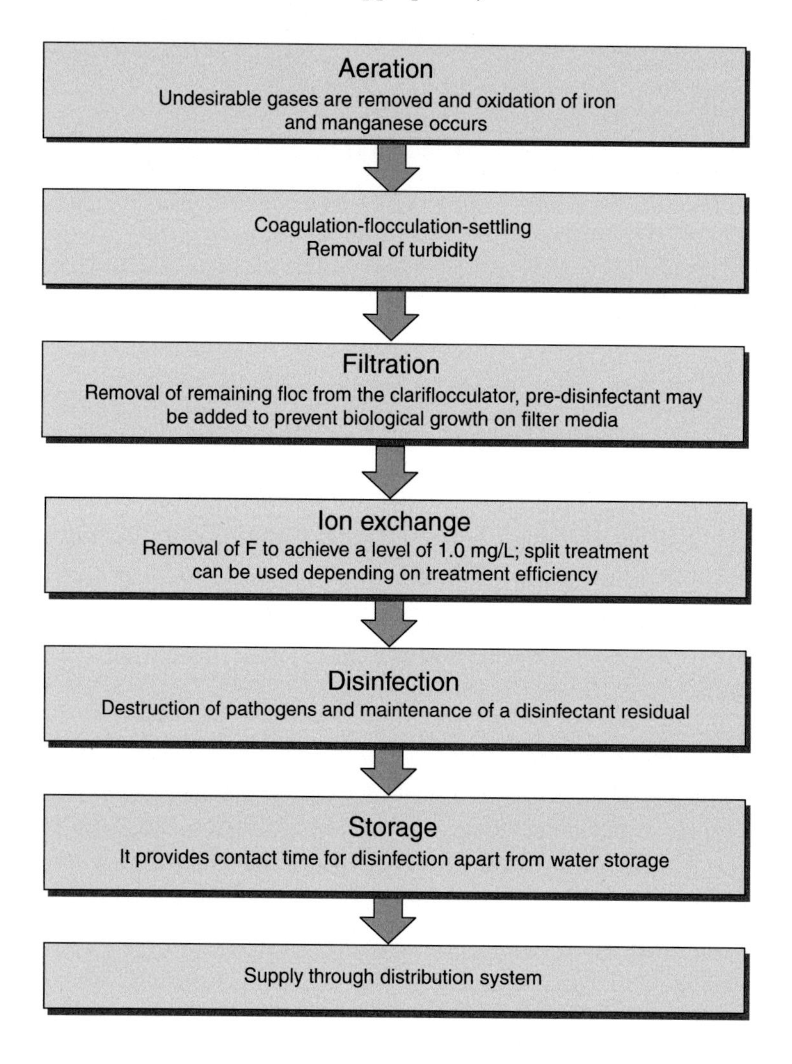

7.2.3 Nitrate Contamination and Its Removal

Nitrate pollution of water resources is a major concern in many parts of the world such as India, UK, Saudi Arabia, North America, Australia, China, Morocco, and Western Iran (Mohseni-Bandpi, Elliott, and Zazouli 2013). The problem of nitrate pollution of shallow groundwater aquifers is widespread and is increasing over time as shown in several reports (Wang et al. 2013; CGWB 2010). This concern is mirrored in increasing research activity from 1960 onwards as shown by Padilla, Gallardo, and Manzano-Agugliaro (2018).

7.2.3.1 *Nitrate in the environment*

Nitrate in water (surface and groundwater) may be of natural or anthropogenic origin. Natural sources include atmospheric precipitation, local mineral deposits (like potassium nitrate), soil-plant interactions such as decomposition of plant materials, presence of nitrogen-fixing bacteria in the root zone of plants, and animal and human waste deposition on soil. Anthropogenic sources such as agricultural, commercial, or industrial activities further contribute to nitrate in water resources. Fertilizer in surface runoff, human and animal wastes, and septic tank discharges all percolate through the soil into shallow groundwater aquifers contributing to nitrate in groundwater. Industrial sources include chemical manufacturing operations and nitrate-containing cutting oils.

Elevated nitrate levels in water are responsible for eutrophication of surface water bodies and the presence of nitrates in drinking water can lead to methemoglobinemia. Nitrate by itself is non-toxic but under reducing conditions such as those found in the stomach, it can be converted to nitrite by gastro-intestinal flora. Infants less than six months of age are at greatest risk of methemoglobinemia, also called the 'blue baby syndrome' due to a number of physiological factors like higher fluid intake per kg of body weight and higher stomach pH which allow survival of nitrate reducing bacteria (Bouchard, Williams, and Surampalli 1992). Hemoglobin (Hb) has ferrous iron which is responsible for oxygen uptake and transport while Methemoglobin (MetHb) which is a transformation product of Hb has ferric iron which precludes (inhibits) oxygen uptake and transport. Several compounds including nitrite can cause the transformation of Hb to MetHb, thus preventing uptake and transport of oxygen in the blood. The maximum contaminant level (MCL) for nitrate in drinking water is 45 mg/L as nitrate, which is equivalent to 10.16 mg/L of nitrate-nitrogen (N). The general wastewater discharge standard for nitrate-N in India is 10 mg/L.

7.2.3.2 *Nitrate removal methods*

Various treatment strategies can be used for removing nitrate from drinking water. Physico-chemical treatment processes like ion exchange (IX), reverse osmosis (RO), and electrodialysis/electrodialysis reversal (ED/EDR) have been used while biological processes include biological denitrification (BD) and microbial harvesting. BD can transform nitrate to gaseous forms such as nitrogen, nitrous oxide and dinitrogen oxide by denitrification (Kapoor and Viraraghavan 1997). Microbial harvesting is the use of algae for removal of nutrients like nitrate. Pre-treatment is often required to avoid scaling of the resin (for IX) or fouling of membranes (for RO and ED/EDR). Common concerns in the application of these removal technologies include sludge and brine management costs and interference from other ions like Ca, Mg and sulphate.

7.2.3.2.1 Ion exchange (IX)

The most common treatment method for nitrate removal is ion exchange (IX). Nitrate is removed from the treatment stream by displacing chloride on an anion exchange resin. Subsequently, regeneration of the resin is necessary to remove nitrate from the resin. Regeneration is accomplished by using a highly concentrated salt solution resulting in the displacement of nitrate by chloride. The result is a concentrated waste brine solution high in nitrate, that requires disposal. If waste brine disposal options are not limiting, IX can be the best option for low to moderate nitrate contamination and removal of multiple contaminants (including arsenic, perchlorate, and chromium). Effective use of IX requires pre-treatment to avoid resin fouling. The most significant drawback of this treatment option is the cost of disposal of waste brine. Also, application of IX may not be feasible for extremely high nitrate levels due to high TDS and large quantities of waste generated.

Conventional IX method (fixed-bed) can give 40 to 65 percent nitrate removal while modified IX methods like magnetic ion exchange (MIX), ion exchange separation (ISEP), and weakly basic anion exchanger (WBAIX) provide 40 to 85 percent removal of nitrates with low waste brine generation (UC Davis 2012).

7.2.3.2.2 Reverse osmosis (RO)

As the second most common nitrate treatment alternative after IX, RO is feasible for both municipal and point-of-use applications. RO efficiently removes multiple pollutants simultaneously including ions, particulate matter and organic compounds. Ions like sodium, chloride, nitrate, arsenic, and fluoride, particulate matter like asbestos and protozoan cysts, and organic constituents like pesticides can be removed by RO. Pre-treatment is necessary before using RO to prevent membrane fouling and scaling, and for increasing the flux through the membrane and thus increasing its life. Water is forced through a semi-permeable membrane under pressure, such that water passes through while the pollutants are not able to pass.

Disadvantages of RO include high capital, operation and maintenance costs, membrane fouling, high pre-treatment and energy demands, and potentially large wastewater volume that requires proper disposal. Brine from an RO process cannot be used for irrigation due to its high TDS.

Key factors to be considered when selecting RO are the pre-treatment requirements, water recovery and power consumption, management of waste concentrate, and the typically higher costs relative to IX.

Conventional RO treatment results in 55 to 70 percent nitrate removal which goes up to 76 percent with blending. For treatment of raw water having high nitrate concentration (200-350 mg/l), RO combined with IX and blending results in 76 to 89 percent nitrate removal (UC Davis 2012).

7.2.3.2.3 Electrodialysis (ED) and electrodialysis reversal (EDR)

ED and EDR have been used for desalination of drinking water for several decades and several plants have been set up to remove nitrate using EDR. ED works by passing an electric current through a series of anion and cation exchange membranes that trap ions including nitrate in a concentrated waste stream. To minimize fouling and the need for chemical addition, the polarity of the system can be reversed as is done in electrodialysis reversal (EDR). By reversing the polarity (and the solution flow direction) several times per hour, ions move in the opposite direction through the membranes, minimizing build-up of scale due to inorganics and deposition of natural organic matter on the membrane surface.

Key factors in the consideration of EDR for removing ions are the pre-treatment requirements, the operational complexity of the system, the limited number of system manufacturers, and the management of waste concentrate.

Unlike conventional RO, EDR is unaffected by silica. EDR costs are similar to RO and in some cases, EDR may be the preferable option as the silt density index (SDI) increases. Further, unlike RO, ED does not remove uncharged constituents in the water (UC Davis 2012). Also, as shown in Table 7.7, water recovery with EDR is much greater than with RO.

Table 7.7 Summary of data for 4 EDR plants run by Ionics (Prato and Parent 1993)

Plant	Number of stages	Water recovery, percent	TDS (mg/L)		Nitrate (mg/L)	
			Initial	Final	Initial	Final
Bermuda	3	90	1614	278	66	8.8
Delaware	3	90	114	11	61	4.5
Industrial	3	80	1753	534	655	128
Italian	2	90	-	-	120	37

Average nitrate removal in a two-stage ED/EDR process was 69.2 percent and average TDS removal was 53 percent. Three-stage EDR resulted in an average nitrate removal of 86.6 percent and average TDS removal 78.3 percent (Prato and Parent 1993). As expected, removal efficiency increases as the number of ED/EDR stages increase, but so do operational costs and energy requirements.

7.2.3.2.4 Biological denitrification (BD)

The first plant for the biological removal of nitrate from drinking water supply became operational in 1981 in the Chateau- Landon in France. Biological denitrification in potable water treatment is common in Europe with recent full scale systems in France, Germany, Austria, Poland, Italy, and Great Britain (Mohseni-Bandpi, Elliott, and Zazouli 2013). These full scale drinking water treatment plants are based on heterotrophic or autotrophic removal processes.

In heterotrophic biological denitrification, strictly anaerobic heterotrophic bacteria use nitrate as the terminal electron acceptor and organic compounds like sugar, glucose, acetone, acetic acid, ethanol and methanol as electron donors. Nitrate is transformed to nitrogen gas while the organic compounds serve as a carbon source. Autotrophic denitrifiers utilize inorganic carbon compounds (e.g., CO_2, H_2CO_3, HCO_3^-) as their carbon source and inorganic compounds like hydrogen gas and sulphide ion serve as electron donors. Substrate and nutrient addition is necessary and post-treatment for removal of microbes and excessive organic carbon are important considerations for this process. Disadvantages of biological denitrification are post-treatment requirements, high capital costs, potential sensitivity to environmental conditions, larger system footprint and high system complexity (Kapoor and Viraraghavan 1997).

Most biological denitrification plants for drinking water treatment are fixed-bed processes and experience in Europe shows high removal efficiencies, often above 90 percent. Autotropic biological denitrification resulted in more nitrate removal (94 percent to 100 percent) compared to heterotrophic bacteria (10-100 percent) (Mohseni-Bandpi, Elliott, and Zazouli 2013).

Table 7.8 Removal efficiencies of different processes for nitrate removal

Treatment Method	Removal Efficiency	Reference
Ion exchange (IX)	Conventional IX – 40 percent to 65 percent Modified IX – 40 percent to 85 percent	(UC Davis 2012)
Reverse osmosis (RO)	Conventional RO – 55 percent to 70 percent With blending – 55 percent to 76 percent	(UC Davis 2012)
RO + IX + blending	76 percent to 89 percent	(UC Davis 2012)
Electrodialysis (ED, EDR)	Two-stage -- 50 percent to 69 percent Three-stage -- 75 percent to 92 percent	(Prato and Parent 1993)
Biological denitrification (BD)	Autotropic – 94 percent to 98 percent Heterotrophic – 70 percent to 97 percent	(Mohseni-Bandpi, Elliott, and Zazouli 2013; UC Davis 2012)

7.2.4 Total Organic Carbon (TOC) and Its Removal

Total organic carbon (TOC) is an important water quality parameter for drinking water plants for the reasons described in Chapter 3. It is a measure of the natural organic matter present in all waters. TOC is not harmful or toxic by itself, but when it reacts with disinfectants like chlorine or chloramine, it can result in the formation of disinfection by-products (DBPs). Disinfection by-products such as trihalomethanes (THMs) and halo-acetic acids (HAAs) are known to be carcinogenic and may also be related to higher incidence of rectal and bladder cancer in humans (Simpson and Hayes 1998). Therefore, maximum contaminant levels (MCL) of THMs have been promulgated in many countries around to the world to protect public health. The USEPA has set an MCL of 80 ppb for total THMs and 60 ppb for HAAs. WHO guidelines and drinking water standards for DBPs in India are summarized in Table 7.9.

Table 7.9 Regulations and guidelines for DBPs in drinking water

Parameter	WHO guidelines	India
Chloroform, ppb	300	200
Bromoform, ppb	100	100
Bromo-dichloromethane, ppb	60	60
Dibromochloromethane, ppb	100	100

Many different processes can be used to remove TOC including coagulation, activated carbon (GAC) filters, and ion exchange.

7.2.4.1 Coagulation

Coagulation is typically followed by flocculation and settling, after which the water is filtered during conventional drinking water treatment. The most common coagulant in conventional water treatment is alum, i.e., aluminum sulphate. Other coagulants include ferric chloride, ferric sulphate, poly-aluminium chloride (PACl) and synthetic organic polymers like Cat-floc. Optimum pH for coagulation with alum is in the range of 5 to 6.5. At doses ranging from 5 to 100 mg Al/L, 97 percent turbidity removal, 25 to 67 percent DOC removals, 44 to 77 percent reduction in UV-254 nm were obtained (Anu Matilainen, Vepsäläinen, and Sillanpää 2010). In a full-scale drinking water treatment plant study, turbidity removal was 95 percent and NOM removal was 47 percent with alum (A Matilainen et al. 2002).

Optimum TOC removals by coagulation are a function of pH to a greater extent than coagulant dose (Bell-Ajy et al. 2000). The optimal pH for NOM removal by coagulation was determined for 16 water utilities in the US and ranged between 5.5 and 6.5 for both alum and ferric chloride. In a laboratory-scale coagulation study with two untreated (raw) river water samples collected in Nov and Mar 2009, turbidity removals were 84.6 percent and 91.4 percent, respectively at optimum alum doses of 10 mg/L and 60 mg/L, respectively (Narayan and Goel 2011). Concomitant removals of NOM were 37.5 percent and 87.7 percent, respectively based on UV-254 nm measurements.

7.2.4.2 Activated carbon

Activated carbon is most commonly used in drinking water treatment plants in either granular or powdered form. A comparison of powdered activated carbon (PAC) and granular activated carbon (GAC) is provided in Table 7.10. Most applications for activated carbon are in water treatment for NOM removal, taste and odor control, and micropollutant removal. Activated carbon is also used as an advanced wastewater treatment process if the wastewater is to be reused for drinking.

Table 7.10 Comparison between GAC and PAC (Crittenden et al. 2012)

	GAC	PAC
Particle size	0.5 to 3 mm	20 to 50 microns
Adsorption capacity	Less than PAC	More than GAC; difference is greater when removal efficiency is >90 percent
Mode of operation	Column or contactors	Added at intake point, rapid-mix tank or slurry contactor (CSTR type) and removed after coagulation–flocculation and settling.
Applications	Taste and odor control in 7 percent water treatment plants in the US	Taste and odor control in 63 percent water treatment plants in the US

PAC is the powdered form of activated carbon where the fine granules are used for seasonal or short-term problem solving. During summer, algal blooms are frequent in water bodies and extracellular exudates from algal cells contribute to taste and odor problems in water supplies. PAC is usually added to the water before the coagulation–flocculation step, and removed after settling along with sludge.

Granular activated carbon (GAC) has a relatively larger particle size compared to PAC and, is typically used in columns or contactors for tertiary level treatment in water treatment plants for the removal of NOM, micropollutants, taste and odor causing compounds. The adsorption efficiency of the GAC decreases over time and eventually needs to be replaced or reactivated (USEPA 1971). Activated carbon can provide 35 to 75 percent TOC removal from raw water. PAC removes about 83 to 91 percent TOC from raw water while GAC removed about 87 percent of TOC .

7.2.4.3 *Ion exchange*

Ion exchange has typically been used for the removal of inorganic pollutants, but recently specialized anion resins have been developed for the removal of organics, such as humic acids. TOC removal in water treatment is also possible through the introduction of styrenic and acrylic strong base anion resins. The large pores in these resins capture large organic molecules. After a brine solution removes the organic compounds from the resin, it can be reused year after year. Ion exchange results in 83 percent to 93 percent removal of the raw water TOC (USEPA 1971).

7.2.4.4 *Reverse osmosis*

The USEPA had recommended RO as the best available technology (BAT) for removal of many inorganic pollutants and organic constituents of water. Inorganic pollutants such as antimony, arsenic and selenium as well as radionuclides such as beta and alpha emitters can be removed by RO. In addition, RO can be effective in the removal of large MW synthetic organics such as pesticides. RO resulted in 65 percent and 66 percent removal of TOC with chlorine dioxide and chlorine, respectively as disinfectants. Beside TOC removal, 82-83 percent removal of THMs was obtained using RO (Rajamohan et al. 2014).

In a recent review of water treatment plant compliance with the Stage 2 Disinfection By-products Rule (2006), the USEPA recommended (i) enhanced coagulation or enhanced softening with GAC; (ii) nanofiltration followed by (iii) GAC 20 with chlorine (USEPA 2016). These treatment methods were able to comply with the MCLs for TTHMs and HAAs.

A summary of TOC removal efficiencies using different treatment processes is provided in Table 7.11.

Table 7.11 TOC Removal efficiency of different treatment processes

Treatment	Removal efficiency	Reference
Coagulation	Alum: 25 to 97 percent DOC removal Ferric salts: 29–70 percent DOC removal Alum: 25 to 47 percent TOC removal	Matilainen, Vepsäläinen, and Sillanpää 2010; (Matilainen et al. 2002)
Reverse osmosis	65 percent to 66 percent TOC removal 82 to 83 percent THM removal	(Rajamohan et al. 2014)

7.3 MUNICIPAL WASTEWATER SYSTEMS

Modern municipal wastewater systems include preliminary, primary and secondary treatment as described in Section 7.3.1. Tertiary treatment is now becoming common due to regulatory requirements for reducing nutrient (mainly nitrogen and phosphorus) levels in treated wastewater. In water-scarce regions where water is to be reused, higher levels of treatment may be necessary and are discussed in Section 7.3.2.

7.3.1 Municipal Wastewater Treatment: Primary and Secondary

A typical conventional municipal wastewater plant includes the following treatment systems (Metcalf and Eddy 2003) and is shown in Figure 7.5.

1. Preliminary treatment systems

The objective is to remove large, floating materials and large particles from the influent. Units include flow meters, screens, grit chambers, microstrainers and comminution (grinders). In some cases, grit chambers are aerated to prevent organic materials attached to the discrete particles from creating nuisance odor conditions and attracting insects. Flotation is sometimes included in preliminary treatment.

2. Primary treatment systems

The objective of primary treatment systems is to remove discrete particles (settleable particles > 10 micron) which are relatively large and have specific gravities greater than water. It generally consists of a primary sedimentation (or settling) tank (PST).

3. Secondary treatment systems

Secondary treatment follows primary treatment of wastewater and is designed for the removal of biodegradable matter that may be present in suspended, colloidal or dissolved form. Various aerobic processes can be used for the removal of biodegradable material and the most common process is the activated sludge process (ASP). Other options include trickling filters, rotating biological contactors, sequencing batch reactors, oxidation ponds, waste stabilization ponds or aerated lagoons. In all cases, a secondary settling tank or clarifier is necessary for removing settleable solids present in the sludge from these processes.

In general, wastewater treated to the secondary level is discharged to inland surface water bodies, sewers or applied to land. Concerns about the presence of pathogens in treated wastewater necessitate disinfection prior to discharge. Disinfection is done using chlorine, ozone or UV radiation.

Sludge from the clarifiers (primary and secondary) is taken for concentration which can be done using clarifier thickening, gravity thickening or flotation. From here, the sludge is taken for digestion which can be done either aerobically or anaerobically. Generally, municipal wastewater sludge is taken to an anaerobic digester where biogas is generated and the sludge volume reduced significantly. The sludge is then dewatered using a variety of treatment options like drying beds, lagoons, vacuum filtration, centrifugation and filter press. The solids are then taken for heat drying or combustion in processes like heat drying, incineration, wet oxidation, or pyrolysis. Finally, the residual solids can be either incinerated or disposed in landfills or used as compost.

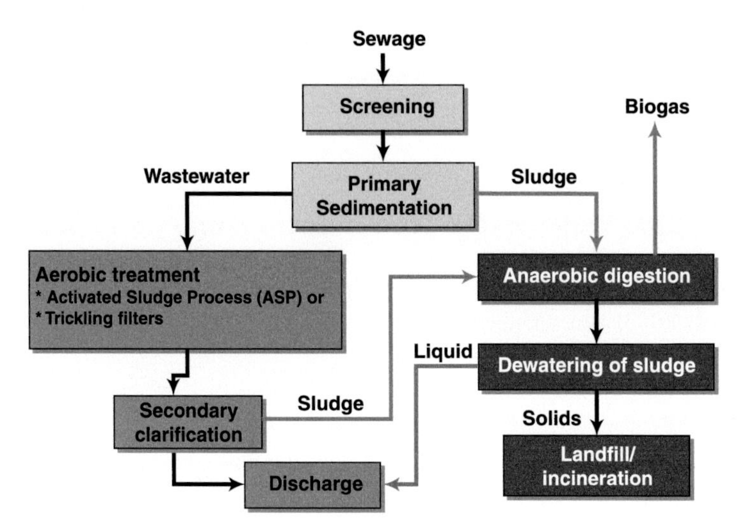

Figure 7.5 Typical flowchart for a conventional municipal wastewater treatment plant.

Examples of some typical wastewater treatment units are listed below along with their functions:

Sump Well: Collects wastewater generated in the city

Pump Well: Pumps wastewater to the treatment units

Equalization Tank: Dampens wastewater flow variations to ensure uniform flow in all units

Screen Chamber: Removes large floating materials which can damage subsequent units by clogging

Grit Chamber: Removes small and heavy suspended solids

Skimming Tank: Removes oil and grease from the surface

Primary Sedimentation Tank (PST): Removes settleable solids (inorganic and organic)

Aerobic Biological Treatment Unit (ASP/ trickling filter/ aerated lagoon/ SBR): Removes biodegradable colloidal and dissolved organic matter by oxidation (aerobic) and nutrients like nitrogen by nitrification-denitrification

Secondary Settling Tank/Secondary Clarifier: Thickens and separates sludge from effluent

Sludge Digester: Reduces volume of sludge generated and produces biogas

Sludge Drying Bed: Dewaters sludge, i.e., dries and reduces sludge volume by separating solids from effluent

7.3.2 Municipal Wastewater Treatment: Tertiary and Higher Levels

Secondary treatment removes only biodegradable organic matter. Nutrients like nitrogen and phosphorus are removed to about 30 percent in secondary treatment which may not be sufficient for preventing eutrophication of receiving water bodies. In such cases, tertiary treatment is essential. Typical values for influent wastewater and expected removals after secondary and tertiary treatment are shown in Table 7.12.

Table 7.12 Typical values for wastewater without treatment and after treatment (Bouwer 1991)

Parameters	Influent, mg/L	Percent removal in conventional treatment plants	Effluent after conventional treatment, mg/L	Tertiary treatment levels, mg/L
Suspended solids (SS)	250	90 to 95	10 to 20	5 to 10
BOD_5	250	85 to 95	20	5 to 10
COD	400	50 to 80	50	20 to 30
Nitrogen	25 to 30	20 to 40	20 to 25	< 5
Phosphorus	10	30	5 to 10	< 1

Tertiary treatment

Nitrogen removal can be achieved using a combination of nitrification-denitrification where ammonia is converted to nitrites and nitrates during nitrification (strictly aerobic) and nitrates are converted to nitrogen gas during denitrification in a strictly anaerobic process. Phosphorus removal can be achieved by adding alum which results in precipitation and removal of aluminum phosphate.

Higher or advanced treatment systems

If the tertiary effluent is to be treated further prior to discharge, it is called advanced treatment. The need for advanced treatment can arise when specific inorganic compounds like heavy metals, heavy metals and organic constituents like surfactants, pesticides, and endocrine-disrupting chemicals must be removed for industrial or other reuse purposes.

Several factors affect the choice of treatment systems such as:

- Contaminants and the extent to which they need to be removed,
- Land availability at the treatment site and its topography,
- Budget involved, and
- Number of skilled personnel available to operate equipment.

Examples of some advanced treatment processes like adsorption, gas stripping, distillation and advanced oxidation processes are provided here.

1. Adsorption

The accumulation of substances in solution on an interface is called adsorption. The substance deposited is called adsorbate which may be in liquid or gaseous phase while the adsorbent is in solid phase. The adsorbents used are activated carbon, synthetic polymeric and silica based adsorbents.

Activated carbon is the most popular adsorbent. It is prepared from a char made of organic materials like almond, coconut or walnut. The organic materials are heated in a retort to red hot temperatures with limited oxygen. This drives out the hydrocarbons. The resulting char is activated using oxidizing gases like steam and CO_2 at temperatures in the range of 800-900 °C (Metcalf and Eddy 2003). Thus, a porous structure with different pore sizes is formed.

Activated carbon can be divided into two classes: powdered activated carbon (PAC) with typical diameter of less than 0.074 mm and granular activated carbon (GAC) with typical diameter greater than 0.1 mm (Metcalf and Eddy 2003).

Once the adsorptive capacity of activated carbon is exhausted, it can be recovered using processes which are collectively termed 'regeneration'. It involves the use of chemicals, steam, solvents and biological processes to oxidize or drive off the adsorbed material. Another process called 'reactivation' involves the same process used for the synthesis of activated carbon. In this process, used carbon is reactivated in a furnace. The adsorbed material is driven off and any new compounds formed on the surface are burned off.

2. Gas stripping

Gas stripping is essentially the mass transfer of a volatile contaminant (or dissolved gas) which is originally in the liquid phase into the gas phase. The liquid which contains the volatile contaminant to be stripped is brought in contact with a gas which does not contain the contaminant to be stripped. The latter is usually air. The basic principle is that the contaminant to be stripped escapes from solution and enters the gas phase to satisfy Henry's Law. Gases such as CO_2, NH_3, H_2S and volatile organic compounds (VOCs) are removed using gas stripping.

3. Advanced oxidation

Advanced Oxidation Processes (AOPs) are used to degrade complex organic compounds in wastewater that are not degraded biologically. These organic compounds are usually oxidized partially to reduce their toxic nature or to make them easily biodegradable. AOPs use hydroxyl free radical ($HO^{\bullet}$), which has the capability to oxidize all reduced compounds which are not oxidized by chlorine, ozone or oxygen. $HO^{\bullet}$ is generated using ozone in combination with UV and/or hydrogen peroxide H_2O_2.

4. Distillation

It is a unit operation which uses vaporization and condensation to separate the solids dissolved in a liquid. It is expensive and therefore used only where other methods cannot be used or where heat is cheaply available. Several types of distillation processes are used, two of which are briefly described below.

In Multiple-Effect Evaporation, a series of boilers is used, each at a lower pressure than its preceding one. After the influent is pretreated, it enters the heat exchanger in the last stage and gets progressively warmed as it moves through all the heat exchangers. On reaching the first-stage, it flows through a set of vertical tubes in a steam-heated thin film and enters the second-stage from the bottom of the first-stage. Thus, it helps in removing a majority of the non-volatile components.

In Multi-stage Flash Evaporation, TSS is removed from the influent. The water is then passed through heat transfer units, each at a lower pressure than its preceding one. As water enters each stage, vapors are formed by flashing. Flashing is boiling caused by reduction in pressure. Finally, the water is pumped out in the lowest-pressure stage. The flashed vapor condenses and gets collected in trays. The latent heat of the vapors is utilized to preheat the water or wastewater.

Water reclamation and reuse

The processing by which wastewater is made reusable is called wastewater reclamation. It is done using various combinations of unit operations and processes. The use of reclaimed water for beneficial uses is called water reuse and is covered in some detail in Chapter 9.

Reclaimed wastewater has been used for many applications that are listed below:

1. The most common use of treated wastewater is for irrigation or horticulture since the effluent from a conventional wastewater treatment plant is high in nutrients like N and P.
2. Landscape irrigation - it encompasses irrigation of parks, golf courses, playgrounds, landscaped areas around offices and residences.
3. Treated wastewater is often used by industries for housekeeping, cooling and process needs.
4. Treated wastewater can be used to recharge groundwater aquifers.
5. It has been used for development of marshes, manmade lakes, golf courses, and storage ponds.
6. Treated wastewater can be used for fire protection, toilet flushing, and air conditioning.
7. In many water-scarce areas, conventionally treated wastewater (up to secondary treatment) is further treated (tertiary and higher levels) for removal of nutrients, heavy metals, pesticides, and other micropollutants and used even for drinking.
8. Treated wastewater can be used to replenish stream flows, conserve ecosystems, recreational uses and to improve stream navigability.

Study Outline

Conventional drinking water treatment

The main objective of any water supply scheme is to provide safe and adequate amounts of potable water to all users in the service area.

The primary objective of drinking water treatment is to provide water that meets all drinking water standards, i.e., there are no health or aesthetic concerns.

The most important pollutants in drinking water are pathogenic microbes and turbidity causing particles (colloids and suspended solids). Pathogens can be reduced or removed by disinfection. Colloids and suspended solids are removed by coagulation and filtration.

If the raw water quality meets all drinking water standards, then the only treatment required is disinfection.

Any water supply scheme has the following elements: water source(s), water collection and transport systems, treatment plant, storage and distribution systems. Besides treated water, the water supply system has another output, which is the wastewater or sludge.

In general, surface water treatment requires the following processes: intake, screening, settling or primary settling, coagulation–flocculation–settling, granular media filtration and finally, disinfection.

In general, groundwater treatment requires the following processes: well point, aeration, softening, granular media filtration and finally, disinfection.

Non-conventional drinking water treatment

Any deviation from conventional drinking water treatment for the removal of specific pollutants other than pathogenic microbes and suspended solids is termed non-conventional drinking water treatment.

Specific pollutants of concern in several parts of the world are arsenic, fluoride, nitrate, NOM, and total dissolved solids (salts). Other than NOM, the other four pollutants are found mainly in groundwater supplies. NOM is present in all waters, surface water and groundwater, at varying concentrations. Arsenic and fluoride are geogenic pollutants while nitrate in groundwater is primarily anthropogenic in origin.

Arsenic in groundwater and its removal

Arsenic is a highly toxic compound and has no known beneficial health effects. It is used as a pesticide and rodenticide.

The two most common forms of arsenic in groundwater are arsenite – As(III) and arsenate – As(V). The speciation of these two forms of arsenic is dependent on pH and redox conditions.

The current drinking water standard for arsenic is 10 ppb or micro-g/L.

The first step in the treatment of arsenic is oxidation of all As(III) to As(V). As(V) can then be removed by any of the following methods: co-precipitation or coagulation with ferric salts, electro-coagulation, ion exchange, adsorption, and membrane filtration.

Fluoride in groundwater and its removal

High fluoride concentrations in water, food, or air can lead to dental and skeletal fluorosis. This is a debilitating disease and can only be prevented not cured. China and India are two countries that are most severely affected by all forms of fluorosis. More than 100 million people are affected in these two countries and several more are affected in other countries.

Defluoridation of groundwater is essential for meeting drinking water standards. In India, the desired fluoride level is 1 mg/L, and 1.5 mg/L is permissible in the absence of an alternative source.

The simplest method for defluoridation of water is addition of alum since aluminum forms strong complexes with fluoride which are removed during settling of floc. However, drinking water standards may not be achievable with this method.

A more expensive but highly efficient method is adsorption using activated alumina or ion exchange using zeolites or synthetic resins.

Membrane filtration processes like ED or EDR, nanofiltration or reverse osmosis can also be used for defluoridation.

Nitrate contamination and its removal

Nitrate is not toxic in itself but in the reducing conditions of the gastro-intestinal tract of humans, it is converted to nitrite which can inhibit oxygen uptake and transport in blood. High nitrate concentrations can lead to methemoglobinemia and infants less than six months of age are at greatest risk.

Another major concern with the presence of nitrate in water sources is its potential for causing eutrophication thereby reducing water quality and depleting dissolved oxygen.

The maximum contaminant level (MCL) for nitrate in drinking water is 45 mg/L, which is equivalent to 10.16 mg/L of nitrogen (N). The general wastewater discharge standard for nitrate-N in India is 10 mg/L.

The most common methods for removing nitrate from groundwater are ion exchange, reverse osmosis, electrodialysis reversal and biological denitrification.

TOC or NOM removal

Total organic carbon (TOC) is a measure of the natural organic matter that is present in all waters.

TOC is not harmful or toxic by itself, but when it reacts with disinfectants like chlorine or chloramine, it can result in the formation of disinfection by-products (DBPs).

Disinfection by-products such as trihalomethanes (THMs) and halo-acetic acids (HAAs) are known to be carcinogenic and are regulated in many countries including India.

Many different processes can be used to remove TOC including coagulation, activated carbon (GAC) filters, and ion exchange.

Municipal wastewater treatment

Modern municipal wastewater systems include preliminary, primary, and secondary treatment.

Tertiary treatment is now becoming common due to regulatory requirements for reducing nutrient (mainly nitrogen and phosphorus) levels in treated wastewater.

In water-scarce regions where water is to be reused, higher levels of treatment may be necessary.

Conventional wastewater treatment

The objective of preliminary treatment systems is to remove large, floating materials and large particles from the influent. Units include flow meters, screens, grit chambers, microstrainers and comminution (grinders).

The objective of primary treatment systems is to remove discrete particles which are relatively large and have specific gravities greater than water. It generally consists of a primary sedimentation (or settling) tank (PST).

The objective of secondary treatment systems is the removal of biodegradable matter that may be present in colloidal or dissolved form. Various aerobic processes can be used and include activated sludge process (ASP), trickling filters, rotating biological contactors, sequencing batch reactors, oxidation ponds, waste stabilization ponds or aerated lagoons. In all cases, a secondary settling tank or clarifier is necessary for removing settleable solids (biomass) present in the sludge from these processes.

In general, wastewater treated to the secondary level is discharged to inland surface water bodies, sewers or applied to land. Disinfection has to be provided prior to discharge to prevent the release of pathogens to the environment. It is done using ozone or UV radiation.

Sludge from the clarifiers (primary and secondary) is taken for concentration and then digestion, which can be done either aerobically or anaerobically. After that the sludge is dewatered, and finally, the residual solids are either incinerated or disposed in landfills or used as compost.

Anaerobic digesters are popular since biogas can be generated and the sludge volume reduced significantly.

Tertiary wastewater treatment

The objective of tertiary treatment is removal of nitrogen and phosphorus and enhanced removal of TSS, COD, and BOD prior to discharge to surface water bodies or over land.

Nitrogen removal can be achieved using biological processes like nitrification-denitrification.

Phosphorus removal can be achieved by adding alum which results in precipitation and removal of aluminum phosphate.

Advanced wastewater treatment

Advanced treatment is provided if the tertiary effluent is to be treated further prior to discharge.

It is needed when specific inorganic (heavy metals) and organic constituents like surfactants, pesticides, endocrine-disrupting chemicals, and other compounds must be removed for industrial or other reuse purposes.

Study Questions

1. What is the primary objective of conventional drinking water treatment? How can it be achieved? Explain with a flowchart.
2. Why should surface water and groundwater be treated differently? Explain with an example of each.
3. What is the difference between conventional and non-conventional drinking water treatment? Explain with examples of each.
4. Why do specific pollutants like fluoride, arsenic and TDS require non-conventional drinking water treatment schemes?
5. When is advanced drinking water treatment needed? Explain with examples.
6. What are the consequences of discharging untreated wastewater to a surface water body or directly over land?
7. When is tertiary and higher level of wastewater treatment needed? What pollutants need to be removed at the higher levels of treatment?

CHAPTER 8

Water Transport and Distribution Systems

Learning Objectives

- *Describe different types of conduits and conditions for their use*
- *Describe different types of distribution reservoirs and conditions for their use*
- *Design distribution networks*

A major issue in the design of water supply schemes is the collection, transport and distribution of water from the source. Collection of water from different sources, i.e., surface water versus groundwater, will differ for obvious reasons and was covered in Chapter 1.

The next major issue is the transport and distribution of water from the source or treatment plant to the community or service area. Different situations require different methods of transport and distribution. Distribution systems can be of three types: gravity, pumped or combined (Peavy, Rowe, and Tchobanoglous 1985).

***Gravity distribution systems*:** If the source is elevated with respect to the community then water can be supplied directly by gravity. Examples are highland lakes or reservoirs that are used as water sources for many cities around the world such as New York and Mumbai.

***Pumped distribution systems*:** If the source is at a lower elevation compared to the community or service area then it has to be pumped directly to the service area.

***Combined distribution systems*:** If the source is at a lower elevation compared to the community or service area, then water can be pumped and stored in surface or elevated distribution reservoirs in the service area or adjacent to the service area resulting in a combined distribution system where gravity and pressure are used. Toronto pumps water out of Lake Ontario, treats the water and stores it in underground reservoirs and elevated reservoirs (called water towers) for further distribution to the service area.

Different types of distribution reservoirs can be built; they can be under-ground, over-ground or elevated. Water is then supplied from these distribution reservoirs to the community or to elevated storage reservoirs built within the community. Thus, combinations of direct gravity supply, pumped and/or elevated supplies are possible depending on the geography of the water source vis-à-vis the community. Definitional sketches of water supply conduits and service reservoirs are provided in Fair, Geyer, and Okun (1966).

The last issue in water supply systems is the analysis of pipe flows and head losses through the distribution system. Hardy–Cross method for analysis of pipe flow in distribution networks is covered in the last section.

8.1 TRANSPORT OF WATER

Structures used to convey water from the source to the treatment plant and/or distribution system are called conduits. Conduits can be of two types: gravity conduits or pressure conduits.

Gravity Conduits

In gravity conduits, water flows freely under gravity. This type of flow is also called open channel flow even if the channel is covered, such as a sewer flowing under less than full capacity. Examples of gravity conduits are canals, aqueducts, flumes, and tunnels. It is necessary to construct these by following the hydraulic gradient. Evaporation, percolation and exposure of water to pollution can lead to high maintenance costs in some cases.

Canals

Channels that are used to convey water, generally by gravity are termed canals. These are almost always concrete-lined structures with different types of cross-sections: parabolic, trapezoidal and V- or U-shaped depending on the nature of flow in the canals and terrain through which the canals pass. Canals are generally constructed ***across*** contours from higher elevation head to lower elevation head such as the Sutlej-Yamuna Link Canal shown in Figure 8.1. Canals can also be constructed ***along*** contours resulting in long, meandering contour canals as shown in Figure 8.2. Narmada Main Canal is a contour canal and the longest irrigation canal in the world. Figure 8.2 shows a map of the Narmada Main Canal and its command area starting from Sardar Sarovar dam in Madhya Pradesh and extending to Rajasthan in the north (not shown in map).[65]

65 http://www.narmada.org/maps/ssp.cmd.area.jpg.

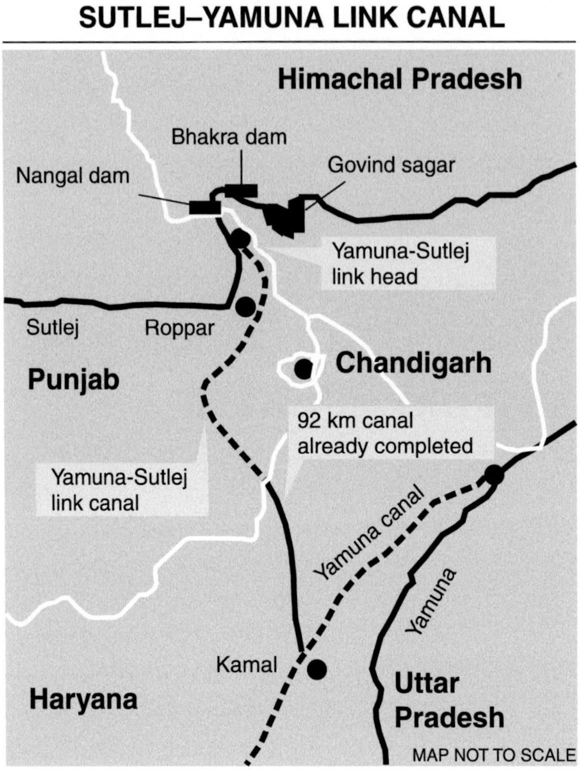

Figure 8.1 Map of the Sutlej–Yamuna Link Canal (Source: Wikipedia).

Aqueducts

Aqueducts are open channels used for conveying water and are generally constructed when the source and service area are both at high elevations and separated by a low-lying area. A schematic of an aqueduct is shown in Figure 8.3 along with photos of the Mathur (*pronounced Mathoor*) Aqueduct in Tamil Nadu, India. Aqueducts are used in terrains where water has to be conveyed over a long distance and pumping costs would be prohibitive. Capital costs of constructing aqueducts may exceed those for pipelines but are offset by the long-term savings in pumping costs.

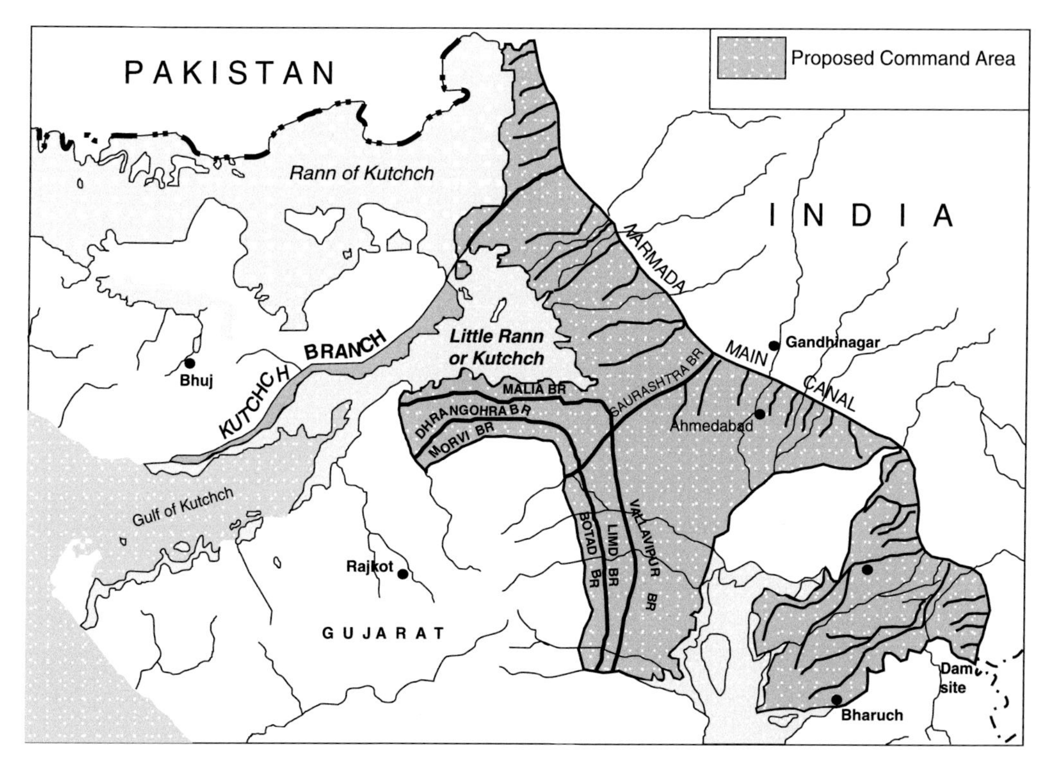

Figure 8.2 Map of the Narmada main canal and its command area.
Note: Map not to scale, and does not represent authentic international boundaries.

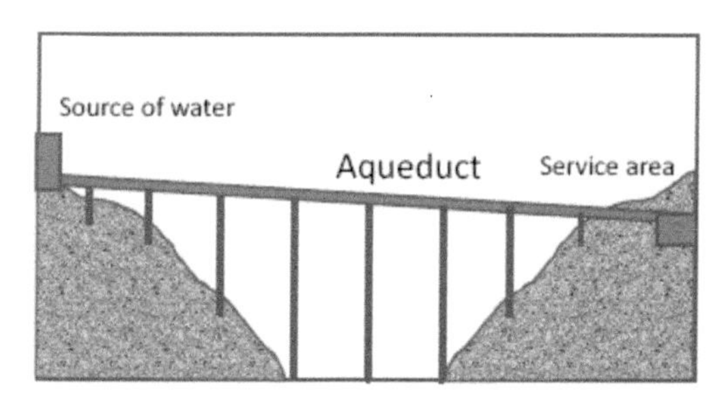

Figure 8.3 Schematic and photos of the Mathur aqueduct conveying water from source to service area (Source: Wikipedia) (see color plate).

The continuity equation relates the flow rate, velocity and cross-sectional area of the conduit.

$$Q = A^*v \tag{8.1.1}$$

Where Q = flow rate, m^3/s

A = cross-sectional area of fluid in conduit, m^2

v = velocity of flow, m/s

The total energy head at any point in a hydraulic system is defined by the following equation (Hammer and Hammer Jr. 2008):

Total energy head, m = Head due to elevation (potential energy), H + Pressure head (potential energy), (P) + velocity head (kinetic energy), $(v^2/2g)$ 8.1.2

H = elevation of the point, m

P = pressure head, m

g = gravitational acceleration, m/s^2

Flow or head loss equations for open channel flow

For uniform, steady, open channel flow, Manning's formula is used and this equation can be rearranged to determine head loss for known flow rates, and specific length and hydraulic radius of pipe.

$$Q = {}^1/_n \, A \, r_H^{2/3} S^{1/2} \tag{8.1.3}$$

where S = slope or hydraulic gradient, or drop in head or head loss per unit length $= -h_L/L$, length of pipe, m/m, r_H = hydraulic radius, m; and n = Manning's coefficient or coefficient of roughness which varies from 0.011 to 0.015. The lower value for Manning's coefficient is applicable for clear water and smooth joints while the upper value is for wastewater and poorly constructed joints (Hammer and Hammer Jr. 2008).

Hydraulic mean radius, r_H is the ratio of the cross-sectional area of the pipe to its wetted perimeter. For a fully-flowing circular pipe, $r_H = d/4$ where d = diameter of the pipe, m. Equations for partially flowing circular pipes and rectangular channels along with methods for sizing of pipes, discharge velocities, and slopes using Manning's equation are provided in various textbooks including Qasim, Motley, and Zhu (2000), Fair et al. (1966).

PROBLEM 8.1.1

A community is to be supplied 10 MLD and a gravity main is to be used for conveying water from a reservoir to the service area which is 5 km away. Determine the size of the gravity main for a velocity of 1 m/s. Determine head loss through the gravity main assuming open channel flow, smooth concrete pipe flowing full. Provide a safety factor of 2.

Solution

$$Q = A*v = 10^4\ m^3/d = 10^4\ (m^3/d)*(1\ d/24\ h)*(1\ h/60\ min)*(1\ min/60\ s) = 0.116\ m^3/s$$

$$A = \text{cross-sectional area of main} = (0.116\ m^3/s)/(1\ m/s) = 0.116\ m^2$$

Calculate diameter of main, d = 0.385 m

Using safety factor of 2, diameter of main to be provided = 0.769. Round off to the upper value of 0.8 m for design purposes.

Since flow in the gravity main is under atmospheric pressure, Manning's formula (equation 8.1.3) is to be used in terms of head loss:

$$S = -h_L/L = (Q*n/A^2\ r_H^{2/3})^2 = [0.116*0.011/0.0116*0.2^{2/3})^2 = 1.03 \times 10^{-3}$$

This means the slope required is 1.03 m/1000 m and head loss over 5 km or 5000 m = 5.17 m.

Pressure Conduits

In these conduits, the water flows under hydraulic pressure which may be higher than atmospheric pressure. They are usually made of wood, iron, steel, reinforced cement concrete, or plastic.

Flow or head loss equations for pressure conduits

The Darcy–Weisbach equation can be used to determine head loss due to frictional losses in pipes under full pressure (Hammer and Hammer Jr. 2008).

$$h_L = f\,Lv^2/d*2g = f\,L\,Q^2/A^2*d*2g = f\,L\,Q^2/12.1*d^5$$

Where $A = \pi d^2/4$, 8.1.4

f = dimensionless friction factor which is a function of the relative roughness of the pipe material and the pipe diameter.

L = length of pipe, m

v = velocity of flow, m/s

g = gravitational acceleration, 9.81 m/s^2

d = diameter of pipe, m

For pressure conduits (pipe flow under pressure), Hazen–William's formula can be used and is more popular in water supply design than the Darcy–Weisbach equation (Hammer and Hammer Jr. 2008).

$$Q = 0.278*C*d^{2.63}S^{0.54} \qquad 8.1.5$$

Where C = coefficient for Hazen–William's formula; C = 130 for concrete and 140 to 150 for plastics, S = slope or hydraulic gradient, drop in head or head loss per unit length =$-h_L/L$, length of pipe, m/m.

PROBLEM 8.1.2

A 10 cm diameter HDPE pipe is to be used for carrying water under pressure. For a flow of 10 L/s and a C value of 150 for a plastic pipe, determine the velocity of flow and head loss through the pipe.

Solution

$$Q = A^*v = 10^{-2}\ m^3/s$$

Cross-sectional area of pipe, $A = \pi r^2 = \pi^*(0.05)^2 = 7.85 \times 10^{-3}\ m^2$

$$v = 1.274\ m/s$$

Head loss through pipe ($S = -h_L/L$) is derived from $Q = 0.278\ C\ d^{2.63} S^{0.54}$

Solving for $S = -[10^{-2}/(0.278^*150^*0.1^{2.63}]^{1/0.54}$

$S = -0.10$ or head loss of 1 m in every 10 m length of pipe.

8.2 DISTRIBUTION RESERVOIRS

Distribution reservoirs are used to store treated water for balancing hourly variations in water demand and for emergencies like fires, break-downs, and repairs. These reservoirs are also called service reservoirs. The main functions of a distribution reservoir are:

1. To balance hourly variations in water demand thus allowing water treatment units and pumps to be operated at a constant rate. This in turn reduces costs and improves efficiency.
2. To maintain constant heads in the distribution mains. In the absence of storage reservoirs, pressure would fall as demand increases, and the pumps may not immediately respond to change in pressure.
3. To allow pumping of water in shifts to meet the whole day's demand without affecting water supply,
4. To supply water during emergencies like fires, breakdowns or repairs, and
5. To reduce the size of pumps, pipelines and treatment units thereby leading to overall economy.

Distribution reservoirs are classified into two types: surface and elevated reservoirs based on their elevation with respect to the ground. In a gravitational type of distribution system, water is stored in the ground, generally, in an underground service reservoir that is at higher elevation compared to the service area and sent from there to the distribution system directly. In a combined gravity and pumping distribution system, water is first stored in ground reservoirs and then pumped into an elevated service reservoir from where it is supplied to the distribution system.

8.2.1 Surface Reservoirs

Surface reservoirs are either circular or rectangular tanks, constructed at ground level or below ground level. They are also called ground reservoirs. They are generally constructed at high surface elevations in the city. If a city has more than one such elevated area, then more than one reservoir can be provided in these areas.

A typical surface reservoir has two compartments so that if one is under repair or cleaning, the other can be used. The two compartments are connected to each other by a shut-off valve or sluice valves.

The overflow pipes are provided with full supply level in order to maintain a constant water level.

Ventilators are provided in the roof slab for free circulation of air. Even though the water stored is treated water, some sludge may settle during storage and will need to be removed. For this reason, the concrete floor is sloped towards the washout pipes which are used to clean the reservoir at suitable intervals.

8.2.2 Elevated Reservoirs

Elevated reservoirs are rectangular, elliptical, or circular overhead tanks at some elevation above the ground level and supported on towers or pillars. They are constructed where pressure requirements necessitate considerable elevation above the ground surface and the use of standing pipes becomes impracticable. Elevated reservoirs are used in areas where a combined gravity and pumping system for water distribution is adopted. Water is pumped into these elevated tanks from surface reservoirs and is then sent to the distribution network. Pumping costs using elevated reservoirs are less than direct pumping costs. It is best to locate elevated reservoirs between the treatment plant and service area to minimize pumping requirements and associated costs (Sincero and Sincero 1996).

These tanks are generally made of reinforced cement concrete (RCC) as they are cheaper, do not corrode and require less maintenance. These tanks can also be made from steel and pre-stressed concrete. Ventilation is provided for free air circulation.

8.3 DISTRIBUTION NETWORKS

Distribution of water here refers to the delivery of treated water from treatment plant to the users in the service area. Every city needs a distribution system to distribute treated water properly. Water can be distributed in a city by gravity, by pumps alone, by pump and reservoir storage or by a combination of gravity, pumping and storage.

A distribution system network consists of a network of pipelines. It essentially comprises of a main supply or trunk feeder, primary feeder, secondary feeder, street mains and service pipes. Trunk feeders feed water to primary feeders which are large pipelines spaced 1 to 2 km apart. Primary feeders feed water to a system of smaller pipelines called secondary feeders which are placed in the city at a spacing of two to four blocks. Water is then passed on from secondary feeders to street mains and subsequently to service pipes. Valves are one of the most important components of a distribution network and are required to control flow.

Distribution networks are of two types: branched or tree system and looped or grid iron system (Sincero and Sincero 1996). Branched systems are an old method of distributing water and are quite common in developing countries (Peavy et al. 1985). They are simple to design and easy to build. They are easily extendable as well. However, they are no longer acceptable in modern water supply systems due to formation of dead ends in such systems. Dead ends lead to build-up of sediment and stagnation of water since pipes at these places are not subject to continuous flow. It is difficult to maintain water quality in the long-term due to bacterial regrowth and increase in chlorine demand at dead ends. Increase in chlorine demand results in inadequate chlorine residual at these points. When repairs are necessary in branched systems, points downstream from the points of repair get no water until repairs are completed. Finally, extensions of the branched systems, i.e., service points that are added after the original system was put in place lead to extremely low pressure at many service points.

In grid iron formation, all pipes are a part of inter-connected loops which prevent formation of dead ends and problems associated with them. The only disadvantage of the grid iron system is the complexity of design which requires various computational methods. The oldest method in use for designing looped systems is the Hardy–Cross method which was published in 1936.

Water pressure requirement

A distribution system should satisfy the pressure requirements of the community, i.e., sufficient pressure should be available at all outlets such as faucets, showerheads, and washbasins. The pressure in the system must also enable water from a fire-fighting hose to deliver 600 liters per minute as per Indian standards. The pressure should be enough to deliver water to the farthest parts of the community. However, high pressures in a distribution system will result in increase in leakage and loss, and can cause bursting of pipes.

Hydraulics of distribution system

The hydraulic analysis of a distribution system helps in determining if the system can meet demand in terms of quantity and pressure (Sincero and Sincero 1996). One of the methods for such analysis is Hardy–Cross method which was published in 1936. This method can be explained using a simple loop which has only one point B at which water is supplied and water leaves the loop (take off) at all other points (A, C, D, E, and F) as shown in Figure 8.4.

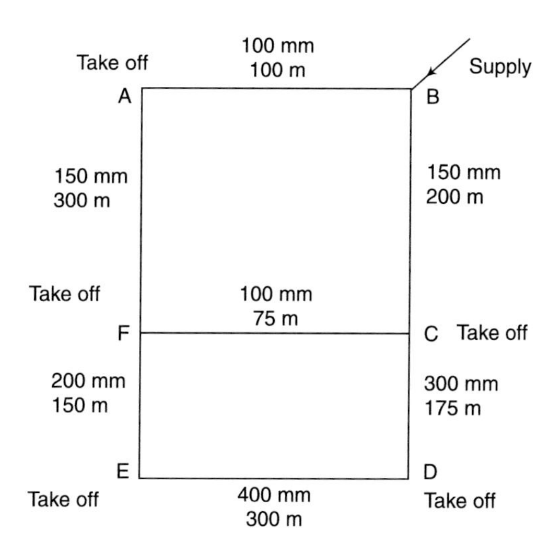

Figure 8.4 Schematic of a distribution network loop.

Let us consider two points: A and B in the above loop. The principle used in the derivation is that head loss in going from B to A = $h_{f\,BA}$ directly, i.e., B to A is equal to the head loss in going from B to A through C and F, $h_{f\,BCFA}$. This is true because otherwise water will only flow through the minimum head loss path and there will be no flow in branch BCFA. Hence, the sum of clockwise head losses in the loop, $\Sigma h_{f\,c}$, has to equal the sum of counter-clockwise head losses, $\Sigma h_{f\,cc}$. In symbol form,

$$\Sigma h_{fc} = \Sigma h_{fcc} \tag{8.3.1}$$

From fluid mechanics,

$$h_f = f\frac{lv^2}{2gd} = f\frac{l}{2gdA^2}Q^2 = kQ^2 \tag{8.3.2}$$

Where h_f is the frictional head loss, f is the friction factor, l is the length of the pipe, v is the velocity, d is the diameter of the pipe, Q is the flow rate, A is the cross- sectional area of flow and k is a constant of proportionality for a specific flow path (or loop) of specified pipe length and pipe diameter. Therefore,

$$\Sigma h_{fc} = \Sigma kQ_c^{\;2} \tag{8.3.3}$$

$$\Sigma h_{f\,cc} = \Sigma kQ_{cc}^{\;2} \tag{8.3.4}$$

Solution to the above problem is arrived at iteratively. Choose Q_c and Q_{cc} such that $\Sigma h_{f\,c} - \Sigma h_{f\,cc} = 0$, so that the system is hydraulically balanced.

In practice, if $\Sigma h_{f\,c} - \Sigma h_{f\,cc} \neq 0$ then Q_c and Q_{cc} values are in error. So, in the loop given above the flows must be apportioned again, so that the difference in head loss is zero. Let us assume that the clockwise flow Q_c is larger by ΔQ. Therefore, to correct the inequality, the counter-clockwise flow should be smaller by ΔQ. The flows are corrected as:

$$\Sigma h_{fc} = \Sigma k(Q_c - \Delta Q)^2$$

$$\Sigma h_{f\,cc} = \Sigma k(Q_{cc} - \Delta Q)^2$$

So that $$\Sigma h_{fc} - \Sigma h_{fcc} = 0$$

i.e., $$\Sigma k(Q_c - \Delta Q)^2 - \Sigma k(Q_{cc} + \Delta Q)^2 = 0 \qquad 8.3.5$$

Expanding the equation using the binomial theorem, and neglecting the second power of ΔQ, we get (Sincero and Sincero 1996):

$$\Delta Q = \frac{\Sigma kQ_c^2 - \Sigma kQ_{cc}^2}{\Sigma 2kQ_c + \Sigma 2kQ_{cc}} = \frac{\Sigma kQ^2}{2\Sigma kQ} \qquad 8.3.6$$

We have taken head loss as proportional to square of flow rate, $h_f = kQ^2$ but actually head loss is proportional to some power 'x' of flow rate, $h_f = kQ^x$. Using this latter equation of head loss, we get ΔQ as $\Delta Q = \frac{\Sigma kQ^2}{x\Sigma kQ}$ and 'x' value is the inverse of the exponent of the friction slope term in the head loss equation. Using the Hazen–William equation, we get $x = 1.85$ (Qasim et al. 2000):

$$h_f = kQ^{1.85} \text{ and } \Delta Q = \frac{\Sigma kQ^{1.85}}{1.85\Sigma kQ} \qquad 8.3.7$$

Several other methods are now available for determining flow and head loss in looped systems.

PROBLEM 8.3.1

Consider a system of parallel pipes as shown in the figure. Determine the flow rates in each branch if the total flow through the system is 500 L/s, entering the system at junction A.

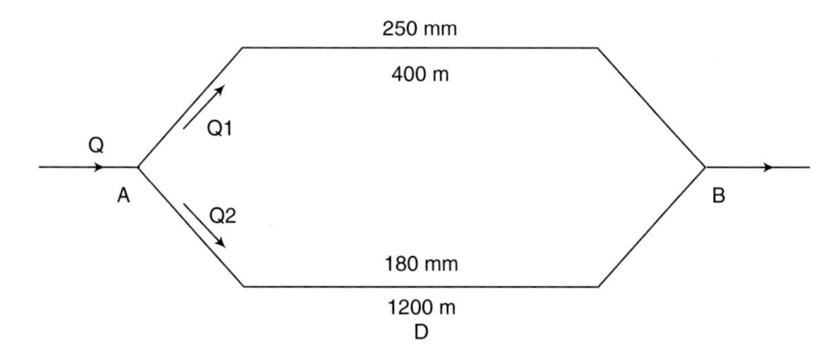

Solution

Assuming Fanning friction factor value as 0.01 and if the flow in branch ACB is Q1 and in branch ADB is Q2 then

$$Q1 + Q2 = Q = 500 \text{ L/sec}$$

Since branch ADB is longer and narrower than branch ACB, it is reasonable to assume that flow in branch ADB is lesser than in ACB. The solution is determined iteratively here.

Trial 1

$$Q = 500 \text{ L/s} = 0.5 \text{ m}^3\text{/s}$$

Pipe ACB - pipe 1

$$k = fL/2gDA^2$$

Assume Fanning friction factor, f = 0.01

Assume as first guess Q1 = 0.4 m^3/s

Length of pipe, L1 = 400.00 m

Diameter of pipe, D1 = 0.25 m

Cross-sectional area, A1 = 0.05 m^2 therefore, k1 = 338.44 s^2/m^5

$$h_f - I = h_{fc} = kQ^n = 54.15 \text{ m}$$

Pipe ADB = pipe 2

Assume Q2 = 0.1 m^3/s

Length of pipe, L2 = 1200.00 m

Diameter of pipe, D2 = 0.18 m

Cross-sectional area, A2 = 0.03 m^2

Therefore, k2 = 5247.35 s^2/m^5

$$h_f - II = h_{fcc} = kQ^n = 52.47 \text{ m for } n = 2$$ (can solve for n = 1.85 but only solution for n = 2 is shown here)

$$h_{fc} - h_{fcc} = 1.68$$

Flow correction, $\Delta Q = \Sigma kQ^2/(2\ \Sigma kQ) = 8.08 \times 10^{-2}\ m^3/s$

The problem can be solved by setting up a spreadsheet with the first guess values of Q1 and Q2 and determining all other values. In the first iteration, the discrepancy between head loss calculated in the clockwise direction – h_f-I and head loss calculated in the counter-clockwise direction – h_f-II is 1.6768. The objective of each iteration was to minimize this error, i.e., the value should approach zero.

Incremental reductions in Q1 were made to reduce the error in head loss calculations. As can be seen in the table below, the error oscillates between positive and negative values for

Q1 = 0.399 m^3/s and Q1 = 0.398 m^3/s indicating that the best solution lies between these values. Further iterations after Iteration 3 were done to find this value and the minimum error was obtained for Q1 = 0.3987 m^3/s.

Iteration number	Q1, m^3/s	Q2, m^3/s	k1	k2	hf – I = $k1Q1^2$	hf – II = $k2Q2^2$	hf – I – hf – II = delta-hf	delta-Q, m^3/s (n = 2)
1	0.400000	0.1000	338.44	5247.35	54.15	52.47	1.6768	0.08
2	0.399000	0.1010	338.44	5247.35	53.88	53.53	0.3517	0.08
3	0.398000	0.1020	338.44	5247.35	53.61	54.59	–0.9832	0.08
4	0.398600	0.1014	338.44	5247.35	53.77	53.95	–0.1811	0.08
5	0.398700	0.1013	338.44	5247.35	53.80	53.85	–0.0477	0.08
6	0.398800	0.1012	338.44	5247.35	53.83	53.74	0.0855	0.08
7	0.398850	0.1012	338.44	5247.35	53.84	53.69	0.1521	0.08
8	0.398750	0.1013	338.44	5247.35	53.81	53.79	0.0189	0.08
9	0.398740	0.1013	338.44	5247.35	53.81	53.80	0.0056	0.08
10	0.398730	0.1013	338.44	5247.35	53.81	53.81	–0.0078	0.08
11	0.398736	0.1013	338.44	5247.35	53.81	53.81	0.0002	0.08
12	0.398735	0.1013	338.44	5247.35	53.81	53.81	–0.0011	0.08
13	0.398737	0.1013	338.44	5247.35	53.81	53.81	0.0016	0.08

PROBLEM 8.3.2

For a series of pipes of known but different lengths and diameters, the overall theoretical diameter can be calculated or equivalent pipe can be imagined as shown in this example. This method is used for analysing complex distribution networks with pipes of different lengths and diameters and reducing them to a single overall diameter and length, i.e., an equivalent single pipe.

There are four pipe segments in series – P1, P2, P3 and P4. P1 has diameter 0.2 m and length of 1000 m, P2 has diameter 0.3 m and length of 500 m, P3 has diameter 0.3 m and length of 300 m, and P4 has diameter 0.4 m and length of 200 m. Total length of pipe, P-total is the sum of the four segments.

Solution

Based on the Darcy–Weisbach formula for head loss through a pipe, $h_f = fLQ^2/A^2{*}D{*}2g$

h_f (P-total) = head loss through each pipe segment = h_f– P1 + h_f– P2 + h_f– P3 + h_f– P4

Since the only difference in each pipe segment is the clubbed parameter ($L/D{*}A^2$) while all other parameters are the same for each pipe segment, a table can be set up for calculating the clubbed parameter for each pipe segment and then solving for the overall diameter of the pipe on a theoretical basis, i.e., the theoretical equivalent pipe.

Pipe segment	Diameter, D (m)	Area, A (m^2)	L, m	L/(D*A^2)
P1	0.200	0.031	1000	5066059.18
P2	0.300	0.071	500	333567.68
P3	0.300	0.071	300	200140.61
P4	0.400	0.126	200	31662.87
P-Total			2000	5631430.34
D^5 =	0.00181			
D =	0.283	m		

Therefore, the equivalent pipe has a diameter of 0.283 m and length of 2000 m.

Study Outline

Transport of water

Conduits can be of two types: gravity conduits or pressure conduits.

Examples of gravity conduits include canals and aqueducts which are at atmospheric pressure while pressure conduits include fully-flowing pipes at higher than atmospheric pressure.

Various equations can be used to determine head loss in different conduits and flowing conditions. The Darcy–Weisbach equation is used for determining head loss due to frictional flow through pipes.

For uniform, steady, open channel flow, Manning's formula is used to determine head loss while Hazen–Williams equation is used for determining head loss in pressure conduits.

Distribution reservoirs

Water has to be provided to the service area after treatment, and distribution reservoirs are required for this purpose. They can be surface reservoirs or elevated reservoirs.

Distribution systems can be based on gravity where water is stored in the ground (often underground) service reservoir and directly sent from there to the distribution system.

If the source of water or treatment plant is at a lower elevation than the service area, water is pumped to elevated reservoirs in the service area which then provide water to the service area. In a combined gravity and pumping distribution system, water is first stored in ground reservoirs and then pumped into an elevated service reservoir from where it is supplied to the distribution system. It is best to locate elevated reservoirs between the source and the service area to minimize pumping requirements and costs.

Distribution networks

A distribution system consists of a network of pipe lines with a main supply or trunk feeder, primary feeder, secondary feeder, street mains and service pipes.

Distribution networks are of two types: branched or tree systems and looped or grid iron systems. Branched systems lead to dead end formation while pipes in a grid iron formation are a part of inter-connected loops which prevent formation of dead ends. The latter is preferred in modern water supply systems.

The hydraulic analysis of a distribution system helps in determining if the system can meet demand, in terms of quantity and pressure. One of the methods for such analysis is Hardy–Cross method which was published in 1936. Several other methods are now available for solving flow and head loss problems in looped distribution systems.

Study Questions

1. What are the conditions under which gravity or pressure conduits are used?
2. What are the functions of distribution reservoirs?
3. Why are elevated reservoirs necessary in some distribution systems? What purpose do they serve?
4. Why is a looped or grid iron distribution system preferred over a branched distribution system?
5. Why is it useful to use an equivalent pipe for a complex distribution system with pipe segments of varying diameters and lengths?

CHAPTER 9

Municipal Wastewater Collection and Disposal

Learning Objectives

- *Compare and contrast separate versus combined wastewater collection systems*
- *Describe different methods for disposal of wastewater and conditions for doing so*
- *Describe options for reusing municipal wastewater and treatment requirements for reuse*

Sewage or municipal wastewater includes human and household wastes, street washing, wastewater from industrial and commercial activities in the municipal area, groundwater and stormwater. There are two major types of sewage collection systems: dry or conservancy systems in unsewered areas and wet carriage systems in sewered areas (Duggal 2007).

The dry or conservancy system is common in rural and semi-urban areas where there are no sewer systems. Night soil (feces and urine, i.e., human excreta) is collected in latrines, privies, septic tanks, cesspits, or twin-pit latrine systems. Anaerobic conditions develop and reduce the solid concentrations in the sewage in these systems, which are removed manually or mechanically at relatively long intervals of 1–5 years.

Wet carriage systems are common in urban centers where sewerage systems have been provided. Wastewater is collected from individual buildings (toilets, sinks, bathrooms, any other water using facilities) and taken for treatment to a sewage treatment plant. The wet carriage system is popular in urban centers all over the world since it does not require manual transport of excreta, it results in better hygiene and reduced risk of waterborne diseases, has less odor and is aesthetically better than the dry systems. The only disadvantages are the high capital and maintenance costs, and excessive water consumption. Only wet carriage systems are addressed in this textbook.

Conventional and advanced wastewater treatment schemes were discussed in Chapter 7, Section 7.3. After treatment, municipal wastewater can be reused or disposed in various ways and are discussed in this chapter.

9.1 MUNICIPAL WASTEWATER COLLECTION SYSTEMS

Wastewater collection systems are generally designed as combined sewer overflow (CSO) especially in old cities with sewerage systems where wet and dry weather flows are collected and treated together. However, a relatively recent trend which started in the twentieth century (Mannina and Viviani 2009) is to design separate sewage collection systems for dry and wet weather flows, respectively. Separate collection systems have two drainage networks, one for stormwater flows and the other for sewage flows, which result in higher capital costs. However, cost-benefit analysis and comparison of the two systems remains a difficult and debatable issue with no clear answers. A comparison of the two types of collection systems is provided in Table 9.1.

Table 9.1 Comparison of various criteria for combined and separate sewage collection systems

Criteria	Combined sewer overflow (CSO) systems	Separate collection systems
Age of sewerage systems	Older cities had combined systems; often sewage flowed into open drains (nallahs) that also collected stormwater.	A more recent trend that can be expensive.
Population density	If the area to be sewered is congested, combined system is better.	If the area to be sewered is not congested, separate system is easier to construct and maintain.
Pumping requirements and costs	If pumping is required, cost is higher since quantity (flow) is greater.	Stormwater drains are expected to follow natural channels and should not require pumping. Sewered flows are much smaller in quantity, therefore pumping cost is less.
Volumetric loading to treatment plant	Higher; can result in washout of processes or process inefficiency.	Lower; greater efficiency if only year-round flows are treated and wet weather (stormwater) flows are by-passed.
Pollutant loading	Lower during wet weather.	Higher in influent to treatment plant.
Capital costs for drainage system	Less, since only one set of drains is required.	More, since two sets of drains are required. Stormwater drains are often open drains while sewer conduits are closed.
Treatment costs	High, since volumetric loading to treatment plant increases.	Low, if only sewage is treated and not stormwater.
Treatment efficiency	Lower, since fluctuations in flow and pollutant loading are greater.	Higher, due to less fluctuation in sewage flow and pollutant loading.
Operation and maintenance (O & M) costs	Higher scour velocities can reduce maintenance costs.	Lower scour velocities in sewage conduits and silting of stormwater drains can increase O & M costs.

Criteria	Combined sewer overflow (CSO) systems	Separate collection systems
Variations in flow to treatment plant	Greater variations due to seasonal impact of wet weather flows; if rainfall is even throughout the year, CSO is better.	If only sewage flows to treatment plant, then there is less variation in volumetric and pollutant loading; if rainfall is uneven throughout the year, separate collection systems are better.
Groundwater recharge by infiltration	Less, since stormwater is collected in the sewer system	More, if stormwater drains are unlined

In countries like India where the year's rain falls over a 3 to 6 month period depending on the location, separate sewage collection systems have proved to be the better option for the following reasons:

1. Stormwater drains should be built along natural drainage channels. Doing so eliminates pumping and wastewater treatment costs. This is a major incentive for urban local bodies that are generally starved of funds. Regular cleaning (especially before the onset of the monsoon season) is essential for ensuring that open stormwater drains are not choked.
2. Dry weather flows which are constant throughout the year need to be treated prior to discharge. Since wastewater treatment is fairly expensive, separate collection systems allow smaller quantities to be treated resulting in lower capital, operation and maintenance costs for wastewater treatment.
3. In existing cities, where most of the area is built-up and paved and natural drainage channels are blocked, pumping of wastewater is often necessary especially during storm events to prevent water-logging. Separate systems allow the dry weather flows to be handled separately and stormwater can be handled (or pumped out) on an 'as needed' basis.

An open, unlined stormwater drain is shown in Figure 9.1a. A closed sewer pipe carrying sewage out of a residential apartment building is laid underground and is under repair as seen in Figure 9.1b.

Figure 9.1 (a) an open brick-lined stormwater drain, and
(b) an underground sewer pipeline under repair (see color plate).

Chandigarh, Gandhinagar, Delhi, Ahmedabad and Mumbai have separate sewer systems while old cities like Paris, Chennai and Kolkata have combined sewerage systems. Water-logging problems that plague Indian cities during the monsoon occur due to choking of open stormwater drains and natural drainage channels. These drains and channels are often encroached upon by unauthorized settlements on both sides reducing their width, while buildings are constructed across natural drains preventing the flow of stormwater (Gupta 2005). It is extremely important to maintain open stormwater drains by cleaning them regularly, especially prior to the onset of the monsoon season. Also, buildings or any other activity such as waste dumping in the drains which can impede the flow of stormwater should be restricted to prevent water-logging in the area.

Stormwater systems can be designed using the Rational Method for determining the rate of runoff (Q_s) or peak stormwater flows (Duggal 2007). This method is applicable to drainage areas less than 64 ha (Iowa Department of Transportation 1995).

$$Q_s = A*I*R/360 \quad 9.1.1$$

where A = area, hectares

I = impermeability factor or runoff coefficient (ratio of runoff to rainfall); impermeability factors for different surfaces are given in Table 9.2.

R = rainfall intensity, mm/h; usually ranges from 25 to 75 mm/h.

Table 9.2 Runoff coefficients or impermeability of different types of surfaces (Duggal 2007; Iowa Department of Transportation 1995)

Surface type	Runoff coefficients
Woods and forests	0.01 to 0.20
Open areas including lawns, and meadows	0.05 to 0.30
Gravel surfaces – compacted	0.45 to 0.60
Gravel surfaces – uncompacted	0.35 to 0.50
Pavements with open joints	0.40 to 0.70
Pavements with cemented joints	0.75 to 0.85
Asphalt pavements	0.85 to 0.90
Paved surfaces or buildings	0.94 to 0.98

PROBLEM 9.1.1

A campus of 60 ha has paved or built-up area of 30 percent while the remaining is either open space or grassy (lawns and wild vegetation) area. The maximum rainfall intensity is 50 mm/h. What is the expected peak stormwater flow? What is the impact of increasing the paved area to 40 percent or 50 percent on peak stormwater flows?

Solution

The impermeability factor, I, has to be calculated for the entire area. Assuming the paved or built-up area has an I value of 0.96 and the open spaces or grassy areas have an I value of 0.1, then I for the entire area with 30% paving is:,

$I = 0.3*0.96 + 0.7*0.1 = 0.358$ which implies that 35.85 percent of the peak rainfall is runoff,

Peak stormwater flow, $Q_s = A*I*R/360$

$= 60*0.358*50/360 = 2.98\ m^3/s$

As the paved area increases, peak stormwater flows will increase proportionately.

For 40 percent paved area, $I = 0.4*0.96 + 0.6*0.1 = 0.444$

$Q_s = 60*0.444*50/360 = 3.7\ m^3/s$

For 50 percent paved area, $I = 0.5*0.96 + 0.5*0.1 = 0.53$

$Q_s = 60*0.53*50/360 = 4.41\ m^3/s$

9.2 WASTEWATER DISPOSAL AND REUSE

After treatment, wastewater can be either discharged or reused. Due to the high costs of wastewater treatment and water scarcity in many regions of the world, the incentive for reusing treated wastewater is increasing. Several disposal and reuse options exist and are described in this section.

9.2.1 Disposal of Municipal Wastewater

Historically, municipal wastewater has been discharged into rivers and streams or used on land for irrigation. Currently, most countries including India have discharge standards for the disposal of treated municipal wastewater. Disposal options include discharge into surface water bodies, over land or into marine or coastal environments. Discharge standards for wastewater differ for the different disposal options as shown in Appendix D. Only general standards that are applicable to all wastewater discharges are summarized in this table. There are several industry-specific standards that are not included here and can be referred to from the CPCB website (www.cpcb.nic.in).

Effluent discharge into surface water bodies

The most common disposal method is discharge of municipal wastewater into inland surface water bodies (generally rivers and streams) followed by land application for irrigation. Conventional (secondary level) treatment is necessary for meeting the discharge standards noted in Appendix D.

which are most stringent for discharge to inland surface water bodies. Based on a CPCB report, all operational sewage treatment plants were able to meet COD standards of 250 mg/L but almost one-third were unable to meet the BOD standard of 30 mg/L for discharge to inland water bodies (CPCB 2013).

Discharge of wastewater to water bodies can be done in two ways as described below (Metcalf and Eddy 2003).

a. Outfalls: Free fall or free flow of sewage (shown in Figure 9.2) which is the most common wastewater discharge method results in foam formation on the water surface and inefficient mixing of wastewater with flowing water. This method is cheaper but it results in unaesthetic (malodorous) and nuisance conditions since it provides a breeding ground for disease vectors like mosquitoes, flies, and other pestilent species. During low flow periods in rivers and streams, sewage flow can result in pools or nallahs within the river bed, exacerbating nuisance conditions.

Figure 9.2 Sewage outfall in Kolkata for discharging sewage to River Hooghly (see color plate).

b. Submerged discharge through diffusers: These diffusers are long pipes that extend far into the river or ocean, are perpendicular to the river bank or beach line and may have single or multiple ports. Nuisance and unaesthetic conditions on the bank can be avoided with submerged diffusers. Depending on the depth of the river, efficient mixing and dilution of wastewater can be accomplished even in low-flow conditions.

9.2.2 Reuse of Treated Municipal Wastewater

Wastewater treatment is expensive and energy-intensive. It is now obvious that treated wastewater is a useful resource that does not have to be discharged to flowing water bodies. Instead it can be used for many non-potable purposes, thus reducing the quantity of potable water needed and its treatment requirements. In many water-scarce regions, wastewater reuse for potable purposes is also being evaluated or implemented. For water-scarce regions or nutrient-impoverished soils, reuse of treated wastewater for irrigation is an excellent solution and has been in practice for more than 5000 years all over the world. In keeping with this paradigm, the current goal in wastewater treatment is zero discharge which also implies that wastewater has to be treated to tertiary or higher levels for specific reuse purposes. Wastewater can be reused at the household level or at a more centralized level from sewage treatment plants run by urban local bodies. Only centralized treatment and reuse applications are discussed here.

Different reuse applications require different levels of wastewater treatment and treatment processes recommended by the USEPA are summarized here (USEPA 2012). Wastewater treated to the primary level (sedimentation only) is not suitable for any reuse purposes. Reuse applications for conventionally (secondary which includes biological processes and disinfection) treated wastewater include the following:

- Non-food crop and restricted landscape irrigation
- Surface irrigation of orchards and vineyards
- Groundwater recharge of aquifers for non-potable purposes
- Wetlands and habitat conservation
- Stream augmentation
- Industrial water use like cooling water and boiler feed water

For higher level wastewater reuse applications, tertiary treatment which includes coagulation, chemical or biological nutrient removal, filtration and disinfection are necessary. These uses include the following (USEPA 2012):

- Food crop irrigation
- Toilet flushing and vehicle washing
- Unrestricted recreational uses
- Landscape irrigation for golf courses and other areas

If wastewater is to be treated to the level of potable water, then advanced treatment processes like activation carbon adsorption, reverse osmosis, advanced oxidation processes and soil-aquifer treatment are recommended (USEPA 2012). Potable water quality level can be achieved by using groundwater recharge of potable use aquifer, and surface water reservoir augmentation. Complete recycling of wastewater after treatment to the level of drinking water standards has been implemented in Namibia since 1968 and Singapore since 1998. Other reuse applications that are common in various parts of the world are aquaculture, and fire-fighting, especially forest fires.

In terms of the volume of wastewater reused, China is the highest, followed by Mexico and USA (Jimenez and Asano 2008). However, only some of the wastewater is treated prior to reuse in developing countries like China, Mexico and India while all of the wastewater is treated prior to reuse in countries like USA, Saudi Arabia and Israel. Total wastewater reused, used for irrigation and treated wastewater (ww) that is reused as a percent of the total water extracted are shown for the top five countries in Table 9.3. Use of treated wastewater as a fraction of the total water extracted is highest for Kuwait, followed by Israel and Singapore.

Table 9.3 Reuse of wastewater in different countries (Jimenez and Asano 2008)

Country	Total treated ww volume reused, m³/d	Treated ww used for irrigation, m³/d	Percent of treated ww used for irrigation	Country	Treated ww reuse/total water extraction, percent
China*	1,48,17,000	12,38,860	8	Kuwait	35.2
Mexico*	1,44,00,000	44,92,800	31	Israel	18.1
USA	76,00,000	9,11,000	12	Singapore	14.4
Saudi Arabia	18,47,000	5,94,521	32	Qatar	13.3
Israel	10,14,000	7,67,123	76	Cyprus	10.4
*Total wastewater reused (treated or untreated)				Jordan	8.1

A summary of some case studies from the USA is provided in Table 9.4 with details of plant capacity, reuse applications and treatment provided.[66]

Table 9.4 Summary of case studies from the USA for reuse of treated wastewater and the treatment provided (Crook 2004)

S.No	City or Region	Country	Capacity	Planned capacity	Application	Treatment processes used
1	York River Treatment Plant, Hampton Roads Sanitation District, Virginia	USA	0.5 MGD		Industrial use in crude oil refinery located close to treatment plant; cooling, boiler feed water, fire-fighting and tree irrigation	Biological oxidation and nutrient removal using an SBR; equalization tank; chlorination, cloth-membrane disk filtration
2	Irvine Ranch Water District, Southern California	USA	Michelson WRP = 15 MGD	Michelson WRP = 33 MGD	Non-potable uses like irrigation. Improved drought-resistance due to reuse	Tertiary level treatment in both plants; after secondary treatment, straining, pressure filtration and disinfection.

(*Contd.*)

66 WRP = Water Reclamation Plant,
MGD = Million US Gallons per Day.

S.No	City or Region	Country	Capacity		Planned capacity	Application	Treatment processes used
			Alisos WRP = 5.5 MGD		Alisos WRP = 7.8 MGD		
3	Montebello Forebay, Southern California	USA	100 mgd			Groundwater recharge; irrigation, industrial uses, recreational impoundments and wildlife habitat	Activated sludge process; tertiary treatment using filtration, nitrification-denitrification, disinfection
4	Monterey County Water Recycling Projects, California	USA	20 mgd			Irrigation	Chemical coagulation, clarification, direct filtration
5	Orange County Water District,	USA			NA		
	1. Water Factory 21		15 mgd		81 mgd	Urban irrigation and other non-potable uses	
	2. Green Acres Project		6 mgd			Urban irrigation and other non-potable uses	
	3. Groundwater Replenishment System		15-40 mgd			Groundwater recharge	
6	St. Petersburg, Florida	USA	65 mgd	64.60%		Urban irrigation and deep well injection	
7	Orlando, Orange County, Florida	USA	NA	68 mgd reused		Irrigation (35 mgd) City's & County's individual reclaimed water systems (7 mgd) Rapid infiltration basins (22 mgd)	

(*Contd.*)

S.No	City or Region	Country	Capacity		Planned capacity	Application	Treatment processes used
8	West Basin Municipal Water District	USA	90 mgd	26.70%	50%	Irrigation (10%)**	Tertiary treatment (Coagulant, flocculation basins, anthracite mono-media filters, and disinfection)
						Cooling water for industries (30%)**	Tertiary treatment (Nitrification, and disinfection)
						Groundwater recharge (26%)**	Tertiary treatment (Decabonation, lime treatment, RO treatment and disinfection)
						Low pressure boiler feed water (24%)**	Secondary treatment, microfiltration, RO treatment, and disinfection
						High pressure boiler feed water (10%)**	Microfiltration, RO, disinfection, and second pass RO

Major advantages of using treated wastewater for various purposes are (Jimenez 2006):

i. Reuse reduces the increasing demand for potable water by providing treated wastewater for non-potable uses in urban areas.

ii. Stream flows or e-flows in rivers and streams can be maintained throughout the year.

iii. Nutrient-rich wastewater can be used to increase agricultural production in water-stressed regions by irrigation, as well as for irrigating lawns and golf courses. Minimum secondary or tertiary treatment is provided depending on the crops to be cultivated. Crops can be either food crops or non-food crops which include fodder and fiber crops. Non-food crops do not require the same level of treated wastewater as food crops, especially those that are to be consumed raw. Nutrient-rich wastewater has proved to be superior in terms of agricultural productivity as compared to groundwater or surface water in five different cities in India (Amerasinghe et al. 2013). Use of treated wastewater also reduces use of fertilizers, improves the organic content of soil and the net income of farmers relying on these irrigation methods.

iv. Use of treated wastewater for various reuse purposes reduces or eliminates their discharge to streams or groundwater reducing pollution of these water sources.

v. Treated wastewater can be used for recharging aquifers or for using aquifers as 'natural filters' for purification of water for potable or other purposes. Israel has been doing this successfully for several decades (Angelakis and Gikas 2014).

Some problems associated with wastewater reuse are listed here.

i. For urban areas, installation and maintenance of a separate wastewater reuse system increases costs and the risk of cross-contamination of water supply lines.

ii. For irrigation purposes, accumulation of salts and heavy metals in soil and cultivated crops can occur in the long-term. Soil salinity increases in the long-term with both freshwater and wastewater irrigation.

iii. Standing wastewater used in paddy cultivation can be a breeding ground for disease vectors like mosquitoes. However in a case study from Tanzania, malaria transmission was lower in the village using treated wastewater despite more malaria vectors. The difference was attributed to higher economic and nutritional status due to higher crop yields in the village using treated wastewater (Jimenez 2006).

iv. The costs of treating and distributing treated wastewater may be too high in some cases thus precluding its use.

v. Excessive application of secondary treated wastewater can lead to leaching of nitrate into groundwater aquifers.

Irrigation or horticulture or landscape irrigation

Treated and untreated municipal wastewater has been used for irrigation around the world for centuries since it provides water and nutrients for plant growth. Effluent from conventional (secondary) wastewater treatment plants is rich in nutrients like N, K and P and can be used for irrigation. Crop yield with sewage farming was found to be 33 percent greater than with well or canal irrigation and many different crops were cultivated (Duggal 2007). In other studies, crop yields with sewage irrigation are 1.25 times to 4 times higher than with freshwater or groundwater, and year-round intensive farming is possible.

However, current concerns regarding pathogens, micropollutants like heavy metals, and synthetic organic compounds in treated municipal wastewater have made this option less acceptable. Use of untreated sewage can lead to skin, diarrhoeal or helminthic diseases for farmers. Further, long-term application of untreated sewage can result in accumulation of heavy metals in the soil and bioaccumulation in the crops (Sengupta 2015; Amerasinghe et al. 2013). A major concern in the use of treated wastewater for irrigation is the possibility of exposure to pathogens for farmers and for those consuming food crops irrigated with wastewater. Ensuring removal of pathogens requires tertiary treatment including disinfection in most cases.

Secondary treated wastewater can be used for three types of vegetation: (i) Forage and field crops, non-food crops like cotton, biofuel crops like jatropha (ii) landscape vegetation for highway medians and border strips, golf courses, parks, recreational areas and (iii) for serviculture, i.e., planted woodlands or reforested areas.

Two methods are commonly used for wastewater irrigation (Duggal 2007; Metcalf and Eddy 2003):

a. Surface irrigation or overland flow: Sewage is distributed in ditches laid carefully along graded slopes. The water percolates into the sides where plants are grown and excess water flows down the slope to a cross-ditch where it is collected for further use or disposal.

   Ridge and furrow method: This is the most common surface irrigation method where sewage is applied to furrows while plants are sown on the ridges. Agricultural drains below the surface may be necessary for collecting infiltrated water.

b. Flooding: Sewage is spread over ploughed land that is without any slope and is enclosed by dykes. The depth of water ranges from 2.5 to 60 cm. Drains may be laid below the surface to collect water that infiltrates through the soil.

In all these systems, wastewater is applied using sprinklers or pipes. Pipes may be perforated or with adjustable discharges. Water is lost mainly by evapotranspiration and infiltration into the subsurface while some of it is used by plants.

Problems related to sewage irrigation are odor problems, and clogging of soil by wastewater solids when the application rate is excessive (Duggal 2007). It is important to let such soil 'rest', i.e., lie fallow for a season, and to apply wastewater intermittently not continuously. Other problems associated with sewage farming are the land area required and the distances to which the sewage has to be transported by pipes or canals.

As is evident from Tables 9.3 and 9.4, irrigation is the most common reuse application for treated wastewater. Concerns about total and fecal coliforms, heavy metals, pesticides and other chemicals and their health impacts have resulted in regulations that require secondary or tertiary wastewater treatment in several countries including the USA and the European Union (Crook 2004; Angelakis and Gikas 2014).

In a case study from Israel, crop yields increased as TDS (measured as electrical conductivity) and the sodium absorption ratio (SAR) decreased with higher levels of treatment (USEPA 2012). Crop yields for watermelon, garlic and corn grain were highest for RO permeate followed by UF: RO blends of 70:30 and 30:70, UF permeate and finally secondary treated effluent. Nutrients and emerging micro-pollutants like antibiotics and other pharmaceuticals were completely removed.

Effluent discharge for groundwater recharge

Another land application method for wastewater disposal is rapid infiltration where wastewater is applied over land in large quantities leading to high percolation rates and low evaporation losses (Metcalf and Eddy 2003). The percolated water is treated 'naturally' as it passes through the subsurface and can be used to recharge groundwater aquifers or to prevent saltwater intrusion into coastal aquifers. Rapid infiltration can also be used as a treatment method where the wastewater passes through the subsurface and is collected at a location down-gradient from the point of application using under-drains or wells for recovery of better quality water as was done in the

Negev Desert in Israel at the Dan wastewater treatment plant (Angelakis and Gikas 2014). Another option is to allow the water to percolate into the subsurface and drain into a surface water body.

Industrial water

The most common industrial reuse application for treated wastewater is for cooling water in thermal power plants and as boiler feed water as shown in Table 9.4. Treated wastewater has been reused by commercial laundries, for vehicle-washing, pulp and paper industries, steel production, textile manufacturing, electroplating and semiconductor industries, water for gas stack scrubbing, meat processing industries, brewery and food and beverage industries (Jimenez and Asano 2008; USEPA 2012). Treatment requirements include nutrient removal, filtration (microfiltration, ultrafiltration or reverse osmosis), and disinfection (Crook 2004).

Habitat conservation and recreational uses

Several locations in the USA, mostly in Arizona, are using treated sewage for restoring habitats of native and migratory birds and other species and also making these places available for recreational purposes (USEPA 2012).

Aquaculture

Natural treatment systems like constructed wetlands in Pune and several other Indian cities and natural wetlands in Kolkata for treating sewage and producing fish and paddy have existed for centuries. The Eastern Calcutta wetlands have been using untreated sewage for paddy, vegetables and fish production with the highest income generation as compared to four other cities in the study (Amerasinghe et al. 2013). No health problems were reported and more than 60,000 people are dependent on these wetlands. Sewage has been used profitably after treatment in Ghana and several Southeast Asian countries like Thailand for aquaculture (USEPA 2012).

Study Outline

Wastewater collection systems

Wastewater collection systems can be categorized as combined sewer outflow (CSO) or separate collection systems. Stormwater and sewage are collected in a single system in CSO. On the other hand, separate collection systems have two drainage networks, one for stormwater flows and the other for sewage flows.

Dry weather flows (sewage only) are constant throughout the year and the choice between CSO and separate collection systems depends on the temporal distribution of precipitation on an annual basis. For cities in regions where precipitation is evenly distributed throughout the year, CSO may be an appropriate option. For cities which receive precipitation during a short period of time (3 to 6 months), separate collection systems may be more appropriate. The choice between the two systems should be made on the basis of a comparison of capital, operation and maintenance costs along with short-term and long-term impacts.

Peak stormwater flows need to be estimated for design of CSO or stormwater drainage systems. The rational method for estimating peak stormwater flows can be used for areas ≤64 ha.

Wastewater disposal and reuse

After treatment, wastewater can be either discharged or reused. The current goal in wastewater treatment is 'zero discharge'. Due to high costs of wastewater treatment and water scarcity in many regions of the world, the incentive for reusing treated wastewater is increasing.

Disposal of wastewater

Wastewater can be discharged into public sewers, into inland surface water bodies or into the sea/ ocean, or applied to land for irrigation. Discharge standards for each of these disposal options are available and the most important parameters are BOD, COD and TSS. Several other water quality parameters are also included in the regulations.

Wastewater can be discharged into surface water bodies through outfalls (free fall of water) or through submerged diffusers.

Reuse of treated municipal wastewater

Different reuse applications require different levels of wastewater treatment and treatment processes. Minimum secondary treatment is required for all reuse applications. For higher applications, higher levels of treatment (tertiary or more) are required.

Some of the major treated wastewater reuse applications are:

- Irrigation for non-food crops, food crops and landscape improvement,
- Groundwater recharge of aquifers for potable or non-potable purposes,
- Wetlands and habitat conservation; aquaculture,
- Stream augmentation for maintaining e-flows or for generating potable water,
- Industrial water uses like cooling water and boiler feed water, and
- Vehicle washing and toilet flushing.

Some concerns associated with wastewater reuse are increase in soil salinity, increase in the population of disease vectors like mosquitoes, high cost of treating and distributing treated wastewater, and leaching of nitrate and other contaminants into groundwater.

Study Questions

1. What is the dry or conservancy system of waste collection? How does it differ from the wet carriage system?
2. What are the different types of wastewater collections systems? Compare their advantages and disadvantages.
3. How can peak stormwater flow be estimated?
4. What is the meaning of 'wastewater discharge standards'? List the water quality parameters that are regulated as part of the general wastewater discharge standards.
5. Why should wastewater be reused and for what purposes can it be reused? What level of treatment is necessary for each of these applications.

Appendix A

A summary of drinking water standards in US and India is provided in the following table along with guidelines provided by the World Health Organization (WHO).

Parameters (in mg/L unless stated otherwise)	USEPA 2011	WHO 2011	IS:10500 1991#	IS:10500 1991*	IS:10500 2012#	IS:10500 2012*
Aluminium	–	–	0.03	0.2	0.03	0.2
Anionic detergents (as MBA	–	–	0.2	1	0.2	1
Arsenic	–	–	0.05	0.05	0.01	0.05
Barium	2	–	–	–	0.7	0.7
Boron	–	–	1	5	0.5	1
Bromate	0.01	–	–	–	–	–
Cadmium	0.005	0.005	0.01	0.01	0.003	0.003
Calcium	–	75	75	200	75	200
Chloramines, as total Cl_2	–	–	–	–	4	4
Chloride	250	200	250	1000	250	1000
Chromium	0.1	–	0.05	0.05	0.05	0.05
Cobalt	–	–	–	–	–	–
Color, Hazen Units	–	–	5	25	5	15
Copper	1.3	3	0.05	1.5	0.05	1.5
Cyanide	–	–	0.05	0.05	0.05	0.05
Fluoride	4	0.6-0.9	1	1.5	1	1.5
Free chlorine residual, min	–	–	0.2	–	0.2	1
Iron	0.3	0.1	0.3	1	0.3	0.3
Lead	0.015	0.05	0.05	0.05	0.01	0.01
Magnesium	–	50	30	100	30	100
Manganese	0.05	–	0.1	0.3	0.1	0.3
Mercury	0.002	–	0.001	0.001	0.001	0.001

Parameters (in mg/L unless stated otherwise)	USEPA 2011	WHO 2011	IS:10500 1991#	IS:10500 1991*	IS:10500 2012#	IS:10500 2012*
Nickel	–	–	–	–	0.02	0.02
Nitrate	10	–	45	100	45	45
Nitrite	1	–	–	–	–	–
pH	6.5-8.5	6.5-8.5	6.5-8.5	6.5-8.5	6.5-8.5	6.5-8.5
Pesticides			0	0.001	individual pesticides	
Phenolic compounds	–	–	0.001	0.002	0.001	0.002
Selenium	0.05	–	0.01	0.01	0.01	0.01
Silver	–	–	–	–	0.1	0.1
Sulphate	250	200	200	400	200	400
Sulphide (as H_2S)	–	–	–	–	0.05	0.05
Total alkalinity as $CaCO_3$	–	–	200	600	200	600
Total ammonia-N	–	–	–	–	0.5	0.5
Total dissolved solids	–	–	500	2000	500	2000
Total hardness as $CaCO_3$	–	–	300	600	200	600
Total organic carbon	2	–	–	–	–	–
Turbidity	–	–	5	10	1	5
Zinc	5	5	5	15	5	15

Desirable concentrations; *Permissible levels in the absence of an alternative source.

Appendix B

Surface water bodies in India are classified by the Central Pollution Control Board and a summary of the use-based classification is provided in the table below.

Designated-Best-Use	Class of water	Criteria
Drinking water source without conventional treatment but after disinfection	A	1. Total Coliforms Organism MPN/100ml shall be 50 or less 2. pH between 6.5 and 8.5 3. Dissolved Oxygen 6 mg/l or more 4. Biochemical Oxygen Demand 5 days 20 °C 2 mg/l or less
Outdoor bathing (organized)	B	1. Total Coliforms Organism MPN/100ml shall be 500 or less 2. pH between 6.5 and 8.5 3. Dissolved Oxygen 5 mg/l or more 4. Biochemical Oxygen Demand 5 days 20 °C 3 mg/l or less
Drinking water source after conventional treatment and disinfection	C	1. Total Coliforms Organism MPN/100ml shall be 5000 or less 2. pH between 6 to 9 3. Dissolved Oxygen 4 mg/l or more 4. Biochemical Oxygen Demand 5 days 20 °C 3 mg/l or less
Propagation of wildlife and fisheries	D	1. pH between 6.5 to 8.5 2. Dissolved Oxygen 4 mg/l or more 3. Free Ammonia (as N) 1.2 mg/l or less
Irrigation, industrial cooling, controlled waste disposal	E	1. pH between 6.0 to 8.5 2. Electrical Conductivity at 25 °C micro mhos/cm Max.2250 3. Sodium Absorption Ratio Max. 26 4. Boron Max. 2 mg/l

Appendix C

WATER QUALITY INDEX

There are several methods for evaluating the water quality of rivers and other water bodies. The most common method is the use of a water quality index (WQI) first formulated by the USA–National Sanitation Foundation (NSF) in 1970 based on a survey of 142 scientists and their assessment of 35 important water quality parameters (Canter 1977). The advantage of this method is that it provides a uniform scale for comparison of water quality in any part of the world. Researchers around the world have frequently modified this method for their conditions and its use remains a matter of debate and development.

The NSF–WQI method is based on comparing concentration levels of various water quality parameters with Q-rating curves. The original NSF–WQI has 9 water quality parameters as shown in the table below. The third column in the table provides the weights (Wi) for each of these parameters and Qi values are determined from the curves provided on the next page.

$$WQI = \sum W_i Q_i$$

Where W_i = weightage of i water quality parameter

Q_i = Quality rating of i water quality parameter

Table C1 Q-values for each water quality parameter used in the US-NSF–WQI calculation

DO (% saturation)	Q-value	BOD5 (mg/L)	Q-value	Nitrate-N	Q-value	Ecoli, colonies/ 100 mL	Q-value	pH	Q-value	Total phosphate, mg/L	Q-value	Change in temperature, °C	Q-value	Total Solids, mg/L	Q-value	Turbidity, NTU	Q-value
0	0	0	96	0	98	1.00	98	<2	0	0	99	–10	56	0	80	0	97
10	8	1	92	0.25	97	2.00	89	2	2	0.05	98	–7.5	63	50	87	5	84
20	13	2	80	0.5	96	5.00	80	3	4	0.1	97	–5	73	100	83.5	10	76
30	20	2.5	73	0.75	95	1.00×10	71	4	8	0.2	95	–2.5	85	150	80	15	68
40	30	3	66	1	94	2.00×10	63	5	24	0.3	90	–1	90	200	73	20	62
50	43	4	58	1.5	92	5.00×10	53	6	55	0.4	78	0	93	250	67	25	57
60	56	5	55	2	90	1.00×10^2	45	7	90	0.5	60	1	89	300	60	30	53
70	77	7.5	44	3	85	2.00×10^2	37	7.2	92	0.75	50	2.5	85	350	52	35	48
80	88	8	40	4	70	5.00×10^2	27	7.5	93	1	39	5	72	400	47	40	45
85	92	10	33	5	65	1.00×10^3	22	7.7	90	1.5	30	7.5	57	450	40	50	39
90	95	12.5	26	10	51	2.00×10^3	18	8	82	2	26	10	44	500	33	60	34
95	97.5	15	20	15	43	5.00×10^3	13	8.5	67	3	21	12.5	36			70	28
100	99	17.5	16	20	37	1.00×10^4	10	9	47	4	16	15	28			80	25
105	98	20	14	30	24	2.00×10^4	8	10	19	5	12	17.5	23			90	22
110	95	22.5	10	40	17	5.00×10^4	5	11	7	6	10	20	21			100	17
120	90	25	8	50	7	1.00×10^5	3	12	2	7	8	22.5	18			>100	5
130	85	27.5	6	60	5	>100000	2	>12	0	8	7	25	15				
140	78	30	5	70	4					9	6	27.5	12				
>140	50	>30	2	80	3					10	5	30	10				
				90	2					>10	2						
				100	1												
				>100	1												

Table C2 Water quality data, WQI parameters and their weightages

			Gangotri		Allahabad (downstream of Sangam)	
Water Quality Parameter	**Units**	**Weight (Wi)**	**Observed values at Point 1, Ci**	**Quality Rating (Qi)**	**Observed values at Point 2, Ci**	**Quality Rating (Qi)**
Dissolved Oxygen*	% saturation	0.17	67.00	70.70	89.00	94.40
Fecal Coliform*	MPN, colonies/100 mL	0.16	1.00	98.00	3410.00	15.65
pH*		0.11	7.90	84.67	8.00	45.80
5-day Biochemical Oxygen Demand*	mg/L	0.11	0.00	96.00	4.00	58.00
Temperature change	°C	0.10	0.00	93.00	1.00	89.00
Total Phosphate	mg/L	0.10	0.00	99.00	0.46	67.20
	mg/L	0.10	0.15	98.60	1.61	91.56
Turbidity	NTU	0.08	0.00	97.00	50.00	39.00
Total solids	mg/L	0.07	50.00	87.00	250.00	67.00
Overall WQI		1		90.48		62.56
				Excellent		Medium

*Parameters measured/reported by CPCB for the Ganga River Basin

Data for some water quality parameters for several locations in the Ganga River Basin were available from the Central Pollution Control Board (CPCB) while the rest are estimates. Calculations for two locations are shown here as representative examples of a relatively unpolluted location (Point 1: Gangotri, source of the River Ganga) and downstream of the confluence of Rivers Ganga and Yamuna (Point 2: Allahabad, Sangam). Quality rating values for each water quality parameter were determined by interpolating between values provided in Table C1 and are summarized in Table C2. Based on the calculated WQI values and the corresponding index shown in Table C3, Point 1 has excellent water quality while Point 2 is moderately polluted.

The following score ranges and the corresponding water quality based on National Sanitation Foundation Water Quality Index (NSF–WQI) are shown in Table C3.

Table C3 Water quality and scores based on NSF–WQI

Score ranges	Water quality
0-25	Very bad
25-50	Bad
50-70	Medium
70-90	Good
90-100	Excellent

Appendix D

Table D1 General discharge standards in India for treated wastewater disposal

The Environment (Protection) Rules, 1986 [545]					
GENERAL STANDARDS FOR DISCHARGE OF ENVIRONMENTAL POLLUTANTS (PART A):					
S. No.	Parameter	Standards			
		Inland surface water	Public Sewers	Land for irrigation	Marine coastal areas
		a	b	c	d
1	Color and odor	See 6 of Annexure-I	–	See 6 of Annexure-I	See 6 of Annexure-I
2	Suspended solids, mg/l, Max.	100	600	200	(a) For process waste water-100 (b) For cooling water effluent 10 percent above total suspended matter of influent.
3	Particulate size of suspended solids	Shall pass 850 micron IS Sieve	–	–	(a) Floatable solids, max. 3 mm. (b) Settleable solids, max. 850 microns.
4	pH Value	5.5 to 9.0	5.5 to 9.0	5.5 to 9.0	5.5 to 9.0
5.	Temperature	shall not exceed 5 °C above the receiving water temperature	–	–	shall not exceed 5 °C above the receiving water temperature
6	Oil and grease, mg/l Max.	10	20	10	20
7	Total residual chlorine mg/L max.	1	–	–	1

The Environment (Protection) Rules, 1986 [545]

GENERAL STANDARDS FOR DISCHARGE OF ENVIRONMENTAL POLLUTANTS (PART A):

8	Ammonical Nitrogen (as N), mg/l Max.	50	50	–	50
9	Total Kjeldahl Nitrogen (as NH_3) mg/l, Max.	100	–	–	100
10	Free Ammonia (as NH_3) mg/l, Max.	5	–	–	5
11	Biochemical Oxygen Demand [3 days at 27 °C] mg/l max.	30	350	100	100
12	Chemical Oxygen Demand, mg/l, max.	250	–	–	250
13	Arsenic (as As), mg/l, max.	0.2	0.2	0.2	0.2
14	Mercury (as Hg), mg/l, Max.	0.01	0.01	–	0.01
15	Lead (as Pb) mg/l, Max.	0.1	1	–	2
16	Cadmium (as Cd) mg/l, Max.	2	1	–	2
17	Hexavalent Chromium (as Cr^{6+}), mg/l max.	0.1	2	–	1
18	Total Chromium (as Cr) mg/l, Max.	2	2	–	2
19	Copper (as Cu) mg/l, Max.	3	3	–	3
20	Zinc (As Zn) mg/l, Max.	5	15	–	15
21	Selenium (as Se) mg/l, Max.	0.05	0.05	–	0.05
22	Nickel (as Ni) mg/l, Max.	3	3	–	5
23	Cyanide (as CN) mg/l Max.	0.2	2	0.2	0.2

The Environment (Protection) Rules, 1986 [545]

GENERAL STANDARDS FOR DISCHARGE OF ENVIRONMENTAL POLLUTANTS (PART A):

24	Fluoride (as F) mg/l Max.	2	15	–	15
25	Dissolved Phosphates (as P), mg/l Max.	5	–	–	–
26	Sulphide (as S) mg/l Max.	2	–	–	5
27	Phenol compounds (as C_6H_5OH) mg/l, Max.	1	5	–	5
28	Radioactive materials :				
	(a) Alpha emitter, micro curie/ml.	1.00×10^{-7}	1.00×10^{-7}	1.00×10^{-8}	1.00×10^{-7}
	(b) Beta emitter, micro curie/ml.	1.00×10^{-6}	1.00×10^{-6}	1.00×10^{-7}	1.00×10^{-6}
29	Bio-assay test	90% survival of fish after 96 hours in 100% effluent	90% survival of fish after 96 hours in 100% effluent	90% survival of fish after 96 hours in 100% effluent	90% survival of fish after 96 hours in 100% effluent
30	Manganese (as Mn), mg/L	2	2	–	2
31	Iron (as Fe), mg/L	3	3	–	3
32	Vanadium (as V), mg/L	0.2	0.2	–	0.2
33	Nitrate Nitrogen, mg/L	10	–	–	20

References

Aggarwal, S. C., and S. Kumar. 2003. *Industrial Water Demand in India: Challenges and Implications for Water Pricing.* http://www.idfc.com/pdf/report/2011/Chp-18-Industrial-Water-Demand-in-India-Challenges.pdf

Al Rmalli, S W., P. I. Haris, C. F. Harrington, and M. Ayub. 2005. 'A Survey of Arsenic in Foodstuffs on Sale in the United Kingdom and Imported from Bangladesh.' *Sci Total Environ* 337: 23–30. doi:10.1016/j.scitotenv.2004.06.008.

Amerasinghe, P., R. Bhardwaj, C. Scott, K. Jella, and F. Marshall. 2013. *Urban Wastewater and Agricultural Reuse Challenges in India.* IWMI Report 147. https://www.gwp.org/globalassets/global/toolbox/references/urban-wastewater-and-agricultural-reuse.-challenges-in-india-iwmi-2013.pdf

Amirtharajah, A., and C. R. O'Melia. 1990. 'Coagulation Processes: Destabilization, Mixing and Flocculation.' In *Water Quality and Treatment: A Handbook of Community Water Supplies* (4th ed.). Edited by Raymond D Letterman, 269–365. New York: McGraw-Hill Professional.

Amoah, I. D., A. A. Adegoke, and T. A. Stenström. 2018. 'Soil-transmitted Helminth Infections Associated with Wastewater and Sludge Reuse: a review of current evidence.' *Tropical Medicine and International Health 23* (7): 692–703. http://doi.org/10.1111/tmi.13076

Andey, S. P., and P. S. Kelkar. 2009. 'Influence of Intermittent and Continuous Modes of Water Supply on Domestic Water Consumption'. *Water Resources Management* 23 (12): 2555–2566. http://doi.org/10.1007/s11269-008-9396-8

Angelakis, A., and P. Gikas. 2014. 'Water Reuse: Overview of Current Practices and Trends in the World with Emphasis on EU States.' *Water Utility Journal* 8: 67–78

Anke, M., B. Groppel, H. Kronemann, and M. Grün. 1984. 'Nickel - an Essential Element.' *IARC Scientific Publications* 53: 339–365.

APHA, AWWA, and WEF. 2005. *Standard Methods for the Examination of Water and Wastewater.* New York: APHA.

Apshankar, Kruttika R. 2018. 'Electrocoagulation-Filtration for the Removal of Fluoride or Nitrate from Drinking Water.' Kharagpur: IIT.

Asian Development Bank. 2007. *2007 Benchmarking and Data Book of Water Utilities in India.* https://www.adb.org/sites/default/files/publication/27970/2007-indian-water-utilities-data-book.pdf

Atkins, P., T. Overton, J. Rourke, M. Weller, and F. Armstrong. 2010. *Shriver and Atkins' Inorganic Chemistry.* Oxford: Oxford University Press.

ATSDR. 2005. *Toxicological Profile of Zinc.* https://www.atsdr.cdc.gov/toxprofiles/tp.asp?id=302&tid=54.

ATSDR. 2000. *Arsenic Toxicity: Case Studies in Environmental Medicine.* https://www.atsdr.cdc.gov/HEC/CSEM/arsenic/docs/arsenic.pdf

AWWA. 1990. *Water Quality and Treatment: A Handbook of Community Water Supplies.* New York: McGraw Hill.

Baker, J. R., M. W. Milke, and J. R. Mihelcic. 1999. 'Relationship Between Chemical and Theoretical Oxygen Demand for Specific Classes of Organic Chemicals.' *Water Research* 33 (2): 327–334.

Bell-Ajy, Kimberly, Morteza Abbaszadegan, Eva Ibrahim, Debbie Verges, and Mark LeChevallier. 2000. 'Conventional and Optimized Coagulation for NOM Removal.' *Journal – American Water Works Association* 92 (10): 44–58. doi:10.1002/j.1551-8833.2000.tb09023.x.

Bhandari, A., R. Y. Surampalli, C. D. Adams, P. Champagne, S. K. Ong, R. D. Tyagi, and T. C. Zhang. 2009. *Contaminants of Emerging Environmental Concern.* USA: ASCE.

Bird, R. B., W. E. Stewart, and E. N. Lightfoot. 1960. 'The Equations of Change for Isothermal Systems.' In *Transport Phenomena,* 1st ed. 71–122. New York: Wiley and Sons, Inc.

Bissen, M., M. M. Vieillard-Baron, A. J. Schindelin, and F. H. Frimmel. 2001. 'TiO2-Catalyzed Photooxidation of Arsenite to Arsenate in Aqueous Samples.' *Chemosphere* 44 (4): 751–57. http://www.ncbi.nlm.nih.gov/pubmed/11482665.

Biswas, Anirban, Debasree Deb, and Aloke Ghose. 2014. 'Seasonal Perspective of Dietary Arsenic Consumption and Urine Arsenic in an Endemic Population.' *Environmental Monitoring and Assessment* 186: 4543–51. doi:10.1007/s10661-014-3718-5.

Boehrer, B., and M. Schultze. 2008. 'Stratification of lakes.' Reviews of Geophysics. 46 (2).

Bouchard, Dermont C., Mary K. Williams, and Rao Y. Surampalli. 1992. 'Nitrate Contamination of Groundwater: Sources and Potential Health Effects.' *Journal of American Water Works Association* 84 (9): 85–90. doi:10.1002/j.1551-8833.1992.tb07430.x.

Bouwer, E. J. 1991. *Biological Processes for Wastewater Treatment - Lecture notes.* Baltimore: The Johns Hopkins University.

Brindha, K., and L. Elango. 2011. 'Fluoride in Groundwater: Causes, Implications and Mitigation Measures.' In *Fluoride Properties, Applications and Environmental Management,* 111–36.

Bureau of Indian Standards.1993. *Code of basic requirements for water supply, drainage and sanitation IS 1172:1993.* New Delhi: Bureau of Indian Standards.

———. 2012a. *Drinking Water Specifications IS 10500: 2012.* New Delhi: Bureau of Indian Standards.

———. 2012b. *Indian Standards, Drinking Water - Specification (Second Revision): IS 10500.* New Delhi: Bureau of Indian Standards.

Canter, L. W. 1999. *Environmental Impact Assessment.* Singapore: McGraw Hill International Ed. Civil Eng Series.

Cantor, K. 1982. Epidemiological evidence of carcinogenicity of chlorinated organics in drinking water. *Environmental Health Perspectives* 46: 187–195.

Central Electricity Authority (CEA). 2017. *All India Installed Capacity (in mw) of Power Stations.* http://www.cea.nic.in/reports/monthly/installedcapacity/2018/installed_capacity-03.pdf

Central Ground Water Board (CGWB). 2010. *Ground Water Quality in Shallow Aquifers of India.* http://cgwb.gov.in/WQ/gw_quality_in_shallow_aquifers.pdf

Central Pollution Control Board (CPCB). 2009. *Status of Water Supply, Wastewater Generation and Treatment in Class I Cities and Class II Towns of India.* New Delhi.

———. 2011. *Basin Wise Compiled Data – 2011.* www.cpcb.nic.in.

———. 2013. *Performance evaluation of sewage treatment plants under NRCD.* New Delhi.

Central Water Commission. 2006. *Evaporation Control in Reservoirs.* New Delhi.

Center for Watershed Sciences. 2012. *Drinking Water Treatment for Nitrate With a Focus on Tulare Lake Basin and Salinas Valley Groundwater.* Davis: University of California. http://groundwaternitrate.ucdavis.edu/files/139107.pdf

Chang, E. E., Y. P. Lin, and P. C. Chiang. 2001. 'Effects of bromide on the formation of THMs and HAAs.' *Chemosphere* 43 (8): 1029–1034.

Chatterjee, Amit, Dipankar Das, Badal K. Mandal, Tarit Roy Chowdhury, and Gautam Samanta. 1995. 'Arsenic in Groundwater in Six Districts of West Bengal, India: The Biggest Arsenic Calamity in the World.' *Analyst* 120 (March): 643–50.

Chick, H. 1908. 'An Investigation of the Laws of Disinfection.' *The Journal of Hygiene* 8 (1): 92–158. http://www.jstor.org/stable/4619368

Chou, C. H., and Carolyn Harper. 2007. 'Toxicological Profile for Arsenic.' doi:http://dx.doi.org/10.1155/2013/286524.

Choubisa, S L., Leela Choubisa, and D. Choubisa. 2009. 'Osteo-Dental Fluorosis in Relation to Nutritional Status, Living Habits and Occupation in Rural Tribal Areas of Rajasthan, India.' *Fluoride* 42 (3): 210–15.

Choubisa, S.L. 2001. 'Endemic Fluorosis in Southern Rajasthan, India.' *Fluoride* 34 (1): 61–70.

Colin, J. 1999. *VLOM for Rural Water Supply: Lessons from Experience.* Task No: 162, WEDC. Loughborough: Loughborough University.

Crittenden, J. C., R. R. Trussell, D. W. Hand, K. J. Howe, and G. Tchobanoglous. 2012. 'Principles of Mass Transfer.' In *MWH's Water Treatment: Principles and Design.* New York: J. Wiley and Sons.

Crook, J. 2004. *Innovative Applications in Water Reuse: Ten Case Studies.* Alexandria, VA, USA: WRA Report.

Daigger, G. T., and J. P. Boltz. 2011. 'Trickling Filter and Trickling Filter-Suspended Growth Process Design and Operation: A State-of-the-Art Review'. *Water Environment Research 83* (5): 388–404. http://doi.org/10.2175/106143010X12681059117211

Das, Dipankar, Amit Chatterjee, Gautam Samanta, and Tarit Roy Chowdhury. 2002. 'A Simple Household Device to Remove Arsenic from Groundwater and Two Years Performance Report of Arsenic Removal Plant for Treating Groundwater with Community Participation.' www.unu.edu/env/Arsenic/Das.pdf.

Das, K. K., S. N. Das, and S. A. Dhundasi. 2008. 'Nickel, its Adverse Health Effects and Oxidative Stress.' *Indian Journal of Medical Research* 128 (4): 412.

Datta, D. V., S. K. Mitra, P. N. Chhuttani, and R. N. Chakravarti. 1979. 'Chronic Oral Arsenic Intoxication as a Possible Aetiological Factor in Idiopathic Portal Hypertension (Non-Cirrhotic Portal Fibrosis) in India.' *Gut* 20 (5): 378–84. doi:10.1136/gut.20.5.378.

Delleur, J. W. 2003. 'The Evolution of Urban Hydrology: Past, Present, and Future.' *Journal of Hydraulic Engineering, ASCE* 129 (8): 563–573.

Duarte, António A. L. S., Sílvia J. A. Cardoso, and António J. Alçada. 2009. 'Emerging and Innovative Techniques for Arsenic Removal Applied to a Small Water Supply System.' *Sustainability* 1 (4): 1288–1304. doi:10.3390/su1041288.

Duda-Chodak, Aleksandra, and Urszula Blaszczyk. 2008. 'The Impact of Nickel on Human Health.' *Journal of Elementology* 13 (4): 685–696.

Duggal, K. 2007. *Elements of Environmental Engineering.* New Delhi: S. Chand.

Dutta, Paritam K., S. O. Pehkonen, Virender K. Sharma, and Ajay K. Ray. 2005. 'Photocatalytic Oxidation of Arsenic(III): Evidence of Hydroxyl Radicals.' *Environmental Science and Technology* 39 (6): 1827–34. doi:10.1021/es0489238.

Dwivedi, Sanjay, R. D. Tripathi, Sudhakar Srivastava, Ragini Singh, and Amit Kumar. 2010. 'Arsenic Affects Mineral Nutrients in Grains of Various Indian Rice (Oryza Sativa L.) Genotypes Grown on Arsenic-Contaminated Soils of West Bengal.' *Protoplasma* 245 (1-4): 113–24. doi:10.1007/s00709-010-0151-7.

Edwards, Marc. 1994. 'Chemistry of Arsenic Removal during Coagulation and Fe-Mn Oxidation.' *Journal of American Water Works Association* 86 (9): 64–78.

Environmental Protection Authority, Victoria. 2002. *Disinfection of Treated Wastewater. Guidelines for Environmental Management.* https://www.epa.vic.gov.au/~/media/Publications/730.pdf

Escobar, M. E. O., N. V. Hue, and W. G. Cutler. 2006. 'Recent Developments on Arsenic: Contamination and Remediation.' *Recent Research Developments in Bioenergetics* 4: 1–32. http://www.cabdirect.org/abstracts/20073215205.html.

Fair, G., J. Geyer, and D. Okun. 1966. *Water and Wastewater Engineering.* New York: J. Wiley and Sons.

Fawell, J., K. Bailey, J. Chilton, E. Dahi, Lorna Fewtrell, and Yasumoto Magara. 2006. *Fluoride in Drinking-Water.* IWA Publishing. https://www.who.int/water_sanitation_health/publications/fluoride_drinking_water_full.pdf

FICCI. 2011. *Water Use in Indian Industry Survey.* New Delhi:FICCI Water Mission. http://ficci.in/Sedocument/20188/Water-Use-Indian-Industry-Survey_results.pdf

Fields, Keith A., Abraham Chen, and Lili Wang. 2000. *Arsenic Removal from Drinking Water by Coagulation-Filtration and Lime Softening Plants.* US Environmental Protection Agency. Columbus, Ohio: Batelle.

Freeze, R., and J. Cherry. 1979. *Groundwater: Englewood.* New Jersey: Prentice Hall.

Gardoni, D., A. Vailati, and R. Canziani. 2012. 'Decay of Ozone in Water: A Review.' *Ozone: Science and Engineering* 34 (4): 233–242. http://doi.org/10.1080/01919512.2012.686354

Garg, S. 2001. *Water Supply Engineering.* Delhi: Khanna Publishers.

Geyer, H., G. Politzki, and D. Freitag. 1984. 'Prediction of Ecotoxicological Behaviour of Chemicals: Relationship Between N-octanol/Water partition Coefficient and Bioaccumulation of Organic Chemicals by Alga Chlorella.' *Chemosphere* 13 (2): 269–284.

Goel, S. 2015. 'Antibiotics in the Environment – A Review.' In *Emerging Micro-Pollutants in the Environment,* 19-42. Washington, DC: American Chemical Society Books.

Goel, S., and E. J. Bouwer. 2004. Factors influencing inactivation of Klebsiella pneumoniae by chlorine and chloramine. *Water Research* 38 (2): 301–8. http://doi.org/10.1016/j.watres.2003.09.016

Goel, S., R. M. Hozalski, and E. J. Bouwer. 1995. 'Biodegradation of NOM: Effect of NOM Source and Ozone Dose.' *Journal of American Water Works Association* 87 (1): 90–105.

Goetz, A., and N. Tsuneishi. 1951. 'Application of Molecular Filter Membranes to the Bacteriological Analysis of Water.' *Jour. American Water Works Association* 43 (12): 943–969.

Graton, L., and H. J. Fraser. 1935. 'Systematic Packing of Spheres: With Particular Relation to Porosity and Permeability.' *The Journal of Geology* 43 (8): 785–909.

GRBEMP. 2013. *Emerging Contaminants in Ganga River Basin: Pesticides, Heavy Metals and Antibiotics.* Kharagpur: IIT.

Greenlee, Lauren F., Desmond F. Lawler, Benny D. Freeman, Benoit Marrot, and Philippe Moulin. 2009. 'Reverse Osmosis Desalination: Water Sources, Technology, and Today's Challenges.' *Water Research* 43 (9): 2317-2348.

Gregory, R., and T. F. Zabel. 1990. 'Sedimentation and Flotation.' In *Water Quality and Treatment: A Handbook of Community Water Supplies,* 4th edition, 367–453. New York: McGraw Hill.

Gupta, K. 2005. 'The Drainage Systems of India's Cities.' *Waterlines* 23 (4): 22–24.

Hammer, M., and M. Hammer Jr. 2008. *Water and Wastewater Technology,* 6th ed. New Delhi: Prentice Hall India.

Hendricks, David. 2011. *Fundamentals of Water Treatment Unit Processes: Physical, Chemical and Biological.* IWA Publishing.

Henry, J. G., and G. W. Heinke. 1996. *Environmental Science and Engineering.* New Jersey: Prentice Hall.

Hering, Janet G., Pen-yuan Chen, Jennifer A. Wilkie, Menachem Elimelech, and Sun Liang. 1996. 'Arsenic Removal by Ferric Chloride.' *Journal of American Water Works Association* 88 (4): 155–67.

Higgins, Joseph J., M. C. Patterson, N. M. Papadopoulos, R. O. Brady, P. G. Pentchev, and N. W. Barton. 1992. 'Hypoprebetalipoproteinemia, Acanthocytosis, Retinitis Pigmentosa, and Pallidal Degeneration (HARP syndrome).' *Neurology* 42 (1): 194–198.

Iowa Department of Transportation. 1995. *Using the Rational Method to Determine Peak Flow.*

Jimenez, B. 2006. 'Irrigation in Developing Countries Using Wastewater.' *Internation Review for Environmental Strategies* 6 (2): 229–250.

Jimenez, B., and T. Asano. 2008. 'Water Reclamation and Reuse Around the World.' *Water Reuse: An International Survey of Current Practice, Issues and Needs* 14: 3-26.

Johnston, Richard, and Han Heijnen. 2001. *Safe Water Technology for Arsenic Removal.* United Nations University. http://archive.unu.edu/env/Arsenic/Han.pdf.

Kang, Meea, Mutsuo Kawasaki, Sinya Tamada, Tasuku Kamei, and Yasumoto Magara. 2000. 'Effect of PH on the Removal of Arsenic and Antimony Using Reverse Osmosis Membranes.' *Desalination* 131 (1–3): 293–98. doi:10.1016/S0011-9164(00)90027-4.

Kapoor, Anoop, and T. Viraraghavan. 1997. 'Nitrate Removal from Drinking Water - Review.' *Journal of Environmental Engineering* 123 (4): 371–80.

Karak, Tanmoy, Ornella Abollino, Pradip Bhattacharyya, Kishore K. Das, and Ranjit K. Paul. 2011. 'Fractionation and Speciation of Arsenic in Three Tea Gardens Soil Profiles and Distribution of As in Different Parts of Tea Plant (Camellia Sinensis L.).' *Chemosphere* 85 (6): 948–60. doi:10.1016/j.chemosphere.2011.06.061.

Karia, G., and R. Christian. 2006. *Wastewater Treatment: Concepts and Design Approach.* New Delhi: Prentice Hall India.

Kartinen, Ernest O., and Christopher J. Martin. 1995. 'An Overview of Arsenic Removal Processes.' *Desalination* 103 (1–2): 79–88. doi:10.1016/0011-9164(95)00089-5.

Kirillin, G., and T. Shatwell. 2016. '*Generalized scaling of seasonal thermal stratification in lakes.*' Earth-Science Reviews 161: 179-190.

Kolpin, D. W., and M. T. Meyer. 2002. 'Pharmaceuticals, Hormones, and Other Organic Wastewater Contaminants in U . S . Streams. 1999 - 2000: A National Reconnaissance.' *Environmental Science and Technology* 36 (6): 1202–1211.

Koonin, E. V., and Y. I. Wolf. 2008. 'Genomics of Bacteria and Archaea: the Emerging Dynamic View of the Prokaryotic World.' *Nucleic Acids Research* 36 (21): 6688–719.

Krasner, S. W., H. S. Weinberg, S. D. Richardson, S. J. Pastor, R. Chinn, M. J. Sclimenti, and A. D. Thruston. 2006. Occurrence of a New Generation of Disinfection By-products. *Environmental Science and Technology* 40 (23): 7175–85. http://www.ncbi.nlm.nih.gov/pubmed/17180964

Kumar, A., and S. Goel. 2018. 'Evaluation of Water Quality in Rivers Ganga and Yamuna using Water Quality Index (WQI) and Streeter–Phelps Model.' *STIWM* 1–26.

Kundu, R., K. Bhattacharyya, A. Majumder, and S. Pal. 2012. 'Response of Wheat Cultivars to Arsenic Contamination in Polluted Soils of West Bengal, India.' *Cereal Research Communication.* doi:10.1556/CRC.2012.0027.

Levenspiel, O. 1998. *Chemical Reaction Engineering.* New Delhi: Wiley Eastern Ltd.

Lofrano, G., and J. Brown. 2010. 'Wastewater Management through the Ages: A History of Mankind.' *Science of The Total Environment* 408 (22): 5254–5264. http://doi.org/10.1016/j.scitotenv.2010.07.062

Loganathan, Paripurnanda, Saravanamuthu Vigneswaran, Jaya Kandasamy, and Ravi Naidu. 2013. 'Defluoridation of Drinking Water Using Adsorption Processes.' *Journal of Hazardous Materials* 248–249: 1–19. doi:10.1016/j.jhazmat.2012.12.043.

Lozoff, Betsy, and Michael K. Georgieff. 2006. 'Iron Deficiency and Brain Development.' In *Seminars in Pediatric Neurology* 13 (3): 158–165.

Madigan, M. T., J. M. Martinko, and J. S. Parker. 2003. *Brock Biology of Microorganisms*. New Jersey: Prentice Hall.

Madigan, M. T., J. M. Martinko, K. Bender, D. Buckley, and D. Stahl . 2015. *Brock Biology of Microorganisms*. London: Pearson Publishers.

Malliarou, E., C. Collins, N. Graham, and M. J. Nieuwenhuijsen. 2005. Haloacetic Acids in Drinking Water in the United Kingdom. *Water Research* 39 (12): 2722–30. http://doi.org/10.1016/j.watres.2005.04.052

Mandal, Badal Kumar, and Kazuo T. Suzuki. 2002. 'Arsenic Round the World: A Review.' *Talanta* 58 (1): 201–35. doi:10.1016/S0039-9140(02)00268-0.

Mannina, G., and G. Viviani. 2009. 'Separate and Combined Sewer Systems: A Long-term Modelling Approach.' *Water Sci Technol* 60 (3): 555–565.

Martin, S., and W. Griswold. 2009. 'Human Health Effects of Heavy Metals: Briefs for Citizens.' *Environmental Science and Technology* 15:1-6.

Masters, G. M. 1998. *Introduction to Environmental Engineering and Science*. New Jersey: Prentice Hall .

Masters, G.M., and W. Ela. 2008. *Introduction to Environmental Engineering and Science*. New Jersey: Prentice Hall.

Matilainen, A., N. Lindqvist, S. Korhonen, and Tuula Tuhkanen. 2002. 'Removal of NOM in the Different Stages of the Water Treatment Process.' *Environment International* 28: 457–65.

Matilainen, Anu, Mikko Vepsäläinen, and Mika Sillanpää. 2010. 'Natural Organic Matter Removal by Coagulation during Drinking Water Treatment: A Review.' *Advances in Colloid and Interface Science* 159 (2): 189–97. doi:10.1016/j.cis.2010.06.007.

Matschullat, Jörg. 2000. 'Arsenic in the Geosphere - A Review.' *Science of the Total Environment* 249 (1–3): 297–312. doi:10.1016/S0048-9697(99)00524-0.

Meharg, A. A. 2003. 'Arsenic Contamination of Bangladesh Paddy Field Soils: Implications for Rice Contribution to Arsenic Consumption.' *Environmental Science and Technology* 37 (2): 229–34.

Meharg, A. A., P. N. Williams, E. Adomako, Y. Y. Lawgali, Claire Deacon, Antia Villada, Robert C. J. Cambell, et al. 2009. 'Geographical Variation in Total and Inorganic Arsenic Content of Polished (White) Rice.' *Environmental Science and Technology* 43 (5): 1612–17.

Metcalf and Eddy. 1991. *Wastewater Engineering: Treatment, Disposal and Reuse*. 3rd ed. New York: McGraw Hill Inc.

———. 2003. *Wastewater Engineering: Treatment, Disposal and Reuse*. New Delhi: Tata-McGraw Hill.

Mihelcic, J., and J. Zimmerman. 2010. *Environmental Engineering: Fundamentals, Sustainability and Design*. New York: J. Wiley and Sons.

Ministry of Water Resources. 2002. *National Water Policy*. Accessed January 1, 2009. http://wrmin.nic.in/writereaddata/ linkimages/ nwp20025617515534.pdf

Mishra, Bhupendra K., Chandra S. Dubey, and Arnold L. Usham. 2014. 'Concentration of Arsenic by Selected Vegetables Cultivated in the Yamuna Flood Plains (YFP) of Delhi, India.' *Environmental Earth Sciences* 72 (9): 3281–91. doi:10.1007/s12665-014-3232-7.

Mohanta, T., and S. Goel. 2014. 'Prevalence of Antibiotic-Resistant Bacteria in Three Different Aquatic Environments over Three Seasons.' *Environmental Monitoring and Assessment* 186: 5089–5100.

———. 2016. 'Statistical Analyses of Water Quality and Antibiotic Resistance Index for Three Aquatic Environments over Three Seasons.' *Pollution Research* 35 (1): 107–122.

Mohanty, Debasis. 2017. 'Conventional as well as Emerging Arsenic Removal Technologies—a Critical Review.' *Water, Air, and Soil Pollution* 228 (381): 1–21. doi:10.1007/s11270-017-3549-4.

Mohseni-Bandpi, Anoushiravan, David Jack Elliott, and Mohammad Ali Zazouli. 2013. 'Biological Nitrate Removal Processes from Drinking Water Supply—a Review.' *Journal of Environmental Health Science and Engineering* 11 (35): 1–11.

Mondal, Swapan. 2008. *Arsenic Removal from Drinking Water by Coagulation with Aluminum Sulfate and Ferric Chloride*. Kharagpur: IIT.

Morris, G. L. 2015. *Water Supply Intake Structures*. http://drna.pr.gov/historico/oficinas/saux/secretaria-auxiliar-de-planificacion-integral/planagua/proyecto-de-caudales-ecologicos/1ra-conferencia-de-flujos-ambientales-en-rios-de-puerto-rico/Land Figs River Structures.pdf.

Morris, J. 1978. 'Water Chlorination: Environmental Impacts and Health Effects.' In R. Jolley, H. Gorchev, and D. Hamilton (Eds.), *Water Chlorination*. Ann Arbor, Michigan: Ann Arbor Science.

Narayan, Sumit, and Sudha Goel. 2011. 'Enhanced Coagulation for Turbidity and Total Organic Carbon (TOC) Removal from River Kansawati Water.' *Journal of Environmental Science and Engineering NEERI* 53 (1): 39–44.

Navarro Silvera, Stephanie, and Thomas E Rohan. 2007. 'Trace Elements and Cancer Risk: A Review of the Epidemiologic Evidence.' *Cancer Causes and Control* 18 (1): 7–27. doi:10.1007/s10552-006-0057-z.

Nazaroff, W. W., and Lisa Alvarez-Cohen. 2004. *Environmental Engineering Science*. Delhi: J. Wiley, India.

Nordin, B. E. Christopher. 1997. 'Calcium and Osteoporosis.' *Nutrition* 13 (7-8): 664–686.

Odell, L. H. 2016. 'Cut Drinking Water Arsenic Levels Using Best Removal Strategies.' *Opflow* 42 (6): 26–30.

Ouypornkochagorn, Sairoong, and Jörg Feldmann. 2010. 'Dermal Uptake of Arsenic through Human Skin Depends Strongly on Its Speciation.' *Environmental Science and Technology* 44 (10): 3972–78. doi:10.1021/es903667y.

Padilla, Francisco M., Marisa Gallardo, and Francisco Manzano-Agugliaro. 2018. 'Global Trends in Nitrate Leaching Research in the 1960–2017 Period.' *Science of the Total Environment* 643: 400–413. doi:10.1016/j.scitotenv.2018.06.215.

Panikkar, A. 2012. *Water profile of India. FAO*. http://www.eoearth.org/view/article/156948/.

Patel, K. S., K. Shrivas, R. Brandt, N. Jakubowski, W. Corns, and P. Hoffmann. 2005. 'Arsenic Contamination in Water, Soil, Sediment and Rice of Central India.' *Environmental Geochemistry and Health* 27 (2): 131–45. doi:10.1007/s10653-005-0120-9.

Payne, L. R. 1977. 'The Hazards of Cobalt.' *Occupational Medicine* 27 (1): 20–25.

Peavy, H., D. Rowe, and G. Tchobanoglous. 1985. *Environmental Engineering*. New York: McGraw Hill.

Pierce, Matthew L., and Carleton B. Moore. 1982. 'Adsorption of Arsenite and Arsenate on Amorphous Iron Hydroxide.' *Water Research* 16 (7): 1247–53. doi:10.1016/0043-1354(82)90143-9.

Prato, Ted, and Richard G. Parent. 1993. 'Nitrate and Nitrite Removal from Municipal Drinking Water Supplies with Electrodialysis Reversal.' In *Proceedings of 1993 AWWA Membrane Conference, Baltimore, Maryland, USA*.

Qasim, S., E. Motley, and G. Zhu. 2000. *Water Works Engineering: Planning, Design and Operation*. New Delhi: Prentice Hall.

Rahman, Mohammad M., Debapriya Mondal, Bhaskar Das, Mrinal K. Sengupta, Sad Ahamed, M. Amir Hossain, Alok Chandra Samal, et al. 2014. 'Status of Groundwater Arsenic Contamination in All 17 Blocks of Nadia District in the State of West Bengal, India: A 23-Year Study Report.' *Journal of Hydrology* 518: 363–72. doi:10.1016/j.jhydrol.2013.10.037.

Rajamohan, R., V. P. Venugopalan, Debasis Mal, and Usha Natesan. 2014. 'Efficiency of Reverse Osmosis in Removal of Total Organic Carbon and Trihalomethane from Drinking Water.' *Research Journal of Chemistry and Environment* 18 (12): 1–6.

Raju, N. J. 2012. 'Arsenic Exposure through Groundwater in the Middle Ganga Plain in the Varanasi Environs, India: A Future Threat.' *Journal of Geological Society of India* 79 (March): 302–14.

Richardson, S. 2003. 'Disinfection By-products and other Emerging Contaminants in Drinking Water.' *TrAC Trends in Analytical Chemistry* 22 (10): 666–684. http://doi.org/10.1016/S0165-9936(03)01003-3.

Richardson, S. D., M. J. Plewa, E. D. Wagner, R. Schoeny, and D. M. Demarini. 2007. 'Occurrence, Genotoxicity, and Carcinogenicity of Regulated and Emerging Disinfection By-products in Drinking Water: a Review and Roadmap for Research.' *Mutation Research* 636 (1–3): 178–242. http://doi.org/10.1016/j.mrrev.2007.09.001

Rittmann, B. E., and P. L. McCarty. 2012. *Environmental Biotechnology: Principles and applications* (3rd ed.). Chennai: McGraw Hill Education (India) Pvt. Ltd.

Roychowdhury, Tarit, Tadashi Uchino, Hiroshi Tokunaga, and Masanori Ando. 2002. 'Arsenic and Other Heavy Metals in Soils from an Arsenic-Affected Area of West Bengal, India.' *Chemosphere* 49 (6): 605–18.

———. 2002. 'Survey of Arsenic in Food Composites from an Arsenic-Affected Area of West Bengal, India.' *Food and Chemical Toxicology* 40 (11): 1611–21. http://www.ncbi.nlm.nih.gov/pubmed/12176088.

Rude, Robert K. 1998. 'Magnesium Deficiency: A Cause of Heterogenous Disease in Humans.' *Journal of Bone and Mineral Research* 13 (4): 749–758.

Santamaria, A. B. 2008. 'Manganese Exposure, Essentiality and Toxicity.' *Indian Journal of Medical Research* 128 (4): 484–500.

Sawyer, C., P. McCarty, and G. Parkin. 2000. *Chemistry for Environmental Engineering and Science*. New Delhi: Tata McGraw Hill.

———. 2003. *Chemistry for Environmental Engineering and Science*, 5th ed. New York: McGraw Hill. https://4lfonsina.files.wordpress.com/2012/11/chemistry_for_environmental_engineering_and_science.pdf

Saper, Robert B., and Rebecca Rash. 2009. 'Zinc: An Essential Micronutrient.' *American Family Physician* 79 (9): 768–772.

Saxena, A., S. Kumar, and P. Goel. 2014. 'Source Mineral for the Release of Arsenic in the Groundwater of Karanda Block, Ghazipur District, Uttar Pradesh.' *Journal of Geological Society of India* 84: 590–96.

Sengupta, S. 2014. 'Reaction Kinetics and Reactors.' In *Comprehensive Water Quality and Purification: Wastewater Treatment and Reuse*, edited by S. Sengupta and S. Ahuja, 30–46. Waltham: Academic Press.

Sengupta, S. 2015. 'Is Sewage Farming Safe?' *Down to Earth* https://www.downtoearth.org.in/coverage/is-sewage-farming-safe-48566.

Shaban, A., and R. Sharma. 2007. 'Water Consumption Patterns in Domestic Households in Major Cities.' *Economic and Political Weekly*, 2007: 2190-2197.

Sharma, Anshul, Sri Malini Adapureddy, and Sudha Goel. 2014. 'Arsenic Removal from Aqueous Samples in Batch Electrocoagulation Studies.' *International Proceedings of Chemical, Biological and Environmental Engineering* 64: 40-43. 91: 10–13. doi:10.7763/IPCBEE.

Shortt, H. E., C. G. Pandit, and T. N. S. Raghavachari. 1937. 'Endemic Fluorosis in the Nellore District of South India.' *The Indian Medical Gazette* 72 (7) : 396–98.

Signes-Pastor, J., K. Mitra, S. Sarkhel, M. Hobbes, F. Burló, W. T. de Groot, and A. A. Carbonell-Barrachina. 2008. 'Arsenic Speciation in Food and Estimation of the Dietary Intake of Inorganic Arsenic in a Rural Village of West Bengal, India.' *Journal of Agricultural and Food Chemistry* 56 (20): 9469–74. doi:10.1021/jf801600j.

Sincero, A., and G. Sincero. 1996. *Environmental Engineering: A Design Approach*. New Jersey: Prentice Hall.

Singh, Kalpana, Dilip H Lataye, Kailas L. Wasewar, and Chang Kyoo Yoo. 2013. 'Removal of Fluoride from Aqueous Solution: Status and Techniques.' *Desalination and Water Treatment* 51: 16-18: 3233–47. doi:10.1080/19443994.2012.749036.

Skoog, D., D. West, F. Holler, and S. Crouch. 2004. 'Effect of Electrolytes on Chemical Equilibria.' *Fundamentals of Analytical Chemistry*. 267–280. Ontario: Nelson Education.

Smedley, P. L., and D. G. Kinniburgh. 2002. 'A Review of the Source, Behaviour and Distribution of Arsenic in Natural Waters.' *Applied Geochemistry* 17 (5): 517–68.

SOES Jadavpur University. 2015. '*Groundwater Arsenic Contamination in West Bengal - India (20 Years Study)*.' http://www.soesju.org/arsenic/wb.htm.

Sophonsiri, C., and E. Morgenroth. 2004. 'Chemical Composition Associated with Different Particle Size Fractions in Municipal, Industrial, and Agricultural Wastewaters.' *Chemosphere* 55 (5): 691–703.

Streeter, H., and E. Phelps. 1925. *A Study of the Pollution and Natural Purification of the Ohio River*. http://udspace.udel.edu/handle/19716/1590

Talmage, W. P., and E. B. Fitch. 1955. Determining Thickener Unit Areas. *Industrial and Engineering Chemistry* 47 (1): 38–41. http://doi.org/10.1021/ie50541a022

Tare, V., A. V. S. Yadav, and P. Bose. 2003. 'Analysis of Photosynthetic Activity in the Most Polluted Stretch of River Ganga.' *Water Research* 37 (1): 67–77.

Tate, C., and K. Arnold. 1990. 'Health and Aesthetic Aspects of Water Quality.' *Water Quality and Treatment: A Handbook of Community Water Supplies*, edited by F. Pontius, 63–156. New York: McGraw Hill Inc..

Tortora, G., B. Funke, and C. Case. 2004. *Microbiology: An Introduction*. San Francisco: Pearson Education, Inc.

Tripathi, Preeti, Sanjay Dwivedi, Aradhana Mishra, Amit Kumar, Richa Dave, Sudhakar Srivastava, Mridul Kumar, et al. 2012. 'Arsenic Accumulation in Native Plants of West Bengal, India: Prospects for Phytoremediation but Concerns with the Use of Medicinal Plants.' *Environmental Monitoring and Assessment* 184 (5): 2617–31. doi:10.1007/s10661-011-2139-y.

UNICEF. 2013. *Water in India: Situation and Prospects*.

US Environmental Protection Agency (USEPA). 1971. *Total Organic Carbon Removal from Municipal and Industrial Wastewater*.

———. 1984. *Design Information on Rotating Biological Contactors*.

———. 1999a. *Wastewater Technology Fact Sheet: Chlorine disinfection*.

———. 1999b. *Wastewater Technology Fact Sheet: Ozone Disinfection*.

———. 1999c. *Wastewater Technology Fact Sheet: Ultraviolet Disinfection*.

———. 2001. *Arsenic in Drinking Water*.

———. 2002. *Arsenic Treatment Technologies for Solid, Waste, and Water*.

———. 2012. *Guidelines for Water Reuse: 2012*.

USEPA. 2007. *Basic Information about Copper in Drinking Water*. United States Environmental Protection Agency. http://water.epa.gov/drink/contaminants/basicinformation/copper.cfm

Wagemann, R. 1978. 'Some Theoretical Aspects of Stability and Solubility of Inorganic Arsenic in the Freshwater Environment.' *Water Research* 12 (3): 139–45. doi:10.1016/0043-1354(78)90001-5.

Wang, L., A. S. Butcher, M. E. Stuart, D. C. Gooddy, and J. P. Bloomfield. 2013. 'The Nitrate Time Bomb: A Numerical Way to Investigate Nitrate Storage and Lag Time in the Unsaturated Zone.' *Environmental Geochemistry and Health* 35 (5): 667–81. doi:10.1007/s10653-013-9550-y.

Watson, H. E. 1908. 'A Note on the Variation of the Rate of Disinfection with Change in the Concentration of the Disinfectant.' *The Journal of Hygiene* 8 (4): 536–542. http://www.jstor.org/stable/4619383

Webster, C. 1962. 'The Sewers of Mohenjo-Daro.' *Jour. Water Pollution Control Federation* 34 (2): 116–123.

Westerhoff, P., P. Chao, and H. Mash. 2004. Reactivity of Natural Organic Matter with Aqueous Chlorine and Bromine. *Water Research* 38 (6): 1502–13. http://doi.org/10.1016/j.watres.2003.12.014

Whittaker, R. H. 1969. 'New Concepts of Kingdoms of Organisms.' *Science* 163: 150–160.

Williams, D. M. 1983. 'Copper Deficiency in Humans.' *Seminars in Hematology* 20 (2):118–128. http://europepmc.org/search?page=1&query=ISSN:%220037-1963%22.

Winslow, C.-E. A., W. Mason, F. Hale, D. Jackson, C. Wigley, and M. Tolman. 1916. 'Tests For Bacillus Coli as an Indicator of Water Pollution.' *Journal–American Water Works Association* 3 (4): 927–946.

World Health Organization (WHO). 1989. *Health Guidelines for the Use of Wastewater in Agriculture and Aquaculture.*

———. 2005. *Chlorite and Chlorate in Drinking-Water. WHO Guidelines for Drinking-water Quality.*

Index

Color Plates

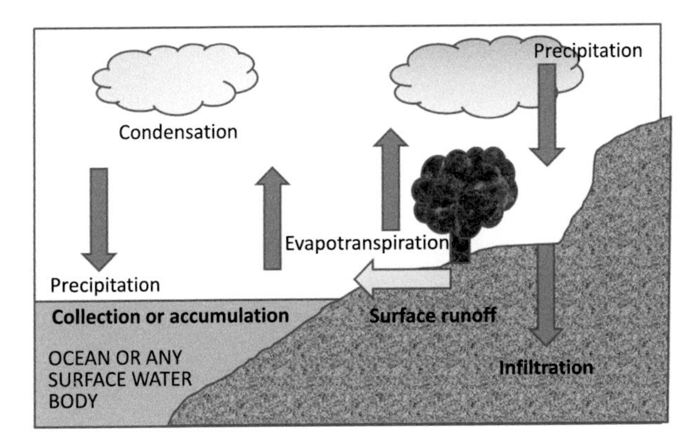

Figure 1.1 Hydrologic cycle and its various components.

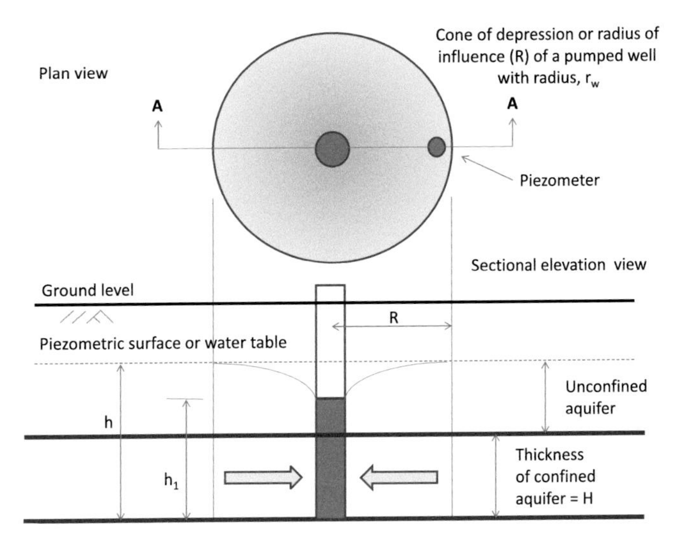

Figure 1.6 Radius of influence of a pumped well in a confined or unconfined aquifer.

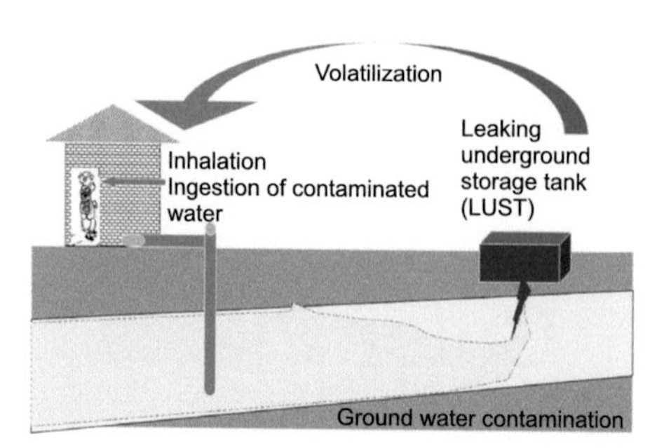

Figure 1.7 Groundwater contamination from leaking underground storage tanks (LUST).

Figure 1.10 Cross-weir intake with canal diversion. The water level is fixed by the weir height, i.e., some damming of water is possible in this structure. (a) Micro-hydroelectric power projects in Ererte, Ethiopia[67] (b) and in Gobecho,[68] Ethiopia.

Figure 1.11 A single point source (large drain hole) discharge of municipal wastewater and multiple drainage points (small drain holes) for discharge of surface runoff and infiltrate from the paved areas above the river embankment are clearly visible in this photo. Location: River Hooghly, Kolkata.

67 https://energypedia.info/wiki/Micro_Hydro_Power_(MHP)_-_Ethiopia,_Ererte.

68 https://energypedia.info/wiki/Micro_Hydro_Power_(MHP)_-_Ethiopia,_Gobecho_I.

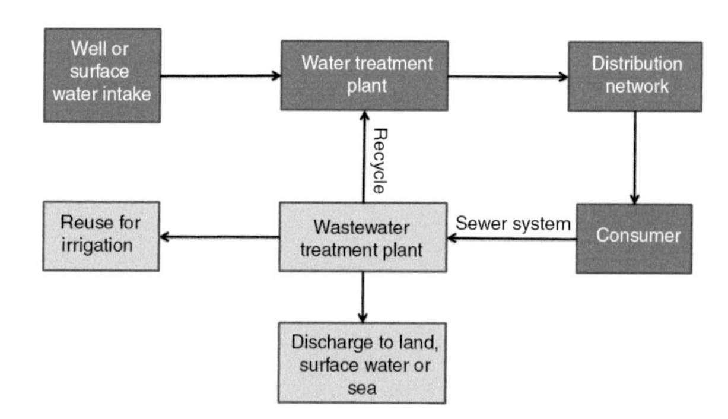

Figure 2.1 Flow of water through a water supply and wastewater system for any urban center.

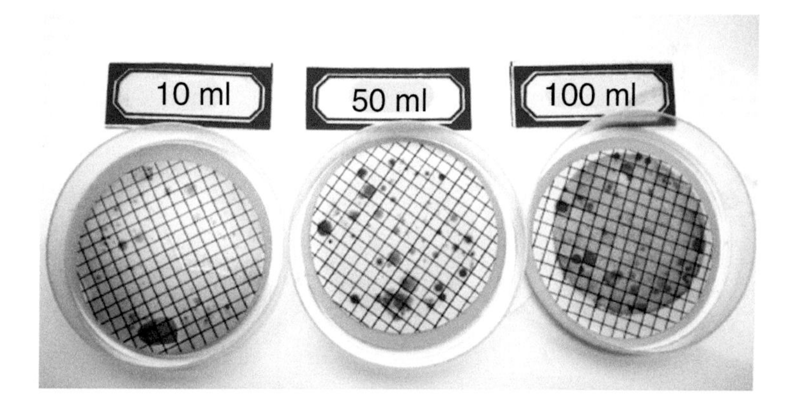

Figure 3.4 Photograph of bacterial colonies detected in different volumes of the same water sample and grown on membrane filters resting on absorbent nutrient pads (Source: B. Mahto).

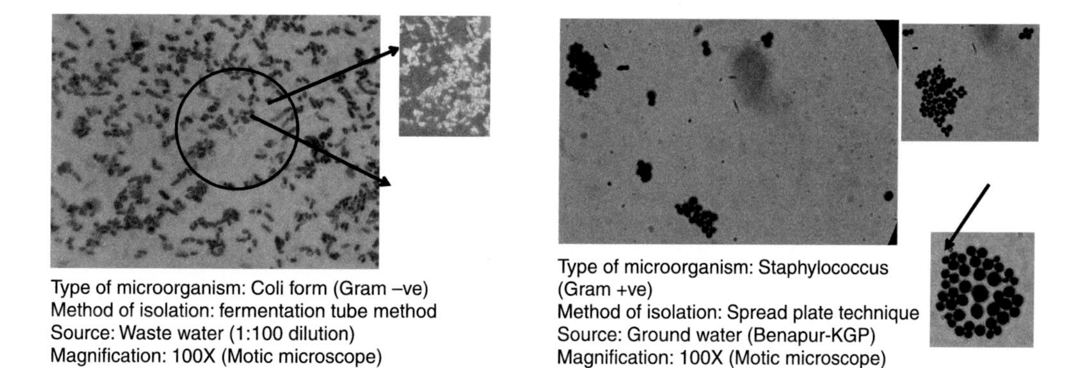

Figure 3.5 Gram-stained bacteria under an optical microscope with 100x objective lens (Source: Iti Sharma).

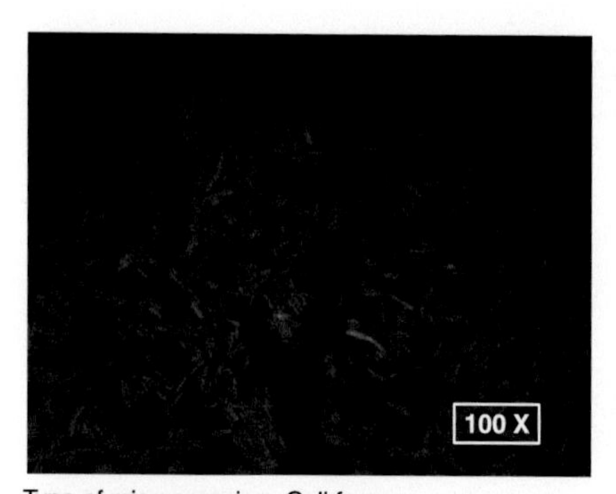

Type of microorganism: Coli form
Date: 3-4-12
Method of isolation: Fermentation tube method
Source: Waste water (1: 100 dilution)
Image: Fluorescence microscope (Dye: Acridine orange), Filter: TRITC (Motic Image Advanced 3.2)

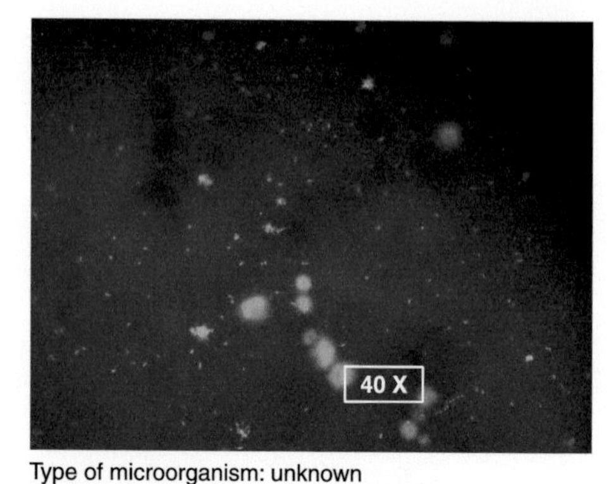

Type of microorganism: unknown
Date: 29-3-12
Method of isolation: Filtration (0.2 μm)
Source: Tap water
Image: Fluorescence microscope (Dye: Acridine orange), Filter: FITC (Motic Image Advanced 3.2)

Figure 3.6 Bacteria observed under epifluorescence microscope based on the AODC method and different filters (Source: Iti Sharma).

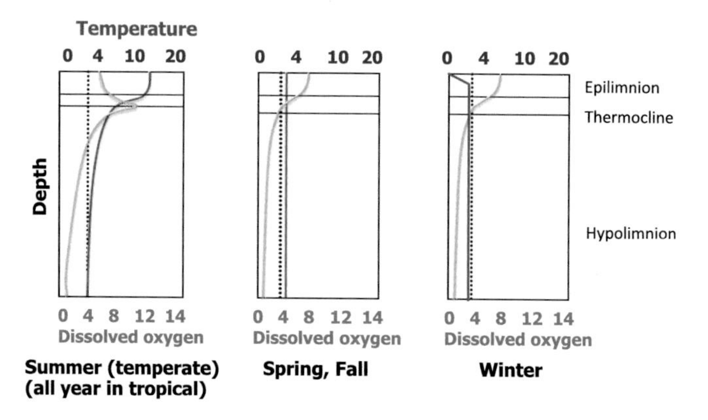

Figure 3.9 Change in Temperature and DO profiles in water with season.

Figure 5.2 Spray or fountain aerators (Gangtok Water Treatment Plant, Source: R. N. Sharma).

Figure 5.3 Cascade aerator (Gangtok Water Treatment Plant, Source: R. N. Sharma).

(a)

(b)

Figure 5.6 Moving screen in a wastewater treatment plant in Pune: (a): Front view, screen is moving from bottom to top; (b). Back view of top of the screen where the screened materials are collected.

Figure 5.10 Series of V-notches in a settling tank to prevent flow of solids into the collection trough or channel.

Figure 5.14b Photo of a clariflocculator in a water treatment plant.

Figure 5.15 Macroscopic and microscopic structures of natural zeolites (Source: Wikipedia).

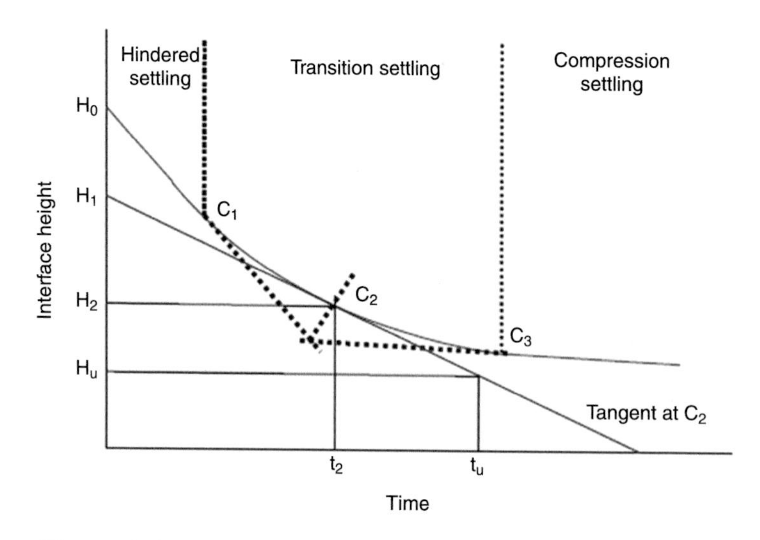

Figure 6.8 Settling column analysis for Type 3 and Type 4 settling.

Figure 6.11 Plug flow digesters in a wastewater treatment plant in India.

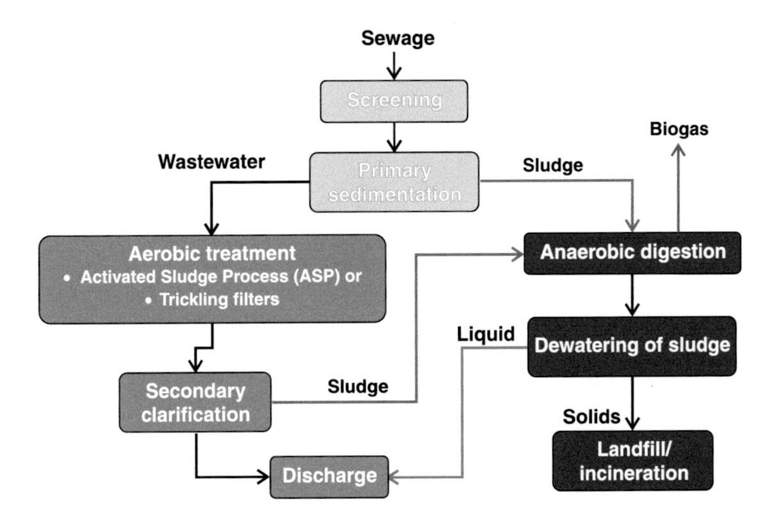

Figure 7.5 Typical flowchart for a conventional municipal wastewater treatment plant.

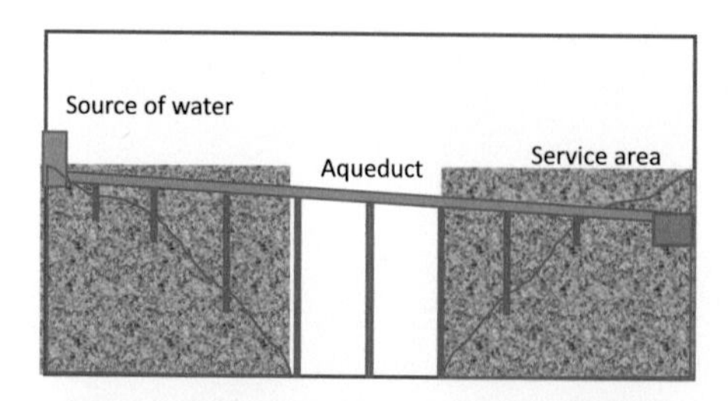

Figure 8.3 Schematic and photos of the Mathur aqueduct conveying water from source to service area (Source: Wikipedia).

Figure 9.1 (a) an open brick-lined stormwater drain, and
(b) an underground sewer pipeline under repair.

Figure 9.2 Sewage outfall in Kolkata for discharging sewage to River Hooghly.

in the United States
r & Taylor Publisher Services